Crop Plant Anatomy

Crop Plant Anatomy

Dr Ratikanta Maiti

Vibha Seeds, India

Dr Pratik Satya

Central Research Institute for Jute and Allied Fibres, India

Dasari Rajkumar

Vibha Seeds, India

and

Allam Ramaswamy

Vibha Seeds, India

www.cabi.org

CABI is a trading name of CAB International

CABI	CABI
Nosworthy Way	875 Massachusetts Avenue
Wallingford	7th Floor
Oxfordshire OX10 8DE	Cambridge, MA 02139
UK	USA
Tel: +44 (0)1491 832111	T: +1 800 552 3083 (toll free)
Fax: +44 (0)1491 833508	T: +1 (0)617 395 4051
E-mail: info@cabi.org	E-mail: cabi-nao@cabi.org
Website: www.cabi.org	

A catalogue record for this book is available from the British Library, London, UK.

Library of Congress Cataloging-in-Publication Data

Crop plant anatomy / Ratikanta Maiti … [et al.].
    p. cm.
  Includes bibliographical references and index.
  ISBN 978-1-78064-019-8 (alk. paper)
1. Plant anatomy. 2. Agricultural productivity. I. Maiti, R. K., 1938-

  QK641.C8653 2012
  580--dc23

                         2012005599

ISBN-13: 978 1 78064 019 8
CABI South Asia Edition ISBN: 978 1 78064 246 8
Commissioning editor: Sreepat Jain
Editorial assistant: Alexandra Lainsbury
Production editor: Fiona Chippendale

Typeset by SPi, Pondicherry, India
Printed and bound in the UK by CPI Group (UK) Ltd, Croydon, CR0 4YY.

Contents

Preface

The field of plant anatomy is one of the oldest life science disciplines. Starting its journey from the inquiries of Greek philosophers, this branch of science has provided a strong support to plant taxonomy, physiology and more recently plant developmental biology and molecular biology. Not very surprisingly, anatomy of crop plants has attracted less attention to botanists and taxonomists, since the major food crops belong to only few taxonomic families. In certain model crops such as rice and maize, the progress in developmental biology can largely be credited to the field of anatomy. However, relating structural anatomy with crop adaptation and productivity has received comparatively less attention from either agriculturists or botanists and few books give proper importance to this field of agricultural science. However, while searching literature, one can find overwhelming support provided by anatomical investigations towards drawing meaningful conclusions from research experiments on crop plants, particularly in the area of physiology and adaptation. Still, the significance of anatomical variation in explaining the adaptation of crop plants towards different abiotic stresses needs to be further investigated. In comparison to many other modern techniques, which target cellular or nucleic acid variation, anatomy provides low cost phenotypic screening options at organ, tissue or cellular level, further helping emerging fields of life sciences such as transcriptomics and system biology as well as providing rapid and reliable phenotypic screening procedures for selection of better performing genotypes.

Increasing crop productivity is the most immediate concern of mankind in today's world. Crop anatomy is intrinsically linked with better genetic potential for productivity of all crops, be it a derivative of primary or secondary metabolic pathway. To a large extent, anatomical structures determine the synthesis of food by crop plants, absorption of mineral nutrients for maintaining the process and channelling of reserve metabolite in certain organs, which are harvested by humans as economic product. Crop plants have thus been selected and manipulated during the course of evolution and selection for storing higher reserve elements. The clue to further improving productivity thus lies in the process of improvement of these physiological processes, which obviously need modifications of structural anatomy at cellular, tissue, organ or whole plant level. Thus in future crop science, anatomy is certain to play a pivotal role in helping crop improvement approaches, both at phenotype and genotype level.

In presenting this book the authors hope to fill a need for a textbook in crop plant anatomy with an applied approach towards crop adaptation and productivity. Not only, however, in our opinion, is there a need for a book for class study and guidance, but also for one which shall serve as a reference text for workers in fields of agricultural science, applied botany, and for teachers and students in other fields of crop ecology and botany.

This book first discusses the significance of anatomy in modern plant science followed by an outline of the basic anatomical structures of angiosperms to give readers a basic idea about plant anatomy. In the second section, structural anatomy of major crop plants has been discussed in detail with an objective to delineate the significance of variations in the anatomical features under different environments. These are followed by two sections, one emphasizing the role of anatomy in adaptation of crop plants and the other on signifying impact of the variations in structural anatomy on crop productivity, both of which are very important for increasing agricultural productivity. Throughout the book we also show how simple, low-cost light microscopy of hand sections can be used for rapid identification of anatomical features and be used for selection of genotypes under different environments, along with citing examples from research publications for further justification. Almost all the anatomical figures presented in the book have been prepared from live samples by the authors using simple light microscopy and low cost digital cameras, which is expected to encourage students and new researchers in the field of agricultural science to explore the tremendous possibility of utilizing anatomical techniques in their research fields. Training on anatomy is mainly undertaken only by laboratory practice. On such practice the authors believe emphasis must be given not only on lectures and text study, but also on extensive practical training. For laboratory teaching the present book should provide a background of facts, terms and history; it may, indeed, be used, in part, as a laboratory guide. The sequence of subjects adopted mainly on the long experience of the first author is giving emphasis of possible application of anatomical traits for adaptation of the particular crop to biotic and abiotic stress factors.

In the treatment of subject matter, emphasis has been placed on adaptability to classroom use from the standpoint of the student beginning anatomical study. Thus the book is, first of all, a textbook in the elements of crop plant anatomy and their significance, providing an introduction to the field. It presupposes an acquaintance only with the fundamental structure and activities of plants, an acquaintance such as is ordinarily obtained from a first course in botany. However, we also understand that while dealing with crop anatomy, which has vast and numerous research applications in the field of plant and crop science, some areas may be neglected, which may not be on purpose. We welcome any suggestions or critical input for further improvement of this book.

We are highly thankful to P. Vidyasagar, Chairman Vibha Agrotech Ltd, for providing facilities in Hyderabad as well as UNALA, ICRISAT, JRI UANL MEXICO, DLAA Poebla, Mexico.

1

Origin and Development of Crop Anatomy

1.1 Introduction

Anatomy (ənăt`əmē) is a branch of biological science concerned with the study of the internal structure of body parts of an organism. The study of gross internal structure of plant organs by the technique of section cutting is called plant anatomy (*ana*=asunder; *temmein*=to cut). Plant anatomy (syn. phytotomy) was originally considered as a section of biology dealing with both external and internal structures of plants. However, during the course of its development as a science, plant anatomy has been separated from morphology, and refers only to the study of the internal structure of the plant. With our increasing capacity to look inside the cell and organelles using advanced microscopic techniques, modern plant anatomy also includes understanding structural composition of cellular organelles, membranes and minute details of cellular structures.

The term plant anatomy is sometimes difficult to distinguish from several other related terms, such as histology, morphology, structural botany etc. For example, a flower structure can be studied by longitudinal or transverse section of the whole flower, which is often referred as flower anatomy. If the flower is not very small, the sections can be studied well by eye observation. However, the same flower can also be studied by separating the calyx, corolla, androecium and gynoecium to describe the flower structure. Thus the difference of anatomy and morphology is somewhat obscure in this case.

Histology, on the other hand, is the study of aggregates of similar cells at organ, tissue or cellular level. While in animals the tissue types and cell types in organs are well differentiated, plant structure is defined by relatively less differentiated cellular organization of meristematic tissues, vascular tissues, photosynthetic tissues or support tissues. In many cases, plant anatomy and plant histology refer to the same studies.

The studies on crop anatomy in the early days were non-specific with a basic interest for discoveries in the field of plant biology, and the emphasis on crop anatomy in relation to agriculture and crop adaptation and productivity was very sporadic. Only in the latter half of 20th century, did the objectives of crop anatomical studies shift towards the adaptation and response of crop species under different agroclimatic environments and stressed situations. Even though several works have been attempted in this important field that merges basic plant biology with agricultural science, no comprehensive documentation is available in this area.

Crop anatomy is not therefore only anatomical studies for the generation of basic biological information about the structure and function of internal organs of crop plants; it is an interdisciplinary science with highly significant applied value for the betterment of agricultural science.

1.2 Plant Anatomy in Ancient Civilizations

Although the subject of anatomy developed only after the discovery of microscopes, several ancient literatures reveal that visual anatomical studies have been carried out in different plant species. Eames and MacDaniels (1925) reported that plant anatomy study was initiated by Theophrastus of Eresus (c. 369–262 BC), who is also regarded as the 'Father of Botany'. He described bark, wood and pith. However, even before the works of Theophrastus, anatomical studies using visual observations were initiated in different civilizations.

Perhaps the earliest well documented studies on plant botany and anatomy were carried out in the early Indian civilization in the Vedic age. People during this period accumulated a great amount of knowledge about botany and anatomy of plants, recognizing the values of different plants in agriculture, medicine, fuel, construction and religious performances (Choudhury, 1971). In *Rig Veda*, the oldest Vedic literature, wood has been termed as '*daru*' to differentiate it from the bark or '*vakala*'. Later, in *Brihadaranyaka Upanisada*, wood anatomy has been described mentioning xylem, pith, bark and fibres. One of the earliest works in the field of plant science is the *Vrksayurveda* (science of longevity of plants) by 'Parasara' (c. 400 BC), which describes flower anatomy including ovary shape and also mentions sexuality in plants (Kanjilal, 1999). It also describes leaf anatomy, mentioning leaf structural compartments storing sap and being covered by a boundary, which may be the cell wall in today's context.

1.3 Resurgence in Anatomical Studies after Invention of the Microscope

After a long dormant period in the development of science, plant anatomy resurged as a major technique in plant science largely due to the discovery of microscopic techniques. During the 17th century, major contributions in the field of plant anatomy were made by Marcello Malpighi (1628–1694) and Nehemiah Grew (1641–1712). However, before their work a significant discovery was made by an English mathematician, Robert Hook (1635–1703). He investigated the internal structure of a thin slice of bottle cork under a crude microscope designed by himself, and described a honeycomb like structure in it, and to each compartment he termed a cell (Latin, *cellula*=a small apartment).

Malpighi is considered as one of the forerunners in the field of animal and plant anatomy. He studied structural similarities in plant and animal body plans and introduced the words 'epidermis' and 'stomata' in his book *Anatome Plantarum* published in 1675. Grew, an English physician, is considered as the father of plant anatomy. He provided detailed anatomical description of flower bud and vascular tissues and many other internal structures of plants that were earlier unknown. He first differentiated between the soft and hard parts of the plant body and described the vertical and horizontal system in plants. He introduced the terms 'parenchyma', 'pith', 'vesicles' and 'cortex' for the first time. He was also the pioneer in comparative anatomy of different plant species, such as a comparison between the stem anatomy of apple, pear, plum or of pine and oak. In spite of the limited resolution of microscopes during that time, Grew generated accurate and vivid descriptions of many plant anatomical features.

The 18th century witnessed some specific developments in the field of plant anatomy, such as the description of vascular bundle elements (Sprengel, 1766–1833), protoxylem (Treviranus, 1779–1864) and cambium (Du Hamel, 1700–1781). In the absence of good resolution, the earlier microscopic observations were less specific. However, a number of theories were developed in this century to explain the functionality of vascular bundles and uptake of water by plants. In this century, Carolus Linnaeus established taxonomy based on structure, with plant anatomy in his epic work, *Species Plantarum* (1753).

In the 19th century, two major developments took place that further established the importance of plant anatomy as an independent science. The compound microscope was further modified to obtain a better look at the cellular and subcellular structures. In 1831, with the help of a compound microscope, Robert Brown described the nucleus in a cell. With the firm establishment of cell theory by Schleiden (a botanist) and Schwann, (a zoologist) in the years 1838–39, the functional significance of anatomical structures were recognized, followed by discovery of cellular components such as protoplasm (Von Mohl, 1846) and cytoplasm (Kolliker, 1862). About the same time Carl Wilhelm von Nägeli (1817–1881) extensively studied the meristem and introduced the terms 'xylem' and 'phloem'. He studied ontogeny of apical meristems and distinguished between primary and secondary

meristems. He also described the structure of starch granules in plant cells.

The second development was progress in understanding of the fertilization process. The anatomy of pistils and ovules helped a lot to understand the process of pollen tube growth, entry and fertilization. E. Strasburger, who was the first to describe fertilization processes of plants in detail, extensively used anatomical drawings to describe the process of fertilization in his book *Die Angiospermen und die Gymnospermen*.

During this period, a number of textbooks on anatomy appeared describing the basic anatomical features of root, stem, leaves, reproductive organs and seeds (Table 1.1). In 1884, the first book on staining technique for plant anatomy was written by V.A. Poulsen. It described many chemicals and their applications in staining various plant parts, such as application of starch to identify and differentiate starch from cellulose, copper-ammonia to detect pectic substances or the use of chromic acid for study of the cell wall. This book served as a very good description of several histochemical techniques that had been used in the 18th and 19th centuries for plant anatomy studies.

Table 1.1. Early works on plant anatomy.

Title	Author	Publication year
An Idea of a Phytological History Propounded Together with a Continuation of the Anatomy of Vegetables, Particularly Prosecuted upon Roots	N. Grew	1673
Anatome Plantarum	M. Malpighi	1675
The Anatomy of Plants begun as a philosophical history of plants	N. Grew	1682
Beiträge zur Wissenschaftlichen Botanik	Carl Wilhelm von Nägeli	1858
The Mechanical Principles underlying the Anatomical Structure of Monocotyledonous Plants (Das mechanische Prinzip im anatomischen Bau der Monokotylon)	S. Schwendener	1874
Structural Botany	A. Gray	1879
Die Angiospermen und die Gymnospermen	E. Strasburger	1879
Physiological Plant Anatomy (Physiologische Pflanzenanatomie)	G. Haberlandt	1884
Botanical Micro-chemistry: An Introduction to the Study of Vegetable Histology	V. A. Poulsen	1884
An Introduction to Structural Botany. Part I Flowering Plants	D. H. Scott	1894

1.4   Plant Anatomy to Describe Species

After the Linnaean plant classification system was widely accepted, plant morphology was considered as a major tool for plant classification. A major initiative of plant anatomists of the 19th century was to compare the anatomical features of members of the same families to verify the Linnaean classification system. This led to a plant classification system based on structural similarity in plants, which, unlike the Linnaean classification system, targeted comparison of many internal characters of plants for classification. Swiss botanist Augustin Pyramus de Candolle advocated heavily for an anatomy-based classification system. He published *Théorie élémentaire de la botanique* in 1813, in which he argued that plant anatomy is the sole basis for plant classification. Specifically, high similarity in anatomical features of two species indicates that they are related. He coined the term 'taxonomy' to describe the structure-based plant classification system. Candolle's arguments provided the empirical foundation for modern plant taxonomy. His views of positional relations of plant structures have been firmly established in several of his works including *Regni Vegetabilis Systema Naturale* (1818–21). Based on this concept he described 161 families, 5000 genera and 58,000 species in *Prodromus Systemis Naturalis Regni Vegetalis* (1824–1873), the first encyclopaedic work on plant classification of that time. Although Candolle's classification system was criticized later for its failure to classify the higher orders of the plant kingdom, it was the first step to a modern plant classification system, which greatly influenced the classification system proposed by George Bentham (1800–1884) and Joseph Dalton Hooker (1817–1911) and later by Charles E. Bessey (1845–1915). This school of thought has been widely accepted in modern plant classification, helping to develop new systems of classification proposed by Hutchinson (1884–1972).

1.5   Progress in Crop Anatomy

The major studies on early plant anatomy rarely included crop species, although there are some exceptions. Since the early attempts of plant anatomical studies primarily concentrated on plant systematics, most of the descriptions of the anatomy of crop species in the early literature were made with the objectives of classification. This is more evident in monocot species, because the majority of the food supply comes from only a few species (rice, wheat, maize, sorghum, sugarcane, millets) in the monocot family. Similarly, a few crop species were considered as model crops in a family, which interested early plant anatomists in describing these crop species. One such example is sunflower (*Helianthus annuus*), belonging to the family Compositae, which has been cited in a number of textbooks to describe general feature of this family. In certain cases, crop species were considered important for special structures, such as tubers of potato, rhizomes of ginger and turmeric, seed structure of oilseeds such as castor, or inflorescence of grass species.

Only in the 20th century has crop anatomy been given some importance. With the rising population, the question of producing more food was given higher priority in the mid-20th century, leading to more emphasis in research on crop science. Since plant anatomy was already established on a solid foundation, anatomy was readily accepted as a tool to investigate basic and applied crop science.

1.6   Modern Crop Anatomy – Emphasis on Structure Function Relation at Cellular Level

Plant anatomy has travelled a long path since the early classification systems. Modern plant anatomy finds numerous applications, one of which is the area of crop anatomy. The scientific studies of anatomy of crop plants for purposes other than plant classification have started recently, and its potential has not been well recognized either in basic science or in applied scientific fields such as agriculture. However, in the late 20th century, the physiological basis of plant productivity has been in the forefront of quests of agricultural scientists and related disciplines. The question of providing food security to billions of people on the earth in the near future is the most important question in front of mankind, where crop anatomy can be used as a handy tool to generate critical inputs in basic as well applied plant sciences for enhancing productivity.

References

Chowdhury, K.A. (1971) Botany: Prehistoric Period. In: Bose, D.M., Sen, S.N. and Subbarayappa, B.V. (eds) *A Concise History of Science in India.* Indian National Science Academy, New Delhi, pp. 371–375.

Eames, A.J. and MacDaniels, L.H. (1925) *An Introduction to Plant Anatomy.* McGraw-Hill Book Co., New York, pp. 321–342.

Kanjilal, D.K. (1999) A note on the Vrksayurveda of Parasara. *Indian Journal of History of Science* 34, 127–131.

2

Relevance of Anatomical Studies in Modern Crop Science

2.1 Introduction

Although plant anatomy is one of the oldest disciplines of plant science, its relevance in modern plant science has rather been underestimated as a basic microscopic technique for dissection of plant tissues. One problem of being an age old science is that major fundamental aspects have been discovered long ago. Thus further scientific progress becomes slower with time, unless the science finds application in new fields. On the other hand, this comes as an advantage to newcomers in the field in that the science is built up on solid foundation. Plant anatomy is built on concrete fundamental discoveries made in the past 350 years. We now have an enormous knowledge on the anatomy of land plants, aquatic plants and lower order species in plant kingdom. Based on this enormous knowledge bank, new research in the field is plunging into finer, in-depth studies of plant developmental structures and plant–environment interactions, as well as new, unforeseen applications of anatomy in plant science.

Unfortunately, the same cannot be said for crop anatomy. It has evolved as a secondary component of crop science helping other scientific disciplines related to crop science and technology. Prominent examples are the science of crop physiology or developmental biology. To a smaller extent, crop anatomy has also been used as a supportive line for plant genetics and crop improvement. Other than these, crop anatomy has primarily been used to justify categorization of crop plants and to generate basic information. Contrary to the general plant anatomy, information on anatomy of crop plants is limited to primarily major crop plants, such as rice, maize or cotton. Little information is available on anatomy of minor crops, such as spices, millets or many legumes.

In the following sections we will discuss both the scopes and relationship of plant anatomy in general and crop anatomy in specific with other disciplines of biological sciences. For clarification, we will chiefly draw attention to crop species. However, we will also resort to examples from non-crop species in certain cases to strengthen our arguments.

2.2 Anatomy in Plant Systematics

Plant systematics is the science of classifying plants in groups and study of the group relations based on certain characters. Historically, morphological characters have predominantly been used in plant systematics.

When plant anatomy was born as a scientific discipline, it was immediately adopted as a tool in plant classification. Wood anatomy was used by Auguste Mathiew in 1858 for description of forest plants in his book *Florae Forestiere* (Naik, 2006). Bureau in 1864 first used anatomical characters for plant systematics at taxa level in the family Bignoniaceae. Along with the morphological descriptions of plants and cytogenetic behaviour of chromosomes, anatomical descriptions (also referred as micromorphology) have been successfully used in classification of several land plants. The basis of classification was similar to morphology in that plants sharing common anatomical features are related. We can group these anatomical evidences in two distinct categories. In the first category, plant anatomy has been used as a primary basis for classification, being supported by other evidence. In most of these cases, anatomy has been used as a primary tool in higher order classification of land plants. One prominent example is the difference in the vascular system of monocotyledonous and dicotyledonous plants. The use of anatomy as the primary basis of classification is more common at genus, family or higher level. However, in certain cases, anatomy has been used as primary evidence in classification at species or subspecies level, mostly when morphological differences are insufficient for classification.

There are only a handful of cases where crop anatomy has been used as primary basis for the purpose of taxonomy and systematic studies. A major reason is that unlike non-crop plants, crop plants have been in closer association with humans, thus their morphological characteristics had been well studied. In many cases, plant classification initiated from these crop plants by noting and comparing their morphological features. Prominent examples are classification of rice, wheat, maize, sugarcane and sorghum in the family Gramineae, based on their similarity in reproductive as well as vegetative features. Only later, anatomy was used to confirm the classification pattern.

The second category, which is more common, includes cases where plant anatomy

has been used as secondary evidence after the primary classes have been established based on morphological characters. At species level characterization, morphological and cytogenetic characters are used predominantly for systematics, followed by confirmation through anatomical observations of different plant parts such as root, stem, leaves or reproductive organs. In many cases, the differences are not very prominent among closer relatives, although their morphological and cytogenetic behaviours are quite distinct. Rice and wheat, the two major food crop species of the world from Gramineae exhibit remarkable dissimilarity in vegetative morphology (many tillers versus few tillers), reproductive morphology (panicle versus spike), chromosome number ($2n=18$ versus $2n=42$), show no crossability with each other and are adapted to different climatic conditions (tropical versus temperate). However, the anatomical features of root, stem, leaves and seed are quite similar in rice and wheat, suggesting remarkable similarity in structural mechanism albeit their other differences. Such anatomical similarity persists in the other members of Gramineae, establishing the importance of anatomy as a key descriptor at family level.

Anatomy is particularly helpful in taxonomic studies of the herbarium material when morphological features are not sufficient to identify the taxonomic status of the plant material. This may happen when the morphological differences are insufficient, or when only parts of plant are available (pieces of leaves or stem).

Out of the major anatomical features of plants, anatomy of leaves has been utilized most in plant classification. Leaf anatomy is the most utilized anatomical descriptor in the families Gramineae, Euphorbiaceae, Zingiberaceae, Musaceae and Ericaceae. Root anatomy and root architecture are also utilized for taxonomic studies, although root characters are less utilized for plant systematics (Scatena *et al.*, 2005; Zobel and Waisel, 2010).

The anatomical variations of stems have been mostly utilized to differentiate the structural aspects of vascular bundles of monocotyledonous and dicotyledonous plants. Among the crop species, stem anatomy has been used for classification of various herbaceous spices in the family Umbelliferae and bast fibre crops in the family Malvaceae. Different anatomical features such as the distribution and thickness of collenchyma, distribution of fibre cells, structure of endodermis and pith cells and various arrangements of cells in vascular bundles are used in taxonomical classification of plants. Besides, anatomical features of wood development are frequently used in phylogenetic studies of forest species. Phylogenetic relationship may be drawn from structure of wood elements such as the lengths, breadth, inclination of vessel tips, and types of pits etc.

Besides the vegetative structures, reproductive anatomy including floral anatomy and seed anatomy are also important in plant systematics. The anatomy of floral arrangement (flower cross-section, described by the floral formula) is the key for Linnaean classification of the plant kingdom.

2.3   Structure–Function Analysis

The plant body develops vertically upward as shoot and downward as root perpendicular to a horizontal axis at ground level. The roots move down to the soil being further from the light source, while the shoots move upward towards the light source for capturing solar energy. Both the shoot and roots develop initially from the same meristematic tissue, but later differentiate and develop specific structures for performing their specialized works. The roots absorb water and nutrients from the soil and supply these to stem and leaves through the vascular bundles. The leaves on the other hand convert solar energy to chemical energy, which is used as food for further growth and overall development of the plant. Thus each plant part has a definite structure for performing definite functions. Each plant part is composed of cells and tissues in a definite arrangement delineated by its anatomical features. Consequently, there

is an intrinsic relationship between the anatomy of the plant parts and their functions. This concept was first outlined by Gottlieb Haberlandt, a prominent German botanist. In his epic work *Physiologische Pflanzenanatomie*, he described the structural anatomy and functional significance of plant tissues, pointing out for the first time that the anatomical features of each plant part reflect the specific function of that plant part. For example, the structural anatomy of the phloem tissues in the vascular bundle determines the extent and speed of solute transport. If the pores of sieve tubes in the phloem are small in size and less in number, then they will definitely transport less food material (sugar and ions) than the phloem tissues having more pores of larger size. Similarly, the structural anatomy of xylem vessels determines the rate of water and nutrient uptake. The number and distribution of tracheary elements are major determinants of water movement through xylem.

Study of anatomy has been used as a key tool in understanding several physiological processes of plants, such as water and nutrient uptake by xylem, photosynthesis and formation of sugars, collection and loading of foods in the phloem, solute transport by phloem from source to sink, cellular respiration, transpiration and many other physiological processes, including the formation and deposition of secondary metabolites. It is the plant's structural anatomy that allows the establishment and performance of physiological activities.

Variations in structural anatomy of crop plants from their wild relatives have been a major reason for selection of the particular crop species over their weedy relatives. In the grass family, the crop species have been selected for better seed production potential (rice, wheat, barley, oat, rye, maize, sorghum, pearl millet), which is determined by higher photosynthesis and channelling of food material from source to sink (vegetative to reproductive organs). The number of seeds in maize cobs is much higher than that of its wild relative teosinte. Similarly, rice produces a higher number of tillers and bears more grains in a panicle than the wild and weedy rice species. The thick noble cane varieties of sugarcane, on the other hand have been preferred for better vegetative growth and sugar production than the thin cane varieties and the wild *Saccharum* species, where food channelling occurs from vegetative (leaf as source) to vegetative (stem as sink) tissues. In both cases, robustness in structural anatomy has helped in better adaptation of these species as preferred crops.

2.4   Anatomy in Crop Ecology

Ecology is defined broadly as the science of interaction of an organism or a group of organisms to its surrounding environment. Ecological adaptation is an essential requirement of survival of any species. Plant species are adapted to different ecological conditions in nature. When a species is introduced in a new ecology, or the existing ecology changes, plants try to adapt to the new environment by modification of their own body plan to survive and reproduce in the altered environment. Such changes are induced by modification of genome constitution through mutation and recombination, higher order chromosomal changes, or changes at phenotype level. Plants with genetic constitution befitting to the environment survive to produce progeny adapted to the environment. The adaptation is manifested by changes in morphology, anatomy, growth or reproduction or a combination of these.

Plants when adapted under different environments show difference in anatomical structures. By interpreting such anatomical changes, one can not only understand the specific responses of anatomical structures of root, stem, leaf, flower and seed of a species under different ecology, but also study the common anatomical modifications taking place in several related and unrelated species under a particular ecology. For example, mesophytic plants surviving under optimum environmental conditions (light, water, nutrient) exhibit several common anatomical features. When many of these plants are adapted to xerophytic conditions

(less water, higher temperature), a number of common anatomical modifications are observed, such as a drastic reduction in the number of stomata on upper leaves, the density of trichomes on leaves forming a microclimate, development of cuticles on leaves, thickening of laminar cells, development of bulliform cells etc. On the other hand, under hypoxia (induced by waterlogging stress), the formation of roots having aerenchyma is a common mechanism for survival. Aerenchyma formation is under the control of ethylene accumulation in the collar zones, which is formed under hypoxic, anoxic or other stress conditions such as heavy soil prohibiting penetration of roots. Similarly, halophytic plants undergo special anatomical modifications, such as enlargement of cellular vacuoles, the formation of salt glands for storing NaCl, thickening of hypodermis and surface hairs to remove excess salt. Hydrophytic plants adapted to flooded conditions produce aerenchyma in tissue functioning as a supply of oxygen under anoxic condition.

Crop species have been adapted to different environments mostly due to migration of human population. The major adaptations of crop plants under extreme environments are many times common, such as aerenchyma formation under submerged condition in rice and maize. A very prominent example of adaptation to change in ecological conditions due to human migration is maize, which originated in the New World (Central and South America), but has been adapted to different environments of the Old World within a very short time period of few hundred years. Due to its special C_4 photosynthesis system, it has adapted easily to the tropical as well as humid subtropical conditions of African and Asian countries.

The study of ecological crop anatomy enables us to understand the basis of adaptation of crop species in different ecological conditions, including extreme environments such as drought, waterlogging, heat, cold stress, or biotic stress conditions. In the third section of this book, we will discuss how adaptation of crop plants under these conditions has affected the anatomical structures of crop plants, or how anatomy has helped in adaptation of some of the crop plants in extreme weather conditions.

Global warming and increase in CO_2 concentration is a major concern of present day agriculture. The ecological conditions for growth and survival of plants and animals in general are undergoing rapid change and are predicted to worsen in near future. Crop species in the future generation are going to face changed ecological conditions, which will affect crop productivity. Understanding structural and physiological changes to sustain in the future environment is a major research area in crop science. Experiments under artificially enriched CO_2 and observation on plants that sustain better under higher CO_2 concentration show that plants undergo distinct morphological, anatomical and physiological changes under such extreme conditions. Such modifications are more obvious on leaf structures, such as modification in stomatal density, alterations in thickness of leaf palisade parenchyma or increase in leaf thickness. Chapter 12 of this book describes possible effects of climate change on crop anatomy and consequences on crop productivity.

2.5 Anatomy in Developmental Studies

Plant anatomy is an indispensable tool in studying growth, development and differentiation. From a tiny seed, the angiosperms develop into a full plant having root, shoot, leaf and reproductive organs. Even during seed development, a single fertilized diploid cell develops and differentiates into a multicellular embryo bearing a cellular structure dedicated for shoot and root development. The anatomical features of the body parts of seed plants are well described in many plant anatomy textbooks. A brief description of anatomy of plant body parts is also provided in Chapter 3 of this book, which will help the readers to understand general plant anatomy as well as anatomy of crop plants.

Plant cells have higher plasticity than animal cells, i.e. the cells can differentiate from uncategorized cell mass to specific cells, de-differentiate from specific cell types to undifferentiated cellular mass and again re-differentiate into specific cell types. This makes the plant cell truly immortal, providing the conditions for growth and survival of the cells are met. In contrast, the animal cells differentiate into specific cell types, but cannot de-differentiate into non-specific cells once the cell fate is determined. The only exception to this category is the stem cells, which can be maintained for indefinite period and can be subjected to differentiate into other cell types. But the capability of re-differentiation is limited to plant cells only. This is the primary basis of vegetative reproduction of plants through stem cuttings, a very common form of propagation for horticultural plants. The ability to re-differentiate comes very handy under stressed conditions, such as formation of aerial roots under submerged condition for harvesting oxygen from the atmosphere.

However, although all plant cells are totipotent, not all the cells in a growing plant are actively dividing. The actively dividing cells are located in the meristematic regions leading to primary growth of the plant. The secondary growth results from the thickening of the differentiated tissues by formation of new layers of differentiated tissues or by maturation and thickening of the cells or deposition of specific materials in the intercellular cavities. Anatomical observation of the meristematic regions is the primary tool to investigate the growth pattern of the dividing tissues and pattern of formation of specific tissues. Anatomy of growth pattern of apical meristems (root and shoot) clearly differentiates the growing regions from the differentiated cells and also identifies distinct layers (L1, L2, L3 etc.) in the apical meristematic regions. These layers are not only structurally different, but also have different functional significance. Different layers form different types of tissues.

2.6 Anatomy in Plant Evolution

Plants have emerged on the planet for millions of years. During the course of evolution plants have advanced from simple structures to more complex organ differentiation. Land plants have evolved from aquatic plants by changing their internal structures for adherence to the soil by modifications in morphology and anatomy of root structures. Likewise, the water transport system of land plants evolved in a different way from those of aquatic plants. Consequently, the evolution of xylem vessel systems selected separate paths in aquatic and land plants. The land plants had to establish the plant stand by mechanical supports, leading to higher root number, root volume and anchorage. The body plan of land plants has also been modified to develop more structural tissues in definite orientations to provide higher mechanical support to the aerial parts. Moreover, since these plants are not in constant association with water, a number of advanced water loss protection mechanism evolved in these plants under different adaptation conditions. Anatomical investigations of terrestrial and aquatic plant body systems have generated numerous evidences for such modifications, which help us to understand the course of evolutionary progress in these two plant systems. The basic evolutionary differences are also very clear in aquatic, semi-aquatic and terrestrial crop species. The aquatic and semi-aquatic crop species such as lotus or water lily have tender stem structures with little mechanical support compared to land crop species. In contrast, both the monocotyledonous and dicotyledonous crop species derive mechanical support from collenchyma and sclerenchyma tissues, and many at times also from parenchyma tissues like tracheary elements. The variations in anatomical configuration of these elements are observable at higher order levels of crop classification (difference in plant architecture of pulse crops and beans), family level (growth habit of leguminous pulse crops like pea and lentil and leguminous green manure crops like *Sesbania* and *Crotalaria*, variation in fibre formation in cotton and jute), genus level (variation in development of bast fibres in *Hibiscus cannabinus* and

Hibiscus rosa chinensis), or at species level (stress tolerant versus susceptible genotypes).

2.7 Anatomy in Anthropology and Palaeobotany

Humans have used plant products as food, fuel, fibre or for construction purposes since prehistoric times. Anatomical investigations of fossil remains of plant parts used for household purposes by ancient civilizations or early human establishments provide important clues about the habitat and culture of that time. Archaeological and anatomical investigations of wood and charcoal samples from the excavated sites can also help in estimating the climatic conditions based on the study of growth rings. The cell walls of late-wood cells are preserved better than that of early-wood cells in remains of wood samples. Such anatomical studies are also helpful in understanding evolution of structural compartments of cells and tissues during evolution. For example, studies of carbonized samples of *Taxodixylon* species aged at 20 million years revealed that the fibrillar structures of cellulose are well conserved throughout the evolutionary period. Anatomical observation of charred remains of fossil plants also helps in identification of nearest relatives of these prehistoric plants in the present-day plant kingdom.

Since crop species have been domesticated early in civilization, investigations on remains of crop plants or seeds generate valuable information regarding the civilization. An important area of anatomical investigations is the assessment of remains of clothes. Early humans primarily used grass species for the purpose of clothes. With advancement of civilization, human population learned to extract fibre from plants and make durable clothes. Fibres from flax were used in producing clothes of very high value in the Egyptian civilization, which was considered as a royal commodity and beyond the reach of general people. The archaeological remains from the pyramids of clothes made of flax fibre support the royal status of this crop.

2.8 Anatomy in Understanding Genetics and Molecular Biology

Genetics is the science of heredity and variation of characters and genes responsible for their expression. Genetic analysis at molecular level is one of the most important areas of present-day biological research, aimed at the identification of genes responsible for various expressions, understanding of the structure and function of individual genes and genomes as well as gene expression in different tissues at different growth phases of organisms or under various external stimulations caused by environment or biotic agents. The expression and variation of anatomical traits are under control of various genes expressing during development. Once these genes are identified, they can be used to study anatomical structure of related species, decipher evolutionary relationship at species, genus or higher order level, to understand various biological mechanisms operating in the plant, or to target alteration of plant anatomy through plant breeding or direct gene transfer.

While the above examples show how molecular genetics can be used to understand the plant anatomy at molecular level and exploit the findings for better crop improvement, anatomy itself has been immensely helpful to gene identification and to understand gene function. There are two broad categories where anatomy has been used as a handy tool to study genetics. The first is the identification of genes through anatomical analysis of phenotypic mutants. Mutations are created by external agents or by error in DNA replication and repair mechanisms. Numerous developmental mutants and segregating progenies have been subjected to anatomical analysis, particularly in *Arabidopsis*, maize and rice to identify genes responsible for these traits.

The second category is the study of *in situ* gene expression in different plant parts.

Originally developed in understanding developmentally regulated gene expressions during embryogenesis in model animals like *Xenopus, Caenorhabditis* and *Drosophila*, the expression study has also been extensively used in the study of gene expression in different tissues and developmental stages in the model plant species *Arabidopsis*.

Transgene integration is routinely confirmed by the expression of GUS gene (β-glucouronidase), which develops a blue colour. Development of blue colour in anatomical sections of germinating seeds or other plant parts of transgenic plants confirms expression of the inserted marker gene, indicating that the transgene construct has been integrated in the host genome.

Mutations in floral anatomy in dicots like *Arabidopsis* and *Petunia* as well as monocots like maize and rice have been proved to be crucial for deducing the genetics of floral development and identification of homeotic genes in plants. Information from analysing the flower shape mutants tells us how the development of sepals, petals, stamens and pistils in flowers are genetically regulated, which is commonly known as the ABC model of flower development, based on the three groups of genes A, B and C, involved in fate determination of cells during development of the flower.

Anatomical studies are also extensively associated with gene expression analysis at protein level. The most common process of detection is immune-localization of proteins by using specific antibody and differential screening of protein expression in wild type and mutant genotypes. This helps in understanding the gradient of protein expression that is developmentally regulated, identifying protein signals in response to biotic and abiotic stimuli, and localization of expression of protein products of a specific transgene. Numerous studies have been conducted in *Arabidopsis*, maize, rice, pea, soybean and wheat using protein immune-localization in anatomical sections of different plant parts.

2.9 Anatomy for Understanding Programmed Cell Death in Plants

Programmed cell death (PCD) is a process of directed cessation of cellular activities and disintegration of cellular materials involved in various processes during the plant life cycle from embryogenesis to leaf and fruit senescence. In many cases, the dead cells preserve their structural integrity, while in other cases like endosperm cells the cellular materials are utilized in the development of other cells. Anatomical studies have helped in comparing the process of PCD in normal and PCD-defective mutants.

Several maize mutants produce distorted embryo or endosperm. A maize mutant type known as long cell (*lc*) exhibits abnormal cell elongation during embryogenesis by expanding only in the longitudinal axis. Anatomical observation using TUNEL assay showed that these mutants are incapable of PCD and follow a generalized cell death pattern (Graziano *et al.*, 2003).

2.10 Anatomy for Studying Postharvest Storage Life of Foods and Vegetables

Higher shelf-life, or postharvest storage life is a desirable character in many crop species, particularly when the plant products are harvested fresh (vegetables, fruits or flowers). Longer shelf life helps in better preservation of fresh foods, thereby reducing postharvest loss and cost of management. In the postharvest condition, fresh vegetables and fruits deteriorate due to rapid loss of water, change in temperature and damages caused by insect pests and diseases. High temperature causes water loss, cellular degradation and enhances ripening. Anatomical investigations help in understanding the extent of shelf life in various vegetable and fruit crops. It is a quick and efficient way to identify genotypic difference in shelf-life and desiccation tolerance. Since temperature plays an important role in maturation and degradation of fruits, the anatomical observations of fruit structures provide valuable information on the effect of temperature during storage and transport.

2.11 Anatomy for Understanding Digestibility of Fodder Grasses

Differences in cell wall composition of leaf blade tissues of fodder grasses affect the digestibility of the grasses in ruminants. A major part of the cell wall of the leaf is digested by aid of microbes present in the ruminant digestive system. The rate of microbial degradation depends on the anatomical structure of the leaf blades, and varies from genotype to genotype. Bermuda grass is a common fodder used for livestock animals. Studies have shown that certain varieties (e.g. 'Coastcross-1') are digested more rapidly by rumen microbes than other varieties.

2.12 Anatomy for Understanding Expression of Various Cellular Enzymes

Histological analysis of enzymatic activities is a common tool in both animal and plant science, having various applications from basic research to applied anatomy. *In situ* localization of enzymatic actions in plant cells are extensively used to understand host–pathogen interactions, the role of different genes and enzymes in developmental and differentiation processes, the identification of key enzymes regulating biological processes, and the detection and diagnosis of viral diseases. Enzyme expression studies are very helpful to understand the breakdown of storage reserves during germination, tissue speciation, fruit ripening, leaf senescence and many other related processes.

2.13 Anatomy for Understanding Effect of External Stimulations on Crop Growth

Plant responses to several external and internal stimulations are often clearly described by anatomical studies. Several plant hormones such as auxin, gibberellins (GA), cytokinins, ethylene, abscisic acid (ABA) and other chemicals trigger many physiological and developmental processes in plants. In crop species, these processes ultimately influence productivity and economic importance of the crop species. For example, seed dormancy is a protective mechanism evolved in plant species having sexual reproduction, so that the seed can germinate under favourable seasonal and climatic conditions. In many crop species seed dormancy is considered to be a hindrance as the dormancy has to be broken for cultivation. Seed dormancy is regulated by the balance of GA and ABA. Moreover, during seed maturation, germination of the embryo is undesirable as the germinating embryo will get no nourishment and will die. On the other hand, a mature seed of a crop should germinate as per the desire of the cultivator. In maturing seeds germination is prevented by a higher concentration of ABA, while in mature seed, the concentration of ABA reduces and the concentration of GA increases, promoting germination of seed under favourable conditions. Anatomical observations have been very helpful in elucidating the roles of these two hormones during different stages of seed maturation and germination.

The effects of external application of gibberellic acids are well established in various crop plants. In rice, it induces cell division and formation of air cavities in the vascular bundle, increases lignification and reduces the size of the vascular channels. During flowering, application of gibberellic acids hastens the panicle exertion, which is commercially utilized in hybrid rice seed production (Satya and Singh, 2010). In the root, GA_3 inhibited the activity of the cambium oriented towards the phloem, thus reducing root growth.

References

Bureau, E. (1864) Monographie des Bignoniacees. Dissertation, Paris, pp. 164–169.

Graziano, E., Bastida, M., Stiefel, V. and Puigdomenech, P. (2003) Longcell, a mutant from maize producing a distorted embryo and generalized cell death. In: Nicolas, G., Bradford, K.J., Côme, D. and Pritchard, H.W. (eds) *The Biology of Seeds, Recent Research Advances*. CABI, Wallingford, UK, pp. 25–32.

Naik, V.N. (2006) *Taxonomy of Angiosperms*. Tata McGraw-Hill, India. 21st Reprint.

Satya, P. and Singh, A.K. (2010) Hybrid rice: concepts, methodologies and recent developments. In: Maiti, R.K. and Sarkar, N. (eds) *Advances in Rice Science*. New Delhi Publishers, India, pp. 155–182.

Scatena, V.L., Giulietti, A.M., Borba, E.L. and Van Den Berg, C. (2005) Anatomy of Brazilian Eriocaulaceae: correlation with taxonomy and habitat using multivariate analyses. *Plant Systematics and Evolution* 253, 1–22.

Zobel, R.W. and Waisel, Y. (2010) A plant root system architectural taxonomy: a framework for root nomenclature. *Plant Biosystems* 144(2), 507–512.

3

Techniques of Crop Anatomy Study

3.1 Introduction

The minute internal structure of the plant body is identified mainly from thin sections taken by hand or microtome and from maceration in which the individual cells are freed from one another. Plant microtechnique consists of three procedures (Fig. 3.1): (i) preparation of plant tissues for microscopic study; (ii) use of the microscope and related equipment for critical study and interpretation of the materials; and (iii) detailed descriptions of each part. There are different steps to study the anatomy of plant parts, which are mentioned below.

3.2 Collection of Desired Plant Materials and Subdivisions

For anatomical study, desired plant organs, such as stems, leaves, floral organs, should be cut with a sharp knife or scalpel and transported between wet blotting paper. In the case of dried herbarium specimens, the specimen can be softened for sectioning to prepare slides. This helps in determining the distinguishing features of vascular arrangement (Hyland, 1941).

3.3 Fixation and Storage

For future anatomical study, the specific organs need to be fixed with reagents to stop the life processes without disintegration of cell structures and organelles. Different fluids are used for fixation, such as glacial acetic acid, 1% chromic acid, formalin (30–40%); the most common fixative is FAA (formalin acetic alcohol).

3.4 Section Cutting

Section cutting is carried out by hand with the use of a sharp razor blade and the assistance of a microtome. The section must be very thin and should not be oblique. Different types of section cutting are used to study plant organs, such as a: (i) transverse section (TS): the plant organ root, stem, petiole, leaf etc. is cut transversely (perpendicular 90°) to its axis; (ii) longitudinal section (LS): the plant organ is cut longitudinally (180°) to its axis; (iii) radial longitudinal section: the plant organ is cut transversely, then cut through the radius; and (iv) tangential longitudinal section: the plant organ is cut tangentially. Sections are made by cutting perpendicularly to the radius.

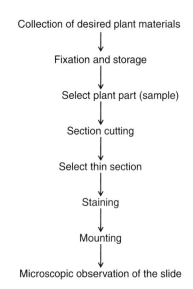

Collection of desired plant materials
↓
Fixation and storage
↓
Select plant part (sample)
↓
Section cutting
↓
Select thin section
↓
Staining
↓
Mounting
↓
Microscopic observation of the slide

Fig. 3.1. Basic steps in microscopic observation of plant parts.

Paraffin blocks are prepared for sectioning by microtome. Excellent paraffin sections as thin as 2 μm can be cut at –15 to –20°C in a caryotome (Greenwood and Berlyn, 1968).

3.5   Processing and Embedding

Fresh cut sections need to be passed through a series of dehydration processes before the final preparation of slides. The procedure is as described below.

1. Fresh sections are kept in a small conical watch glass in water.
2. 30% ethanol for 20 min + light green (1% in water).
3. 50% ethanol + 2–3 drops of safranin (1% in 50% ethanol) for 25 min.
4. 70% ethanol for 30 min.
5. 100% ethanol for 15 min.
6. 100% ethanol + xylol (1:1) for 20 min.
7. Pure xylol 5 min.
8. Clove oil 5 min.
9. Mounting in Canada balsam, DPX.

For studying lipid materials, acetone is used as the dehydrating and infiltration agent and ester wax is used for the tissue embedding medium (Chayen and Gahan, 1959).

3.5.1   For preparation of sectioning by microtome

The tissue part, leaf, stem or floral part is passed through a series of dehydration procedures using different grades of alcohol, starting from 30% to 100% ethanol, and ultimately in xylene. When the particular tissue is completely dehydrated, it is then embedded in the required grade of paraffin solvent.

3.5.2   Preparation of whole mounts and macerations

Dry preparation of plant samples is done through herbarium preparation; for wet preparation, museum jars are used. Chemical and mechanical separation of cells is done by maceration. The best known preservation fluid is 70% ethanol, although formaldehyde is also an excellent preservative. The most useful concentration for bulk preservation contains 5 parts formaldehyde (35–40% strength) in 95 parts water.

3.5.3   Slide of whole mount

Temporary and semi-permanent slides are used. It is better to mount material in 10% glycerin. More durable slides of algae and fungi are mounted in lactophenol solutions.

3.6   Preparation of Slide

Preparation of the slide includes staining and mounting the selected section. Different tissues or cells take different types of stains based on the chemical composition, e.g. sclerenchyma takes safranin and appears pink in colour.

3.7   Stains Used for Microscopy

The most commons stains that are used are aniline blue, fast green, safranin, cotton blue, methylene blue or crystal violet.

3.7.1   Mounting media

Media used for mounting vary between glycerine 10%, glycerine jelly, lactophenol, erythrosine or Canada balsam (or DPX mounting medium), depending whether they are for temporary or permanent preparations.

3.7.2   Safranin

Safranin is a very common vital stain. It is prepared by mixing safranin (0.25 g) with 10 ml of ethanol (95%) and 100 ml of distilled water.

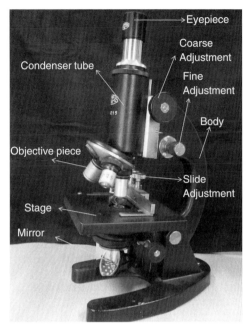

Fig. 3.2. Parts of a compound microscope.

3.8   Microscope

A microscope is used to study the anatomical structure of plant parts. Several instruments, such as a dissecting microscope, compound microscope, phase contrast microscope and electron microscope, can be used in the field of anatomy for observing the internal parts of the tissues or cells. An electron microscope is an advanced piece of equipment used for observing the internal part of cells. The function of a microscope is to produce an enlarged image of an object, which is accomplished by a system of lenses.

3.8.1   Compound microscope

A compound microscope has a metallic body, a condenser tube, a concave mirror for the light source, an objective piece, an eyepiece, a stage for positioning the slide, and a slide adjustment with both coarse and fine adjustment (Fig. 3.2).

3.8.2   Types of microscopy

- Dark field: the object under study appears self-luminous on a dark background.
- Phase contrast: light passes through a thin unstained biological specimen; there is very little amplitude differential between the background light and the light diffracted by the specimen.
- Fluorescence microscope: fluorescent substances can be used in this technique. First the specimen is treated with fluorescent dye. The light source should have a low wavelength; UV rays are used.
- Electron microscopy: the electron microscope was discovered by Knoll and Ruska in 1931. In this type of microscopy, electrons are used to illuminate a specimen to create an enlarged image. The specimen should be 500 nm/0.5 μm in thickness. The condenser lens is placed between the electron source and the specimen and is responsible for focusing the electron beam on the specimen. There are three types of electron microscope:
  - Scanning electron microscope (SEM): this is mainly used to magnify the

surfaces of solid specimens. The beam current is very low (10^{-10}–10^{-12} amp), so even delicate biological surfaces can be examined with a minimum of heat damage.

- Scanning transmission electron microscopy (STEM): it uses a focused incident probe across a specimen that has been thinned to facilitate detection of electrons.
- Reflection electron microscopy (REM): an electron beam is incident on the surface. The reflected beam of elastically scattered electrons is detected.

3.9   Maceration of Wood/Fibres

Wood is a hard substance and the cells are closely attached to each other. For studying the components of wood the cells must be free from one another. For this a technique called maceration is used to separate the cells. Place a piece of wood in a glass test tube. Add a mixture of concentrated HNO_3 (10%) and chromic acid (10%). Heat the material in a test tube on a burner for 10–15 min and then cool the test tube in running water at normal temperature for 20 min. Remove the piece of wood and wash with water and place the wood on a slide and press it to separate the elements. Cover with a cover slip and observe under the microscope. Different types of vessels, parenchyma and fibres can then be examined.

3.10   Novel Technique for Studying Epidermal Layer from a Leaf

It is difficult to visualize the epidermal system because it is very thin in most plants. The surface view of the epidermal tissue system of a leaf is studied by the epidermal peeling or impression method.

3.10.1   Method

Epidermal pealing can be done directly on the leaf of the plant in the field, so we can obtain the epidermis with the stomata open and in a natural condition. For this a simple technique can be used:

1. Take a small amount of xylene (C_6H_4($CH_3)_2$) in a glass Petri dish.
2. Gradually dissolve thermocol (a polystyrene commonly available in shops) in the xylene and stir with a glass rod. Progressively add more thermocol to the solution and stir with a glass rod until it turns into a honey-like solution (gummy).
3. Apply the solution by finger once only on the lower and upper surfaces of the leaf at different places.
4. Allow the solution to semi-dry.
5. Put a transparent cellophane tape on the semi-dried area of the leaf.
6. Remove the tape gently and paste on to the labelled microscope slide.
7. Observe the slide under low and high power for the epidermal tissue system.

3.11   Estimation of Pollen Viability

Different methods for the estimation or determination of pollen viability are available. Three generally used techniques are: (i) the iodine potassium iodide (IKI) technique; (ii) tetrazolium; and (iii) the Evans blue method.

3.11.1   Iodine potassium iodide technique

This is a very simple and quick method for the determination of pollen viability of a plant. Take 1:2 ratio of iodine and potassium iodide solution and mix in a container. Take the anthers and squeeze with a needle. Place the pollen on a glass slide and add 1–2 drops of IKI, and leave for 10–15 min. Examine the slide under the microscope for staining. Fertile pollen takes stain and appears dark blue in colour. Count the number of stained pollen grains in the microscopic field. Unfertile pollen grains appear yellow in colour. Take at least 20 readings, then take the average percentage of stained pollen grains to give the percentage of pollen viability.

3.11.2 Tetrazolium test

2,3,5 Tetrazolium chloride seems to be a better alternative for the test. Immerse the pollen grains in the above chemical and immediately cover with black paper and incubate in the dark for 30 min to avoid oxidation. Viable pollen becomes stained and can be counted.

3.12 Microchemical Tests

Microchemical tests reveal with more or less accuracy the chemical nature of important structures of protoplasm. For studying histochemistry the section should be thick. Certain chemical tests are useful in detecting basic substances, e.g. the iodine potassium iodide test for starch, Osazone test for sugars, Fechling's solution test for reducing sugars, Millon's reagent test for proteins, and Sudan III for fats and oils.

3.12.1 Cellular differentiation

In different plant tissues, the basic structure of the plant cell presents modifications, both in the cell membrane, the form of cells and the contents, which show the microscopic characteristics of medicinal plants and are important in the identification and the detection of adulterants. During the differentiation of the membrane there are chemical modifications that originate changes in their physical properties, among which are the accumulation of cellulose or hemicelluloses and the incrustation of the wall for lignin or cutin and suberin.

3.12.2 Reactions for cellular membrane

The following colour reactions of cell membranes vary according to the different concentrations of cellulose, hemicelluloses and pectins present in the wall.

- Chloro-zinc-iodide reaction: the blue colour is observed in the presence of cellulose, and a yellow colour with pectin.

The membranes that contain these substances in distinct proportions also appear as a blue, pale violet or brown colour.

- Iodine reaction: gives a light blue colour reaction with cellulose and deep blue when hemicelluloses are present.
- Ammoniacal solution of copper oxide reaction: cellulose is dissolved in the solution, but hemicellulose is not dissolved. The dissolved cellulose when treated with sulphuric acid gives a blue-coloured precipitate of cellulose, which bleaches to white. Commonly used to test cellulosic fibres.
- Phlorglucinol with HCl reaction: in the membrane, cellulose is observed as a red colour.

3.12.3 Reaction for lignified membranes

Lignin is a hard substance, which impregnates the cell membrane of the tracheids, fibrous vessels etc., and chemically is a complex phenyl propanic polymer (C8–C3). For the detection of the lignified membrane the following reactives are utilized:

- Aniline acid sulfate: with this stain the membranes are stained shining yellow.
- Phlorglucinol with HCl reaction: the lignified membrane is stained rose or red.
- Chloro-zinc-iodide: gives a yellow colour in the lignified membrane.

3.12.4 Reactions for suberized and cutinized membranes

Suberin and cutin are substances that are mainly found formed by the highly polymerized fatty acids. Suberin is found in the cells of phellem and in the endodermic cells, while cutin forms a secondary deposit on the cells of cellulose (membrane), but may also be cutinized. Suitable reactives for suberin and cutin are:

- Sudan-glycerine: coloured red on heating. The reagent is prepared with 0.01 g of Sudan III in 5 ml OH and 5 ml glycerine.

- Chloro-zinc-iodide: this gives a yellow-brown colour.
- Dilute aqueous paint: stains the membrane red.
- Concentrated sulphuric acid: does not dissolve suberin or cutin.
- Oxidizing agents: at room temperature, chromic acid does not give a result.

3.12.5   Reactions for chitinous membranes

Chitin constitutes the greater part of the membranes of insects and crustaceans. It does not give a reaction with the reactives utilized for cellulose and lignin. On heating with 50% potash at 160–170°C for 1 h, it is converted into ammonical chitinose and acids; this may dissolve in acetic acid at 3%, the chitinose precipitating when a small excess of alkali is added. The chitin gives a violet colour when it is treated with a solution at 50% potassium iodide followed by 1% sulphuric acid.

3.12.6   Determination of callose in sieve plate

Callose forms a film in the mature sieve plates, which forms a stopper and blocks completely the sieve plate. Thus the sieve tubes may be identified by the presence of callose. Callose is also deposited in pollen tubes, anthers and also in reaction to wounding. Reactions for the determination of callose are:

- Solution of coralin: stains the callose red.
- Aniline blue: stains the callose blue.
- Chloro-zinc-iodide: stains callose brownish-red.
- Solution of nitrate of ammonical copper: does not dissolve callose.
- Solution of potash: in 1% dissolves the callose in cold water.

References

Chayen, J. and Gahan, P.B. (1959) An improved and rapid embedding method. *Quarterly Journal of Microscopic Science* 100, 275–277.

Greenwood, M.S. and Berlyn, G.P. (1968) Feulgen cytophotometry of pine nuclei: effects on fixation role of formalin. *Stain Technology* 43, 111–117.

Hyland, F. (1941) The preparation of stem sections of woody herbarium specimens. *Stain Technology* 16, 49–52.

4

General Anatomy of Crop Plants

4.1 Introduction

The study of the gross internal structure of plant organs by the technique of section cutting is called plant anatomy. Anatomy is the branch of science that deals with the internal structures and organization of organisms and is also known as internal morphology. Theophrastus is known as 'the father of anatomy'.

Anatomy is a recent branch when compared with morphology and taxonomy. The rapid development of the study of anatomy took place only after the development of magnification through microscopes. The specialized study of cells has been treated as an independent branch called cytology. The study of anatomy now includes minute internal structures such as tissues and tissue systems and gross internal structures such as stele, vascular skeleton etc. The study of tissues and tissue systems is called histology; therefore the term histology should not be used as a synonym of anatomy.

4.2 Fundamental Parts of the Plant Body

The plant body consists of two main systems, the root system and the shoot system. Both of these systems have a common axis.

The root system is generally underground and positively geotropic. The chief functions of the root system include fixation of the plant in the soil, absorption of mineral water and conduction of mineral water to the root system.

The shoot system is the aerial part of the plant and is negatively geotropic. The main axis of the shoot system is called stem. The stem bears branches, leaves, buds and flower etc. The stem is distinguished into nodes and internodes. The leaves develop over the stem at the nodes only. The main functions of the shoot system include displaying the foliage on branches, the conduction of mineral water from the root system and food material to the root system, and gives mechanical strength to the plant body (Fig. 4.1).

The axis consists of two parts: the portion that is normally aerial is known as the 'stem' and the portion that is subterranean is called the 'root'. The leaves are characteristic of the stem and do not occur on the root.

4.3 Development of the Plant Body

A vascular plant begins in existence as a morphologically simple unicellular zygote ($2n$). The zygote develops into the embryo and thereafter into the mature sporophyte.

Fig. 4.1. Morphology of flowering plant (cotton) showing different parts.

The development of the sporophyte involves division and differentiation of cells, and an organization of cells into the tissues and tissue systems. The embryo bears a limited number of parts, generally only an axis bearing one or more cotyledons. The cells and the tissues of this structure are less differentiated. However, the embryo grows further, because of the presence of the meristems, at two opposite ends of the axis, the future shoot and root. After the germination of the seed, during the development of the shoot and root, new apical meristems appear, which cause vegetative growth and the reproductive stage of the plant is attained.

The first-formed plant body is known as the *primary plant body*, since it is built up by means of the first or *primary growth*. The tissues of the primary plant body are known as primary tissue; for example, the first formed xylem is called primary xylem. In most vascular cryptogams and monocotyledons, the entire life cycle of the sporophyte is completed in the primary plant body. The gymnosperms, most dicotyledons and some monocotyledons show an increase in thickness of the stem and root by means of secondary growth. The tissues formed as a result of secondary growth are called secondary tissues. The bulk of the plant increases because of secondary growth; in particular, the vascular tissues are developed, which provide new conducting cells and additional support and protection.

A special meristem, the cambium, is concerned with the secondary thickening. The cambium arises between the primary xylem and the primary phloem, and lays down the new xylem and phloem adjacent to these.

In addition, a cork cambium or phellogen commonly develops in the peripheral region of the axis and produces a periderm, a secondary tissue system assuming a protective function when the primary epidermal layer is disrupted during the secondary growth in thickness.

The primary growth of an axis is completed in a relatively short period, whereas the secondary growth persists for a longer period, and in a perennial axis the secondary growth continues indefinitely.

4.4   Outline of the Cell Structure

The plant body consists of a number of small, microscopic, box-like compartments called cells (Fig. 4.2), which are the fundamental basis of the life of all organisms (except viruses). The cells are the universal elementary units of organic structure. A plant cell may be defined as a microcosm having a definite boundary, or the cell-wall within which occur complicated chemical reactions. A cell devoid of this chemical reaction is considered to be a dead cell. There are several parts that constitute cells (Figs 4.3, 4.4), which are tabulated below.

4.5   Tissue

In unicellular and colonial forms each cell performs all the life activities. As the evolution proceeds, the complexity of the organism increases and specialization of cells

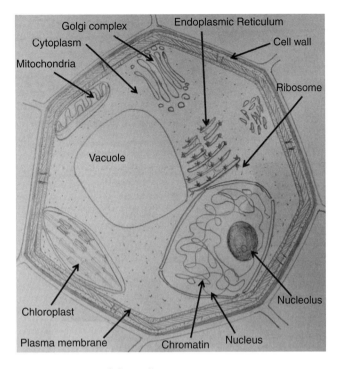

Fig. 4.2 Diagrammatic representation of plant cell.

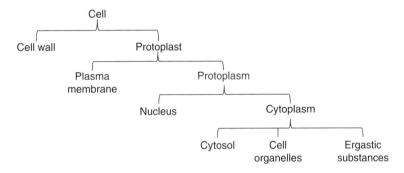

Fig. 4.3. Basic outline of plant cell structure.

begins. Cells performing similar functions organize into groups. A group of such cells, which are similar in origin, structure and function, is described as tissue. In fact these tissues are formed in response to the basic division of labour.

In complex organisms cells organize into tissues, which in turn organize into tissue systems. Various tissue systems constitute the organ and different organs form the organism.

Various types of tissues are found in the plant body (Fig. 4.5). Based on the activity, all the tissues of plants are classified into mainly two types, meristematic tissues and permanent tissues.

1. Meristematic tissues are groups of immature cells, which are primarily concerned with the production of new cells.

2. Permanent tissues are groups of differentiated cells, which have lost the power of

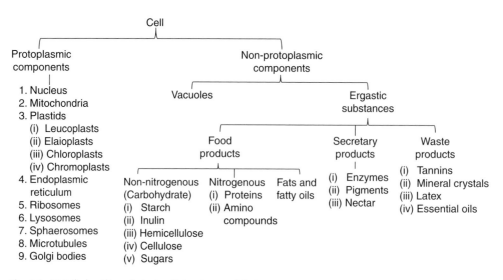

Fig. 4.4. Detailed outline of plant cell structure and their components.

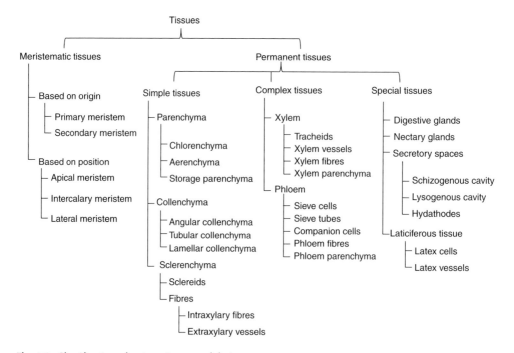

Fig. 4.5. Classification of various tissues and their components.

division and acquired a definite shape to perform a specific function.

Based on the origin, position, function, stages of development, and cell division and growth, meristematic tissue is classified into various types (Fig 4.6).

4.6   Tissue Systems

Various types of tissues in the plant body organize into tissue systems. All the tissues of a plant that perform the same function together form a tissue system.

Sachs (1875) classified the tissue systems into three types (Figs 4.7, 4.8): (i) epidermal tissue system; (ii) ground tissue system; and (iii) vascular tissue system.

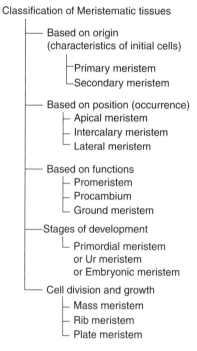

Classification of Meristematic tissues

— Based on origin
(characteristics of initial cells)
 ┬Primary meristem
 └Secondary meristem

— Based on position (occurrence)
 ├ Apical meristem
 ├ Intercalary meristem
 └ Lateral meristem

— Based on functions
 ├ Promeristem
 ├ Procambium
 └ Ground meristem

—Stages of development
 └ Primordial meristem
   or Ur meristem
   or Embryonic meristem

— Cell division and growth
 ├ Mass meristem
 ├ Rib meristem
 └ Plate meristem

Fig. 4.6. Types of classification of meristematic tissue.

4.7 Anatomy of Plant Organs

Different organs of the plant body consist of several tissues forming tissue systems. These tissues differ from one part to other, i.e. in the arrangement and amount of the tissues.

4.7.1 The root

Introduction

The root system generally develops beneath the surface of the soil. The roots are generally divided into two categories: (i) primary, normal roots; and (ii) adventitious roots. Primary roots originate from the embryo and generally persist throughout the plant's life. Adventitious roots arise secondarily from stem, leaf or other tissues and may be either permanent or temporary.

Functions of root

Primary roots help the plants in anchoring to the soil and absorb water and soluble substances. The functions of adventitious roots vary. Absorption takes place for the most part by diffusion though the cell walls of the root hairs. Water is also absorbed through the epidermis of the root in the region distal to the zone of root hairs. Generally, the older roots are incapable of

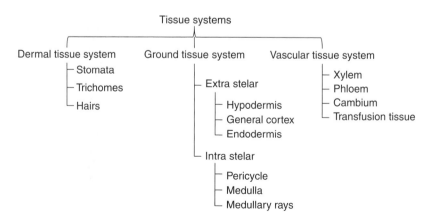

Tissue systems

Dermal tissue system   Ground tissue system   Vascular tissue system
├ Stomata   ├ Xylem
├ Trichomes   ├ Extra stelar   ├ Phloem
└ Hairs   ├ Cambium
                          ├ Hypodermis   └ Transfusion tissue
                          ├ General cortex
                          └ Endodermis

                        └ Intra stelar
                          ├ Pericycle
                          ├ Medulla
                          └ Medullary rays

Fig. 4.7. Tissue systems of a flowering plant.

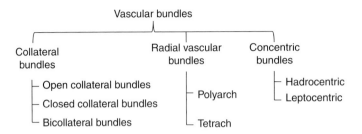

Fig. 4.8. Types of vascular bundles.

absorption and serve only for conduction, support and storage.

Anatomy of dicotyledonous root

Dicot plants show a taproot system. In the root hair region it shows the primary structure; behind the root hair region secondary growth is initiated.

The initial organization of the primary dicot root is simple when compared to that of the stem. Transverse section of the root shows three main parts: the epidermis, cortex and stele (Figs 4.9, 4.10).

1. The epidermis of the root is also called the rhizodermis, epiblema or siliferous layer, and is the outermost region of the root. It is uniseriate, having a single layer of compactly arranged thin-walled living cells. Both cuticle and stomata are absent. Epidermis gives protection and absorption of water and minerals. Epidermal cells produce unicellular root hairs, which are useful in increasing the surface area for absorption of water and helps in absorbing water from soil. Generally root hairs are short lived. Epidermis ruptures after secondary growth.
2. The cortex is the middle region, lying between the epidermis and the stele (Fig. 4.10). Cortex is well developed and is larger and thicker than the stele. It is multiseriate and relatively homogeneous. It consists of general cortex and endodermis.

(i) The general cortex is extensively developed in the root. It consists of loosely arranged thin-walled parenchyma cells with prominent intercellular spaces. The cells are usually round or oval. The cells are colourless and store starch.

This region helps in lateral conduction of water and salts. After secondary growth the general cortex ruptures.

(ii) The endodermis is very distinctive in roots. It is single layered, having compactly arranged barrel-shaped cells. Endodermal cells are characterized by the presences of Casparian strips or Casparian bands. Casparian bands are made of suberin. In transverse section, Casparian bands appear as lens-shaped thickening on the radial walls of the endodermis. Endodermis is also called the starch sheet layer, as starch grains are present in the endodermal cells. The root endodermis acts as a barrier between the cortex and the stele. Endodermis ruptures due to secondary growth.

3. The stele consists of: (i) pericycle; (ii) vascular strands; and (iii) pith.

(i) The pericycle is the outermost region of the stele. It is uniseriate and parenchymatous. Cells of the pericycle retain meristematic activity. Lateral roots arise endogenously from the pericycle and help in absorption of water. During secondary growth, pericycle forms phellogen or cork cambium and a small amount of vascular cambium.

(ii) Vascular strands are composed of the vascular tissues, xylem and phloem. Xylem conducts water and salts, whereas phloem transports food materials. Xylem and phloem are arranged in the form of separate strands on different radii. They alternate with each other. Hence the vascular strands in the root are described

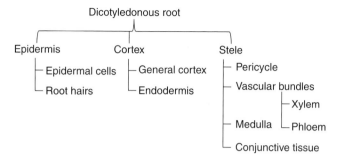

Fig. 4.9. Outline of the dicot root.

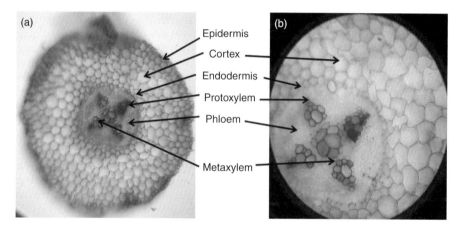

Fig. 4.10. (a) Transverse section of young dicotyledon (cotton root 40×) showing epidermis, multilayer cortex and single layer endodermis. (b) Sector enlarged (100×) showing xylem strands (tetrarch condition).

as separate and radial. Xylem is exarch having protoxylem towards the pericycle and metaxylem towards the centre and scattered secondary xylem vessels. Cambium is absent and hence vascular bundles are described as closed type.

During secondary growth pericycle becomes meristematic and divides, gives rise to cork cambium or phellogen, and produces a few brownish layers of cork cells or phellem towards the outside, and the phelloderm on the inside. Secondary xylem and secondary phloem are formed due to the activity of the cambium ring. After secondary growth the secondary vascular tissues form a continuous cylinder and usually the primary xylem becomes embedded in it. Primary xylem is located at the centre and primary phloem elements are generally seen in the outer region (Fig. 4.11).

(iii) Pith or medulla: the non-vascular tissue present between the xylem and phloem strands is called conjunctive tissue. It may be totally parenchymatous or part of it may be converted into sclerenchyma. During secondary growth the cambial cells that originate from the pericycle lying against the group of protoxylem function as ray initials and produce medullary rays. These rays are transversed through cambium in the xylem and phloem; this is a characteristic feature of roots.

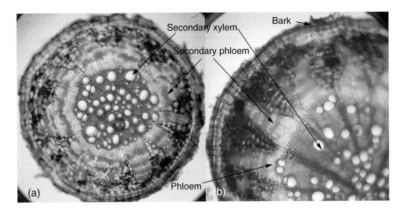

Fig. 4.11. (a) Transverse section of matured dicotyledonous (cotton 40×) root showing secondary growth. Secondary xylem vessels are larger in size. (b) Sector enlarged (100×) illustrating bark and the cortex is compressed due to increase in secondary structures.

Anatomy of monocotyledonous root

Most of the monocot roots do not show secondary growth. The root is almost circular in outline and shows mainly epidermis, cortex and stele (Figs 4.12, 4.13).

1. The epidermis is the outermost region of the root. It is uniseriate, having a single layer of compactly arranged thin-walled living cells (Fig. 4.13). Both cuticle and stomata are absent. Most epidermal cells produce tubular outgrowths called root hairs. The epidermis of the root gives protection and also helps in absorption of water and minerals. Root hairs are especially useful for absorption of water.

2. The cortex is well developed and it is larger and thicker than the stele. It is multiseriate and relatively homogeneous and has parenchyma cells. It is separated into the general cortex and endodermis.

(i) The general cortex consists of loosely arranged thin-walled parenchyma cells with prominent intercellular spaces. The cells are usually round or oval. The cells are colourless and store starch, and also help in lateral conduction.

   As the epidermis disintegrates in mature roots, exodermis is formed for protection. The outermost layers of the cortex become thick walled and suberized. They are arranged compactly and give protection. This region is called the exodermis.

(ii) The endodermis is the innermost layer of the cortex. It is very distinctive in monocots compared to dicots. It is single layered with compactly arranged barrel-shaped cells. Endodermal cells are characterized by the presence of lignosuberin thickening, called the Casparian band. When endodermal cells become thick-walled, thin-walled passage cells are left opposite to protoxylem regions. They facilitate the entry of water and salts into the stele from the cortex.

3. The stele is the central part of the root. The stele consists of pericycle, vascular strands, medulla and conjunctive tissue.

(i) The pericycle is the outermost region of the stele. It is uniseriate, having a single layer of thin-walled parenchymatous cells, which are compactly arranged without intercellular spaces. The cells of pericycle become sclerenchymatous in mature roots. Later roots arise endogenously from the pericycle and it does not give rise to vascular cambium or cork cambium.

(ii) Vascular strands comprise vascular tissues of xylem and phloem. Xylem conducts water and salts and phloem transports food materials. Xylem and phloem are arranged in the form of separate strands on different radii and alternate with each other. Hence the

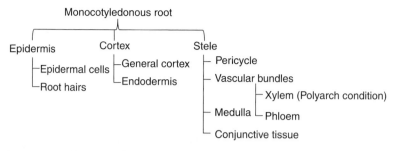

Fig. 4.12. Organization of monocotyledonous root.

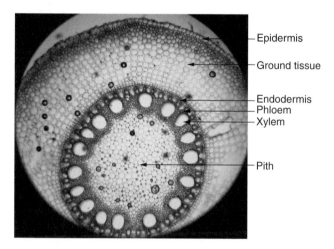

Fig. 4.13. Transverse section of monocotyledonous (maize) root (100×) showing epidermis, multilayered ground tissue (cortex), endodermis, phloem, xylem (polyarch) and central pith.

vascular strands in root are described as separate and radial. Xylem is exarch, having protoxylem towards the pericycle and metaxylem towards the centre or medulla (centripetal development). Xylem is polyarch, having more than eight protoxylem points. Cambium is absent and hence vascular bundles are described as closed type.

(iii) Pith or medulla: pith is absent. The non-vascular tissue present between xylem and phloem strands is called conjunctive tissue. It may be totally parenchymatous or a part of it may be converted into sclerenchyma. During secondary growth the cambial cells that originate from the pericycle lying against the group of protoxylem function as ray initials.

(iv) The medulla is distinct and usually parenchymatous. The medulla stores food materials.

(v) Conjunctive tissue: the non-vascular tissue left between xylem and phloem strands is called conjunctive tissue. It may be parenchymatous or sclerenchymatous.

4.7.2 The stem

Introduction

The part of the axis of the plant that is aerial in nature and consists of branches, leaves and reproductive structures is called the stem. The stem possesses nodes and internodes, which are absent in the root system and differ in vascular system. The difference lies mainly in the arrangement of the xylem and the phloem. In the root, the strands of primary xylem and phloem lie in different radii, separated from one another. In the stem strands lie side by side in the same radius, i.e. they are conjoint, collateral. The xylem of the root is always exarch, whereas that of the stem is exarch, endarch or mesarch. The endarch condition is the most commonly occurring type in present-day plants.

Origin of the stem

During the development of the embryo the first stem, the meristem, organizes. The fully developed embryo commonly bears an axis, the hypocotyl–root axis. The axis consists of one or more cotyledons at its upper extent, called shoot primordium, and root primordium covered by a root-cap at the lower extent. The radicle (embryonic root) is found at the lower end of the hypocotyl and the embryonic shoot is found above the insertion of the cotyledons. The embryonic shoot consists of an axis bearing un-extended internodes and one or more leaf primordia, commonly called the plumule, and its stem part is termed the epicotyl. The origin of shoot organization is found in the hypocotyl-cotyledon stem. Hypocotyl is the first stem unit of the plant and the cotyledons are the first leaves, and is located below the cotyledonary mode.

During the germination of the seed, the root meristem forms the first root.

Anatomy of dicotyledonous stem

A transverse section of primary stem shows three regions, the epidermis, cortex and stele (Fig. 4.14). Dicot stem shows secondary growth.

1. The epidermis is the outermost region surrounding the stem. It is uniseriate, having a single layer of compactly arranged barrel-shaped living cells. The outer walls of the epidermal cells are cutinized. Moreover, the epidermis is covered by a separate layer of cutin, called cuticle, on its outer surface (cuticularized). The cuticle reduces transpiration. Few stomata are present in the epidermis for exchange of gases and transpiration. All the epidermal cells are colourless except the guard cells. Multicellular hairs are present on the epidermis. The epidermis ruptures after secondary growth. In aged plant species such as trees, dead bark tissue occupies the place of epidermis.

2. The cortex is the middle region present between the epidermis and stele. Cortex is smaller and thinner than the stele. It shows a hypodermis, a middle cortex and an endodermis.

(i) The hypodermis is the outermost region of the cortex. It is multilayered, with two to six layers, and collenchymatic. Collenchyma is living mechanical tissue. Hypodermis gives considerable strength, flexibility and elasticity to young stems. Having chloroplasts, it may carry out photosynthesis.

(ii) The middle cortex is also multilayered and parenchymatic. The cells may be

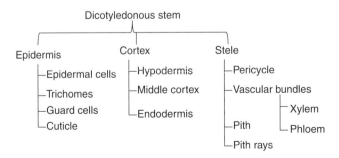

Fig. 4.14. Outline of structure of dicot stem.

round or oval with prominent intercellular spaces. This region also stores food materials temporarily and flavonoids are deposited.

(iii) The endodermis is the innermost layer of the cortex (Fig. 4.15). It is usually distinct in stems without Casparian bands. In young stems the innermost layer of the cortex contains abundant starch, hence this layer is called the starch sheath. The starch sheath layer is homologous to endodermis, but morphologically unspecialized, hence such a layer may be called endodermoid. After secondary growth the endodermis ruptures.

3. The stele is the central region of the stem and is larger and thicker than the cortex. It consists of pericycle, vascular bundles, and medulla and medullary rays.

(i) The pericycle (perivascular region) is the outermost non-vascular region of the stele. It is multiseriate and sclerenchymatic. Sclerenchyma is mechanical tissue and gives strength and rigidity to the stem.

(ii) Vascular bundles comprise vascular tissues of xylem and phloem, in which 15–20 vascular bundles are arranged in one ring called the eustele. Each vascular bundle is wedge-shaped. It consists of xylem towards the centre of the axis and phloem towards the periphery. In the vascular bundle xylem and phloem are present together (conjoint) on the same radius (collateral), with a strip of cambium (fascicular cambium) between xylem and phloem (open). Hence the vascular bundle is described as conjoint, collateral and open type. Xylem is endarch with protoxylem present towards the centre. The xylem consists of many vessels. Xylem vessels are arranged in rows.

(iii) Medulla and medullary rays: this region is distinct, present at the centre of the stem and is usually parenchymatous. The cells are round or oval with intercellular spaces. Medulla stores food materials. The extensions of parenchymatous pith between vascular bundles are called primary medullary rays,

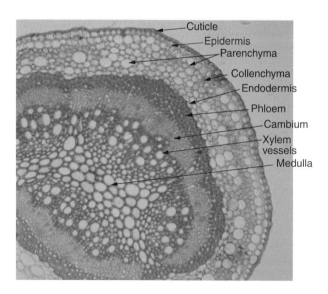

Fig. 4.15. Transverse section of dicotyledonous stem (*Amaranthus* 100×) showing uniseriate epidermis followed by four to five layers of collenchyma and single layered endodermis. The stele is at the centre of the stem, consisting of medulla and vascular bundles. Secondary growth has begun with cambium forming secondary tissues.

which help in lateral conduction and may gives rise to secondary meristem (inter-fascicular cambium).

Monocot stem

Generally monocots lack secondary growth. The stem is usually circular in outline in transverse section and is differentiated mainly into epidermis, cortex and stele (Fig. 4.16).

1. The epidermis is the outermost region of the stem, and is meant for protection. It is uniseriate, having a single layer of compactly arranged barrel-shaped living cells. The outer surface of the epidermis is covered by thick cuticle, which checks transpiration. Few stomata are present in the epidermis for the exchange of gases and transpiration. All the epidermal cells are colourless except the guard cells.

2. The cortex is highly reduced and usually represented by a narrow hypodermal region. In mature stems, hypodermis is two- to three-layered sclerenchymatic tissue and gives mechanical strength. In young stems, chlorenchymatic patches are present just below the stomata region and perform photosynthesis.

3. The stele is the central part of the stem and is larger and thicker than cortex. The stele in the monocot stem, with numerous vascular bundles scattered irregularly in the ground tissue, is called an atactostele, and is the most advanced type of stele. The stele consists of ground tissue and vascular bundles.

(i) Ground tissue: in monocot stems usually there is no clear demarcation between the cortex and pith since endodermis,

pericycle etc. are indistinguishable. The whole region is parenchymatic and is best described as ground tissue (or conjunction tissue).

(ii) Vascular bundles are numerous and are scattered irregularly in the ground tissue (Fig. 4.17). Usually the peripheral bundles are small and closely arranged, and the central bundles are large and are widely arranged. Each vascular bundle is oval in shape and is surrounded by fibrous bundle sheath. It is many cells thick towards outer and inner regions and only a few cells thick at lateral regions. Hence the vascular bundles are described as fibrovascular bundles. Vascular bundles are conjoint, collateral (with xylem and phloem present on the same radius) and closed (i.e. without cambium). The xylem is present towards the centre of the axis and consists of a few vessels, which are arranged in the form of letter 'Y'. The protoxylem often disintegrates and forms protoxylem lacuna. Phloem is present towards the outside, between the metaxylem vessels, and consists of sieve tubes, companion cells and a few parenchymatous cells at the sides.

4.7.3   The leaf and the petiole

Introduction

The leaf is a laterally growing organ on the stem. It performs various metabolisms, which include photosynthesis, respiration

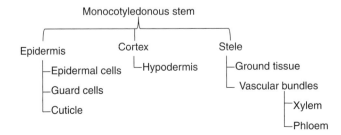

Fig. 4.16. Outline of monocot stem.

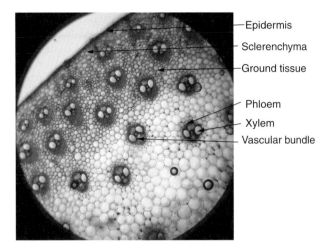

- Epidermis
- Sclerenchyma
- Ground tissue
- Phloem
- Xylem
- Vascular bundle

Fig. 4.17. Transverse section of monocot (maize) stem (100×) showing scattered vascular bundles in ground tissue (parenchyma). Note the peripheral vascular bundles are smaller and inner (central) vascular bundles are somewhat larger in size.

and transpiration. To effectively carry out these metabolisms, leaves have mesophyll consisting of chlorophyll and vascular tissues.

Leaves show variation in anatomy based on the habitat. Usually, leaves show two types of structures: dorsi-ventral leaves and isobilateral leaves.

1. Dorsi-ventral leaves: in this types of leaf morpho-anatomical variations are present in the dorsal and ventral surfaces. The dorsal surface is dark green in colour and is directly exposed to sunlight and the ventral surface is light green in colour. Mesophytic leaves show dorsi-ventral symmetry. These types of leaves are generally present in dicotyledons, e.g. cotton and sunflower.

2. Isobilateral leaves: do not show any variations either morphologically and anatomically between the dorsal and ventral surfaces. The leaf blade is nearly erect and both surfaces usually receive direct and equal amounts of sunlight. The internal structure is more or less similar in both the upper and lower halves. This type of leaf is generally present in monocots and hydrophytes, e.g. leaves of grasses.

Anatomy of dorsi-ventral leaf (dicot)

A typical mesophytic dicot leaf shows a dorsi-ventral nature, i.e. the leaf blade consists of distinct dorsal and ventral surfaces. In transverse section the leaf shows three distinct regions, the epidermis, mesophyll and vascular bundles (Fig. 4.18).

1. The epidermis is mainly for protection and is present on both surfaces of the leaf blade. The epidermis covering the upper surface of the leaf is called the upper epidermis or ventral epidermis or adaxial epidermis. The epidermis present towards the lower side of the leaf is called the lower epidermis or dorsal epidermis or abaxial epidermis. Epidermis is single layered, having compactly arranged barrel-shaped (tabular) cells. The outer walls of the epidermal cells are cutinized. Moreover, the epidermis on its outer surface is covered by a continuous layer of cutin called cuticle. The cutinized outer walls and cuticle reduce the loss of water due to transpiration. Anisocytic stomata are present in both of the epidermal layers, but more stomata are usually present in the lower epidermis. Stomata facilitate the exchange of gases between the leaf and environment. All the epidermal cells except guard cells are colourless. Epidermis contains multicellular hairs.

2. The ground tissue of the leaf, present between the upper and the lower epidermal layers, is called mesophyll (Fig. 4.19). The mesophyll is chlorenchymatic and is concerned mainly with photosynthesis. The mesophyll is differentiated into an upper palisade tissue and lower spongy tissue.

(i) Palisade tissue is present below the upper epidermis (only in upper plane). The palisade cells are thin walled, cylindrical and contain numerous chloroplasts. The cells are compactly arranged in one layer, perpendicular to the upper epidermis. Plants showing xerophytic characters have palisade tissue in two planes e.g. *Neerium* and *Eucalyptus*. Narrow intercellular spaces are present in the palisade tissue. Palisade tissue is the most highly specialized type of

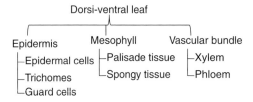

Dorsi-ventral leaf

Epidermis   Mesophyll   Vascular bundle
 ├Epidermal cells ├Palisade tissue   ├Xylem
 ├Trichomes   └Spongy tissue   └Phloem
 └Guard cells

Fig. 4.18. Anatomy of dorsi-ventral leaf (dicotyledon).

photosynthetic tissue. The upper surface of the leaf is dark green in colour due to palisade tissue. Compactly arranged palisade tissue reduces the high transpiration loss of water and offers drought resistance.

(ii) Spongy tissue forms the lower part of the mesophyll present towards the lower epidermis. Cells of the spongy tissue are thin-walled, irregular in shape and arranged loosely with large continuous intercellular spaces. Stomata open into large intercellular spaces called sub-stomatal chambers. The cells of spongy parenchyma contain fewer chloroplasts, hence the lower surface of the leaf is light green in colour.

3. Vascular system: the leaf receives the vascular supply from the stem. The vascular tissue occurs in the form of discrete bundles called veins. Veins are interconnected to form reticulate venation. Vascular bundles (veins) occur between the palisade and spongy tissue (median in position). They are collateral and closed. In the vascular bundle, xylem is present towards the upper

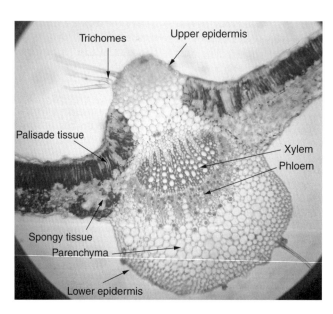

Fig. 4.19. Transverse section of dorsi-ventral leaf (cotton 100×) showing organization of tissues. Stellate trichomes on upper and lower epidermis, palisade towards upper and spongy tissue towards lower epidermis can be observed. Central vascular bundle can be observed, consisting of xylem towards the upper epidermis and phloem towards the lower epidermis.

epidermis and phloem towards the lower epidermis. The vascular bundle is surrounded by parenchymatous bundle sheath. The cells of the bundle sheath are called border parenchyma. The larger vascular bundles are supported by hypodermal parenchymatous strands. These strands are considered to be the sheath extension.

Anatomy of isobilateral Leaf

A typical monocot leaf shows isobilateral nature. The monocot leaf shows three distinct regions, the epidermis, mesophyll and vascular tissue (Fig. 4.20).

1. Both sides (upper and lower) of the leaf are covered by epidermis. The upper epidermis may be easily identified due to the presence of xylem and bulliform cells composed of large and small cells. Under water

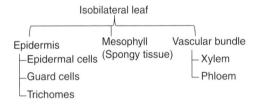

Fig. 4.20. Outline of the transverse section of an isobilateral leaf.

stress, smaller cells lose water and as a result the leaf coils and rolls thereby preventing transpiration. The epidermis is uniseriate and composed of more or less oval cells (Fig. 4.21). The outer walls of the epidermal cells are cutinized and also covered by cuticle. Cuticle reduces the loss of water due to transpiration. Stomata are covered by two dumbbell-shaped guard cells. The graminaceous type of stomata is present in both epidermal layers, but the stomata are usually more concentrated in the lower epidermis. The epidermis is rough in nature due to the presence of silica crystals. Silica crystals offer drought resistance by covering some parts of the epidermis and form a physical barrier. Bulliform cells are present in the upper epidermis.

2. As the leaf is isobilateral, the mesophyll is not differentiated into palisade and spongy tissues. It is composed of compactly arranged, thin-walled, isodiametric chlorophyllous cells, having well developed intercellular spaces among them.

3. The vascular tissue occurs in the form of discrete bundles called veins. The vascular bundles are collateral and closed. Most of the bundles are small in size but fairly large bundles also occur at regular intervals. The xylem is found towards the upper side and phloem towards the lower

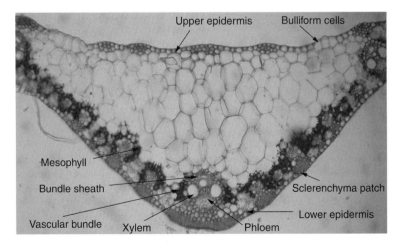

Fig. 4.21. Transverse section of isobilateral leaf showing different parts. Note: sclerenchyma patches are more frequent in lower epidermis. Vascular bundles are situated just above the lower epidermis. Parenchyma occupies at the centre of the leaf.

side of the bundles. In C$_4$ plants such as maize, sorghum and sugarcane, each bundle remains surrounded by a bundle sheath consisting of thin-walled parenchyma cells. The cells of the bundle sheath generally contain large chloroplasts, grana is absent and starch grains are present. This special structure was first identified by Haberlandt in 1882 (Haberlandt, 1914) and it is known as Kranz anatomy. Kranz anatomy is the special character of C$_4$ plants, and is one of the adaptations for drought resistance in plants of semi-arid zones. This type of arrangement is absent in C$_3$ plants such as rice. Kranz anatomy is also found in a few dicots, such as *Amaranthus*.

Anatomy of petiole

The tissue of the petiole may easily be compared with the primary tissues of the stems. Transverse sections of petiole show four regions, the epidermis, hypodermis, ground tissue and vascular bundles (Fig. 4.22).

1. The epidermis is uniseriate, having a single layer of compactly arranged barrel-shaped living cells. The outer walls of the epidermal cells are cutinized, the cuticle helping to reduce the transpiration. Multicellular hairs are present on the epidermis.

2. The multilayered (two to six layers) hypodermis formed of collenchyma cells is found immediately beneath the epidermis (Fig. 4.23). Collenchyma is living mechanical tissue but

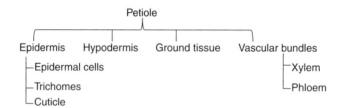

Fig. 4.22. Outline of petiole transverse section.

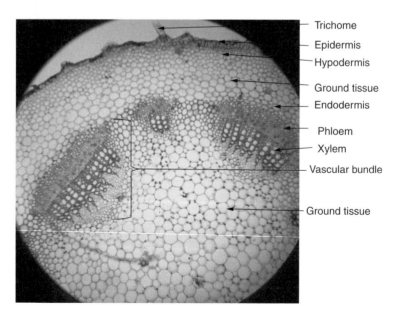

Fig. 4.23. Transverse section of cotton petiole (small portion 100×) showing different parts. New vascular bundles are developing and show similar structure to the stem.

flexible, being composed of cellulose and pectin. The hypodermis gives considerable strength, flexibility and elasticity to young petioles. Having chloroplasts it may carry out photosynthesis. Due to the presence of the thick layer of angular collenchyma, the petiole is strong, stiff and gives strength.

3. The ground tissue occurs just below the hypodermis, and consists of thin-walled parenchymatous cells having well defined intercellular spaces among them. Vascular bundles are arranged in half rings scattered in ground tissue.

4. The vascular bundles are of various sizes in the same petiole. Each vascular bundle is wedge-shaped. In the petiole, the xylem is always found towards the upper side whereas phloem occurs towards the lower side (as in the leaf).

Features of special interest include:

- Generally a groove is present towards the upper side of the petiole.
- Typically the vascular bundles are distributed in a semicircle in ground tissue.
- Generally the central bundle is largest and remains encircled by the endodermal sheath.
- The xylem is always present towards the upper side and phloem towards the lower side.

4.7.4 Anatomy of floral parts

Introduction

The flower is morphologically a determinate branch (stem) with crowded appendages, the internodes being much shortened, or in many cases obliterated. The appendages are modified in function and appearance and the upper ones are often so placed as to appear terminal on the axis. Anatomically, however, the flower is a typical stem with appendages; in no fundamental way does it depart structurally from the normal stem with leaves. It is actually a modified stem.

PARTS OF THE FLOWER AND THEIR ARRANGEMENT   The flower consists of an axis known as a receptacle and lateral appendages. The appendages

are called floral parts or floral organs. The sepals and petals constitute the calyx and corolla, respectively. The calyx and corolla are the sterile parts. The stamens consist of an androecium, where free or united carpels compose the gynoecium (Fig. 4.24). The stamens and the carpels are the reproductive parts.

The flower shows definite growth. In the flower the apical meristem ceases to be active after the formation of floral parts. In more specialized flowers there is a shorter growth period and they produce a smaller and more definite number of floral parts than the more primitive flowers. In still more advanced flowers there are specialized characters, such as whorled arrangement of parts instead of parts within a whorl, zygomorphy instead of the actinomorphic condition, and epigynous instead of hypogynous condition.

The sepals show similar structure to leaves in their anatomy. Each sepal consists of ground parenchyma, a branched vascular

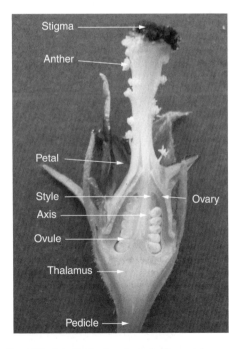

Fig. 4.24. Longitudinal section of a flower (okra – hypogynous flower), showing different parts. Ovules are enclosed in the ovary and are attached to the central axis, which arises from the thalamus.

system and an epidermis. The chloroplasts are found in the green sepals but there is usually no differentiation into palisade and spongy parenchyma. They may contain cells, laticifers, tannin cells and other idioblasts. The epidermis of sepals may possess stomata and trichomes.

The petals also resemble leaves in their internal structure. They contain ground parenchyma and more or less branched vascular system and an epidermis. They may also contain crystal-containing cells, tannin cells, laticifers and certain other idioblasts. They contain pigments containing chromoplasts. Very often, the epidermal cells of petals contain volatile oils, which emit the characteristic feature fragrance of the flowers. In certain flowers the anticlinal epidermal walls of the petals may be convex or papillate. The epidermis may also contain stomata and trichomes.

Usually the stamen contains two lobed four-loculed anthers. The anther is situated on a slender filament that bears a single vascular bundle. In certain primitive dicotyledonous families the stamens are leaf-like and possess three veins, whereas in advanced types they are single veined.

The structure of the filament is quite simple. The vascular bundle is amphicribral and remains surrounded by parenchyma. The epidermis is cutinized and bears trichomes. Stomata may also be present on the epidermis of both anther and filament. The vascular bundle is found throughout the filament and culminates blindly in the connective tissue situated in between the two anther-lobes.

The outermost wall layer of the anther is the epidermis. Just beneath the epidermis there is endothecium, which usually contains strips or ridges of secondary wall material mainly on those walls that do not remain in contact with the epidermis. The innermost layer is composed of multinucleate cells; this is nutritive in function and known as tapetum. The wall layers that are located between the endothecium and tapetum are often destroyed during the development of pollen sacs. On the maturation of the pollen the tapetum disintegrates and the outer wall of the pollen sac now consists of only the epidermis and endothecium. At the time of dehiscence of the anthers the pollen are released out through the stomium.

Structure of the pollen grain

Pollen grains are haploid (x), unicellular and uninucleate male spores, each of which has two wall layers, an outer thick exine and inner thin intine. Exine may be smooth or ornamented. In the exine, one or more unthickened slit-like or circular areas called germ pores are present. The position of germ pores is constant and has taxonomic value (Fig. 4.25).

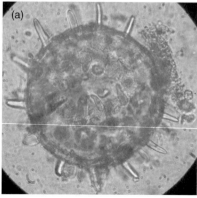 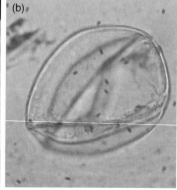

Fig. 4.25. Structure of the pollen grain (400×): (a) okra: spherical in shape containing spine-like structures and ornamentation; and (b) maize: smooth, isobilateral and without any ornamentation.

Each pollen grain divides to form two unequal cells, a vegetative and a generative cell. Pollination occurs at this two-celled stage. Further development of male gametophyte takes place on the stigma and style.

Gynoecium

A single unit of a gynoecium is called a carpel. A flower may possess a solitary carpel or more than one. If two or more carpels are present they may be united or free from one another. When the carpels are united the gynoecium is known as syncarpous; a gynoecium with a single carpel is said to be apocarpous. The apocarpous gynoecium is termed a simple pistil and the syncarpous gynoecium is termed a compound pistil.

The vascular skeleton of the flower

In structure, the petiole is a typical stem, herbaceous or woody, with a ring of vascular bundles or with an unbroken cylinder of vascular tissue. In the receptacle the stele is modified in shape, often expanding and becoming like an inverted or flattened cone or pyramid. From this receptacular stele depart the traces to the various floral organs, traces which in origin, structure, and behaviour are similar to those of leaves. Gaps accompany the exit of traces, and the crowding of the organs thus breaks the cylinder up into a network of strands. The traces pass off successively to sepals, petals, stamens and carpels, according to the manner of arrangement of these parts in the flower.

4.7.5 Anatomy of seed

Introduction

Seeds are the dispersal and propagation units of the Spermatophyta, which includes gymnosperms (conifers and cycads and related species) and angiosperms (flowering plants). Seed is an integumented fertilized ovule consisting of embryo. The constituent parts of a seed include a seed coat, cotyledons, endosperm and embryo. The embryo is a young sporophyte, diploid

($2n$), formed as a result of fertilization. The mature embryo consists of cotyledons, hypocotyl (stem-like embryonic axis below the cotyledons) and radicle (embryonic root). Endosperm is formed due to triple fusion; it consists of food storage tissue, triploid ($3n$) as a result of double fertilization; two-thirds of the genome is of maternal origin.

Classification

Based on the endosperm the seeds are classified into two types.

1. Endospermic seeds: the endosperm is present in the mature seed and serves as a food storage organ. Testa and endosperm are the two covering layers of the embryo. The amount of endosperm in mature seeds is highly species-dependent and varies from an abundant endosperm layer (*Nicotiana tabaccum*) to a single layer (*Arabidopsis thaliana*).
2. Non-endospermic seeds: the cotyledons serve as the sole food storage organs as in the case of sunflower (*Helianthus annuus*). During embryo development the cotyledons absorb the food reserves from the endosperm. The endosperm is almost degraded in the mature seed and the embryo is enclosed by the testa.

Seed coat

The seed coat is the outer protective layer of the seed, developed from the integuments of the ovule, diploid maternal tissue. The seed coat covers the embryo as protection. The seed coat consists of two layers, the testa and the tegman.

1. The testa is the outermost layer of the seed coat, and is formed from the outer integument of the ovule. It is usually hard and impermeable to water and sometimes contain ornamentations. Testa contains macro- and microsclereids. Macrosclereids are arranged very compactly and are dumbbell-, boat-, bone-shaped (osteosclereids). Microsclereids are small in size with a square shape and dentate edges. Hardness of the seed coat depends on thickness of the testa and the arrangement of sclereids. Rate of absorption of water (imbibition) depends on the thickness of the testa. Very hard and thick testa causes seed dormancy in some plants (Fig. 4.26).

2. The tegman is the inner layer of the seed coat. It is formed from the outer integument and is thin and soft. It covers the embryo.

Cotyledons

Cotyledons are also referred as primary leaves. Monocots have one cotyledon and dicots have two cotyledons. Cotyledons are made up of parenchyma tissue in the form of chlorenchyma. It consists of food materials such as starch granules (*Oriza*), proteins (*Pisum*) and oil (*Arachis*).

Endosperm

Endosperm is made up of parenchyma, which in turn consists of storage tissue such as aleurone layers in *Oryza* and starch granules in maize.

Perisperm

Diploid maternal food storage tissue originates from the nucellus. Remaining food materials unused by the embryo are known as perisperm. Usually perisperm consists of large parenchyma cells with stored food materials like oil granules and tannins. It is observed only in some species, e.g. *Coffea arabica* and *Beta vulgaris*.

More special structural features found in seeds

- A raphe is a small ridge-like structure on the seed coat formed from adnate funicle.
- The hilum is the funicular scar on the seed coat that marks the point at which the seed was attached via the funiculus to the ovary tissue.
- The chalaza is the base of an ovule, bearing an embryo sac surrounded by integuments. It is at the non-micropylar end of the seed.
- The operculum is a little seed 'lid', and refers to a dehiscent cap of a seed or a fruit that opens during germination.
- A micropyle is a minute pore formed by integuments towards the distal end of the ovule. It forms a canal or hole in the coverings of the nucellus through which the pollen tube usually passes during fertilization. Later, when the seed matures and starts to germinate, the micropyle serves as a minute pore through which water enters.
- Fibrous: a seed with a stringy or cord-like seed coat, as mace in *Myristica*.
- Funicular: a seed with a persistent elongate funiculus attached to seed coat, as in *Magnolia*.
- The aril develops from the funicle part of the ovule, and is considered as a third integument. In *Myristica fragrans* the aril is a branched ribbon and covers the seed.
- The strophiole is a small outgrowth of the hilum region, which restricts water movement into and out of some seeds, e.g. bean.
- A caruncle is a small cushion-like outgrowth present on the micropylar end of the seed. It is formed by the over

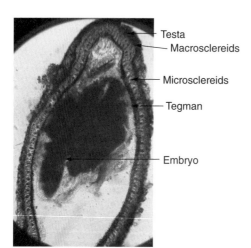

Fig. 4.26. Transverse section of cucumber seed showing inner embryo covered by outer testa and inner tegman. Macrosclereids are almost rectangular in shape, microsclereids are small dot-like structures.

Testa
Macrosclereids

Microsclereids

Tegman

Embryo

growth of the integuments on the micropylar end. The caruncle helps to absorb water during seed germination.

4.7.6 Anatomy of embryo and young seedling

Anatomy of embryo

The morphology and underlying anatomy of the embryo in mature seed varies considerably among species (Figs 4.27, 4.28).

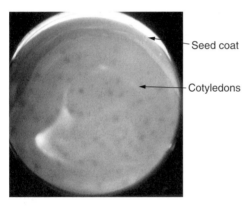

Fig. 4.27. Transverse section of cotton seed (40×) illustrating dark seed coat and cotyledons showing black spots indicating gossypol glands.

Embryo development begins in the embryo sac, following fertilization and division of the zygote. In general, the mature angiosperm embryo consists of an embryonic axis, with either a single cotyledon (monocotyledons) or a pair of cotyledons. There are apparent exceptions, e.g. in some species of Apiaceae and Ranunculaceae the cotyledons are fused at maturity. In angiosperms the number of cotyledons rarely exceeds two, though individuals of some species may develop three or more cotyledons (*Magnolia grandiflora*). Gymnosperm embryos frequently have more than two cotyledons, sometimes as many as 12 (polycotyledonous), e.g. *Pinus, Pinea*, depending on species.

The embryonic axis bearing the cotyledons commonly shows a polarity from the earliest stages of embryogenesis, with the proximal end towards the micropyle being the radicle containing the root meristem, and at the distal end of the shoot the meristem (plumule), sometimes with recognizable leaves. The radicle is usually adjacent to the micropyle, through which it emerges on completion of germination. The first shoot segment above the cotyledons is the epicotyl, and the region of the axis between the radicle and the point of insertion of the cotyledon is the hypocotyl. The cotyledons are often well developed

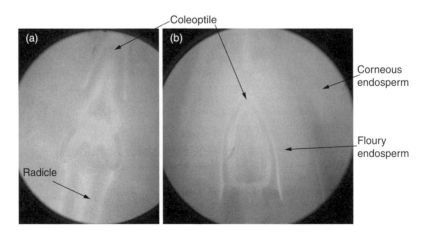

Fig. 4.28. Transverse section of maize (40×) seed showing coleoptile and radicle (a) and corneous and floury endosperm (b).

and serve as food storage organs in non-endospermic seeds.

The embryo contains thin-walled parenchyma cells. The cells are round or polyhedral in shape with intercellular spaces, and often packed with starch or aleurone. Some isolated areas of embryo bear procambium and a few mature protoxylem and protophloem.

Anatomy of cotyledon

Cotyledons consist of traces and are varying in number. Lower taxa of both monocotyledons and dicotyledons have only two traces. Four traces are common, with transition associated with the fusion of the middle pair. Sometimes three vascular bundles are common. The larger numbers of traces are most likely rare (e.g. *Canna*).

Many monocotyledons consist of two strong lateral traces that supply the sides of the sheathing leaf base. The median vein that supplies the cotyledon tip bends downward to the scutellum. The lateral veins continue downward to the sheathing wings. The tip of the median bundle remains unbranched in small and cylindrical forms. Two or three traces are present in most of the cotyledons. All the vascular bundles continue as veins toward the tip.

Cotyledons of angiosperms containing an even number of traces indicate the primitive condition. Cotyledons of the primitive taxa (e.g. Ranales, Liliales and Helobiales) have two strong traces, often with pairs of lateral traces. Even-numbered traces are a characteristic feature of gymnosperms. Odd-numbered traces of angiosperms are derived from even-numbered traces.

The vascular system of the cotyledon of monocots contains two major bundles or two with additional vascular bundles, and dicots have generally three major bundles.

Families shows a constant number of traces (Table 4.1).

Table 4.1. Families showing number of traces.

Number of traces	Family
2	Most of the lower monocotyledonous families
2	Liliaceae
2	Zingiberaceae
2	Amaryllidaceae
1, 2–3	Iridaceae
1 to several	Araceae
Several	Cannaceae
2	Ranales

Mesocotyl

Mesocotyl consists of hypocotyl and the adnate part of the cotyledon and forms a compound structure. It has the vascular tissues of hypocotyls and vascular bundles of cotyledon. Adnation of the cotyledon neck to the hypocotyl is frequently found and is observed externally and internally. In some genera, the presence of a longitudinal ridge on the axis of the embryo is considered as external evidence for fusion. Presence of a vascular bundle longitudinally in the cortex of hypocotyl is taken for internal evidence. The vascular supply of the scutellum is like an inverted V. In the cross-section of the mesocotyl, vascular bundles appear inverted, the xylem external to the phloem due to the bending down of the tip of the cotyledon.

In a monocotyledonous embryo the scutellum is considered as the cotyledon, and mesocotyl is considered as the first internode of the stem. The coleoptile represents the first leaf and the hypocotyl is represented only by a plate of tissue.

Cotyledonary sheath is the cap-like or sheath-like structure that encloses the plumule and the basal part of the cotyledon. Generally this sheath is called the coleoptile. Anatomical studies reveal the coleoptile is the basal part of the cotyledon.

Most of the grasses have small, very short plate-like hypocotyl and received little attention anatomically. In some monocots it is long and prominent and in the embryo.

The hypocotyl of dicots is long compared to monocots and is cylindrical or plate like. Poorly differentiated hypocotyl occurs in orchids, saprophytes and parasites.

Anatomy of seedling root

Diarch or tetrarch condition of the vascular cylinder is observed in the primary root of the seedling. In primary roots the monarch condition is rarely found. Monocots generally consist of polyarch condition and dicots with triarch, tetrarch or pentarch condition (Fig. 4.29). The tetrarch condition is considered as the basic type, since it is associated with arborescent taxa. Diarch is observed in most of the herbaceous taxa. The Ranunculaceae contains the diarch condition and the woody families show the tetrarch condition in roots.

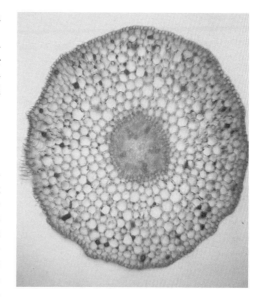

Fig. 4.29. Transverse section of seedling root (cotton 100×) having five regions of protoxylem (pentarch condition). Tissues are not very differentiated and parenchyma occupies majority parts.

References

Haberlandt, G. (1914) *Physiological Plant Anatomy*. Macmillan, London.
Sachs, J. (1875) *Text Book of Botany, Morphological and Physiological*. Clarendon Press, Oxford.

5

Cereals

Cereals are the carbohydrate rich food grains, which are consumed largely by humans and provide the daily need of nutrients for survival and livelihood. Examples of cereals include rice (paddy), wheat, maize, sorghum, pearl millet etc. This book is mainly concentrated on the anatomy (including general botany) of the following field crops: rice, maize, wheat, sorghum and pearl millet.

5.1 Rice

5.1.1 Introduction

In terms of agricultural production, rice (*Oryza sativa* L.) is the second most important crop after wheat. But it is the most important cereal grain crop of the world with respect to human consumption, as it feeds over half of the world population including more than 75% of the poorest people. More than 90% of the rice is grown in Asia, supplying primary nutrition of 60% of the world's population. Botanically, rice is an annual grass of the Gramineae family. The genus *Oryza* includes more than 20 wild species and two cultivated diploid species ($2n = 2x = 24$), *Oryza sativa* (Asian rice) and *Oryza glaberrima* (African rice). Ecogeographically, three distinct subspecies

of rice are recognized, *indica, japonica* and *javanica. Indica* rice is adapted to tropical and subtropical humid climates, while *japonica* is adapted to temperate regions. The *javanica* rice is limited to tropical regions of South-east Asia.

5.1.2 Origin, distribution and adaptation

Rice was first domesticated in the region of the Yangtze River valley. Archaeological and historical evidence points to the region of the foothills of the Himalayas in the north and hills in the north-east of India to the mountain ranges of South-east Asia and south-west China as the primary centre of origin of *O. sativa*, and the delta of the River Niger in Africa for that of *O. glaberrima*, the African rice. The earliest remains of cultivated rice in India have been found in the north and west and date from around 2000 BC. Perennial wild rices still grow in Assam and Nepal. Rice appears to have spread to southern India after its domestication in the northern plains around 1400 BC. It then spread to all the fertile alluvial plains watered by rivers. Cultivation and cooking methods are thought to have spread to the west rapidly and, by medieval times, southern Europe saw the introduction of rice as a hearty grain.

Today, the majority of all rice produced comes from China, India, Pakistan, Indonesia, Bangladesh, Vietnam, Thailand, Myanmar, the Philippines and Japan. Asian farmers still account for 92% of the world's total rice production. Rice is grown in all parts of India, northern and central Pakistan. Basmati rice cultivated in the northern plains of the Punjab region is famous all over the world for its smell and quality.

Rice is well adapted to diverse environmental conditions and climates from sea level up to high elevations. It is grown in north-eastern China at 53°N of the equator and in New South Wales at 35°S in Australia and is widely adapted below sea level at Kuddanad (Kerala) to above 2000 m in Kashmir. Most of the rice areas lie between the equator and 40°N latitude and between 70° and 140°E longitude. The highest yields are recorded between 30° and 45°N of the equator.

O. sativa is the dominant rice species, cultivated in Asia, Africa, Europe, north-Central and South America and Oceania. *O. glaberrima* is another cultivated species limited to Africa. *Indica* rices grow widely in tropical regions and are adapted to cooler areas and largely grown in temperate climates such as central and northern China, Korea and Japan. A number of biological, environmental and socio-economic conditions cause the existence of the large differences in rice productivity around the world.

5.1.3 Utilization

As a cereal grain, it is the most important staple food for a large part of the world's human population, especially in East, South, South-east Asia, the Middle East, Latin America and the West Indies. Rice is cooked by boiling or steaming, and absorbs water during cooking. In some countries parboiled rice is popular. Raw rice may be ground into flour for many uses, including making many kinds of beverages such as amazake, horchata, rice milk and sake. Rice flour does not contain gluten and is suitable for people on a gluten-free diet. Rice may also be made into various types of noodles. Raw, wild, or brown rice may also be consumed by raw-foodist or fruitarians if soaked and sprouted.

5.1.4 Morphological description

Vegetative characters

Rice is an herb having a plant height of 2–4 m. The plant generally takes 3–6 months to complete its life cycle. Most cultivated species are hydrophytic and a few species are mesophytic. Rice has a fibrous root system, adopted for mostly hydrophytic conditions. Seminal roots help in the establishment of the seedling and are short lived. In perennial rices, short underground shoots densely covered with sheathed leaves are present. Annual species have long but slightly branched adventitious roots, whereas a main root and subterranean shoots are completely absent. The stem is referred to as a culm or stalk. The culm is composed of nodes and internodes, and is enclosed within the sheath, but a small portion of the culm just below the panicle is exposed after heading. The number of elongated internodes ranges from three to eight. Internode elongation is associated with growth duration. Tillers develop from the leaf axils at each unelongated node of the main shoot or other tillers during vegetative growth. Tillers should not be lodging; tiller stems with strong mechanical tissue (sclerenchyma) may impart non-lodging. Tiller stems may be screened microscopically for the variations in the intensity of sclerenchyma (Maiti, unpublished). Rice leaf is composed of the leaf sheath, the blade, the ligule and the auricles. The sheath is elongated, with a ribbon-shaped leaf base enrolling the culm. The blade is narrow, flat and longer than the sheath. At the base of the leaf blade, a white band structure called the collar is present. A pair of hairy and sickle-shaped auricles is situated at the junction between the collar and the sheath. Leaf sheaths may possess cilia. Like stalks, leaf sheaths also have a clearly marked or ribbed venation and become keel-like by the protrusion of the mid-vein in the upper part of the sheath beneath the leaf lamina. Leaf lamina consists of parallel venation. The ligule is oblong or blunt and 1–5 mm long in most of the species. In *O. latifolia*, the ligule is short but breaks up at the apex into a row of hard, sometimes woolly hairs.

Floral characters

The rice panicle consists of the base, axis, primary and secondary branches, pedicel and glumes, and bears hundreds of spikelets, the characteristics of many cereals (Fig. 5.1). The panicle axis is extended from the panicle base to the apex and may contain from eight to ten nodes. Panicles are developed from the nodes of primary and secondary branches on the top of which spikelets are located. Articulation is present in all rice species between lower and upper empty glumes. This is fairly distinct in most species because the spikelets fall easily on ripening of seeds. However, it is less distinct in the cultivated species *O. sativa*, where during the ripening of fruits the spikelets do not drop off but remain on the branches. The spikelet is borne on the pedicel, a short stalk or secondary branch. Two rudimentary glumes are present on the upper part of the pedicel. A pair of sterile lemma and the rachilla are situated between the rudimentary glumes and the spikelet. The flower is enclosed in the lemma and palea with awns. Flowers are sessile, bisexual, ligulate, zygomorphic, epigynous and trimerous. In every flower, lemma, palea, two lodicules, pistil and stamens are present. There are three stamens, long anthers, dithecous, basally fixed; the stamens have about 2–2.5 mm long, linear-oblong anthers. The anthers dehisce through a longitudinal slit. The pistil is composed of the stigmas, styles and ovary. The ovary is unilocular, unicarpellary, with a single ovule on basal placentation, silky-like style, and bifid feathery, plumose stigma. Two fleshy bodies called the lodicles are situated at the base of the ovary. The lodicles swell with water, thereby separating the lemma and palea at flowering. Anthesis involves a series of events. At the beginning of anthesis, the tips of the lemma and palea begin to open, the filaments elongate, and anthers begin to emerge from the lemma and palea. The filaments elongate more to push the anthers out of the lemma and palea. Anthesis lasts about 1–2.5 h. Anther dehiscence take place just before the extension of the lemma and palea, with many pollen grains falling on the sticky stigmas. Rice is a self-pollinated crop and the plants are pollinated by wind (anemophily). The complete process from pollination to fertilization takes from 18 to 24 h. The fruit is a caryopsis and generally called grain. Fruits are small, compressed from the sides, 1–8 mm long, oblong, elliptical or orbicular, depending on the shape and size of the spikelet. Small fruits of rice are initially yellowish, and then turn dark brown. Seeds are endospermic.

5.1.5 Anatomical description

Root anatomy

In transverse section, the root organization shows three distinct parts, the epidermis, cortex and stele.

Root epidermis has a single layer of compactly arranged thin-walled living cells. Both the cuticle and stomata are absent. Epidermal cells produce unicellular root hairs, which are useful in increasing the surface area for

Fig. 5.1. Panicle inflorescence showing spikelets. Each spikelet consists of sessile, bisexual, ligulate, zygomorphic, epigynous, and trimerous flowers. The boot leaf contributes the major amount of photosynthates to maturing seeds.

absorption of water from the soil. Root hairs are generally short lived. The epidermis is present as a layer of sclerenchyma imparting strength to the root system.

Cortex is parenchymatous and in the central region, present between epidermis and the stele. The cortex is poorly developed. Numerous air spaces form aerenchyma in the cortical region for gaseous exchange (Fig. 5.2).

The centre region is occupied by vascular tissue forming stele. The vascular system is poorly developed because the rice plant is adapted for a hydrophytic habitat. Xylem and phloem are arranged in the form of separate strands on different radii. Xylem consists of very small xylem vessels. Sclerenchyma is almost absent; cambium and pith are absent.

In hydroponically grown rice, roots are observed to be bounded by an epidermis, exodermis, and a fibrous layer (Clark and Harris, 1981). The exodermis has a suberin lamella along its inner tangential wall.

Significance of variations in root anatomy

The structural anatomy of rice root is adapted more towards a hydrophobic environment.

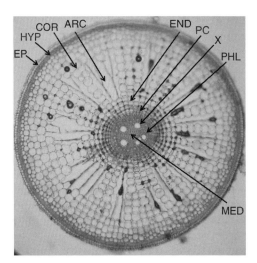

Fig. 5.2. Transverse section of root (100×) illustrating epidermis (EP), sclerenchymatic hypodermis (HYP), cortex (COR), aerenchyma (ARC), endodermis (END), pericycle (PC), xylem (X), phloem (PHL) and medulla (MED).

The root system in rice is poorly developed compared to other cereal crops and root hairs and root cap are absent, but in some genotypes there may be a root packet. A high proportion of air chambers in the root is suitable for aeration of internal tissues under anaerobic submerged conditions, which help in gaseous exchange. In some genotypes there may not be air chambers in the root. There is a reduction in mechanical tissue and vascular tissue, but that reduction may be less or more in some genotypes. The presence of sclerenchyma below the epidermis is desirable for imparting strength to the tissue system of roots.

Traditional upland *indica* cultivars possess a larger diameter of root stele and xylem vessels compared to modern upland varieties. *Japonica* rice has a significantly larger stele diameter relative to the root diameter (Kondo *et al.*, 2000).

Asian cultivated rice (*O. sativa* L.) has genetic diversification of root characteristics, but this variation has not been elucidated completely with reference to the genetic background. To elucidate the variations in root anatomical and morphological traits among diverse varietal groups of cultivated rice, Uga *et al.* (2009) analysed four anatomical traits (stele transversal area, root thickness, total transversal area and number of late metaxylem vessels) and two morphological traits (ratio of deep rooting and root length index) in 59 accessions. An earlier principal-coordinate analysis study using data on 179 restriction fragment length polymorphisms categorized these accessions into three varietal groups: 13 *japonica*, 21 *indica*-I, and 25 *indica*-II. Based on a principal-components investigation of the six traits, the *japonica* group had extensive variation in root anatomy compared to the two *indica* groups. In particular, *japonica* upland rice was characterized by a larger stele and xylem structures (Uga *et al.*, 2009).

Stem (culm) anatomy

Rice is a monocot and lacks secondary growth. The internal structure of the stem in transverse section generally consists of epidermis, cortex and vascular tissue.

Epidermis is the outermost region of the stem and is meant for protection. It is uniseriate, having a single layer of compactly arranged barrel-shaped living cells. The outer surface of the epidermis is covered by a thin cuticle. A few stomata (graminaceous type) are present in the epidermis for the exchange of gases necessary for adaptation to the hydrophytic system.

Cortex is highly reduced and is usually represented by a narrow hypodermal region. In mature stems hypodermis is sclerenchymatic and gives mechanical strength. In young stems chlorenchymatic patches are present just below the stomata regions and perform photosynthesis. The cortex shows a lot of air chambers (Fig. 5.3). Vascular bundles are present just below the epidermis.

Stele is the central part of the stem and is larger than the cortex. The stele consists of ground tissue and vascular bundles. There is no clear demarcation between the cortex and pith since endodermis, pericycle etc., are indistinguishable. The ground tissue also shows air chambers. The central part of the stem is parenchymatic and is best described as ground tissue (or conjunction tissue).

Vascular bundles are numerous and are scattered irregularly in the ground tissue. Usually the peripheral bundles are small and closely arranged, and the central bundles are large and are widely arranged. Each vascular bundle is oval in shape and is surrounded by fibrous bundle sheath. It is many celled thick towards outer and inner regions and few celled thick at lateral regions, hence the vascular bundles are described as fibrovascular bundles. Vascular bundles are conjoint, collateral (with xylem and phloem present on the same radius) and closed (i.e. without cambium). The xylem is present towards the centre of the axis and consists of few vessels which are arranged in the form of letter 'Y'. The protoxylem often disintegrates and forms protoxylem lacuna. Phloem is present towards outside, between the metaxylem vessels. It consists of sieve tubes, companion cells and a few parenchymatous cells at the sides. The stele consists of numerous vascular bundles scattered irregularly in the ground tissue, and is called an atactostele. It is the most advanced type of stele.

Significance of variations in stem anatomy

Rice stem is less defined in structure than in sorghum or wheat stem. It adapts well to anaerobic conditions by formation of chambers (aerenchyma) in the cortex region for the exchange of gases. Variation in sclerenchyma structure arises from differences in the thickening of the cell wall in the outer cortical parenchyma and the number of sclerenchyma cell layers, and contributes to both biotic and abiotic stress tolerance. Since *japonica* rice has higher thickening of sclerenchyma than *indica* rice, it is more tolerant to stem borers and desiccation. Rice exhibits thinner cuticle and less wax deposition on the epidermis compared to other cereal crops, although genotypic variation for cuticle thickness is observed among and within subspecies.

Leaf anatomy

The rice leaf comprises the leaf blade, leaf sheath, and the joining region of these two parts containing appendages such as ligule and auricle. Leaf anatomy in transverse

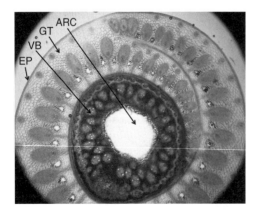

Fig. 5.3. Transverse section of culm (stem 100×) showing epidermis (EP), vascular bundles (VB), ground tissue (GT) and central aerenchyma (ARC).

section shows epidermis, mesophyll and vascular bundles (Fig. 5.4).

The epidermis is uniseriate and composed of more or less oval cells. Bulliform cells are observed in the adaxial surface of the leaf blade, which helps in rolling the leaves under low water levels. The outer walls of the epidermis are covered by a thin layer of cuticle. Graminaceous type of stomata is present in both epidermal layers. Stomata are covered by two dumb-bell shaped guard cells. Stomata are usually more frequent in the lower epidermis. Epidermis is rough in

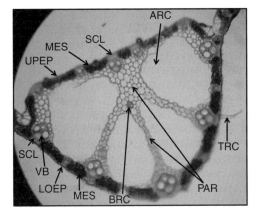

Fig. 5.4. Transverse section of leaf (100×) showing upper epidermis (UPEP), lower epidermis (LOEP), sclerenchyma patches (SCL), vascular bundles (VB), aerenchyma (ARC), parenchyma (PAR), mesophyll (MES) and brachisclereid (BRC).

nature due to the presence of silica crystals. The upper epidermis may be easily identified due to the presence of xylem and bulliform cells. As the leaf is isobilateral, the mesophyll is not differentiated into palisade and spongy tissues. The mesophyll tissue of rice leaf is composed of lobed chlorenchyma with a length:breadth ratio of approximately 2:1. Large air chambers (aerenchyma) are present in the mesophyll tissue, which is a characteristic feature of a hydrophytic plant. Papilla and trichomes cover the entire leaf surface except on the adaxial surface of the leaf sheath.

The vascular system is poorly developed because the rice plant is adapted for a hydrophytic habitat. The vascular bundles are collateral and closed. Most of the bundles are small in size but fairly large bundles also occur at regular intervals. Xylem is found towards the adaxial side and phloem towards the abaxial side of the bundles (Fig. 5.5).

There are distinct anatomical differences between the early leaves (first to third) and leaves that appear later. The first leaf of rice lacks a leaf blade and is very small in size compared to other leaves. The fifth and later leaves have a strong midrib, which provides mechanical strength to the leaves. The leaves also differ in the capability of photosynthetic activities; early leaves are photosynthetically less efficient than later leaves.

Development of leaf blade and leaf sheath in rice is well coordinated. The leaf

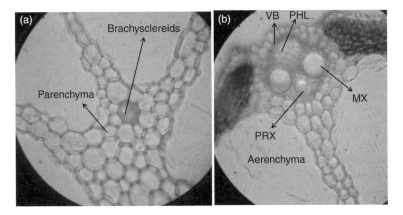

Fig. 5.5. (a) Transverse section of leaf (400×) showing the aerenchyma and vascular bundle; (b) vascular bundle (400×) showing proto- and metaxylem regions.

develops from the shoot apical meristem by rapid cell division. Leaf sheath elongation starts only after leaf blade elongation is completed. Air spaces known as lacuna are formed in the leaf sheath and midrib of the leaf blade.

Significance of variations in leaf anatomy

Rice and wheat leaves are distinct from other C₃ monocot leaves in terms of higher photosynthetic capacity, which has primarily resulted from the extensive selection and breeding for higher productivity. Rice leaves differ from wheat in allocation of leaf nitrogen to rubisco (ribulose-1,5-bisphosphate carboxylase/oxygenase), the principal photosynthetic enzyme, and rice also has a higher mesophyll cell area than wheat. In the nitrogen responsive cultivars, these areas increase consistently along with increase in intercellular air space favouring photosynthetic activities. Air chambers are present in rice mesophyll cells, which help in oxygen supply as well as re-fixation of CO_2 generated from photorespiration. These air chambers are large, regular and are separated by partitions of photosynthetic tissues. Higher adaptation of rice than wheat in hot and humid areas is attributed to better photosynthetic activity in rice leaves due to better spatial arrangements of chloroplasts in mesophyll cells, higher mesophyll cell number and availability of air spaces to re-capture CO_2 released through photorespiration (Makino, 2011). The outer layer of chlorenchyma cells in rice is covered by chloroplast and stromules (stroma-filled appendages), which is absent in wheat (Sage and Sage, 2009). The general features of rice leaves are also conserved largely in the wild relatives, which suggests that selection did not bring much change in the leaf structure; rather, rice has been selected for having leaf structure with high photosynthetic capacity.

The strength in rice leaves is provided by the large midrib, sclerenchyma strands and often by the presence of sclereids. There is a great reduction in vascular tissue, small and big vascular bundles are arranged in an alternate manner. In small vascular bundles, xylem elements are lacking. Phloem is well

developed when compared with xylem. These characters exhibit better adaptation to an aquatic nature.

Anatomy of rachilla

Transverse section of the rachilla shows ridges and grooves in outline (Fig. 5.6). Single-layered epidermis covers the internal tissues. There are four to five layers of sclerenchyma present beneath the epidermis in the region of ridges and grooves. Mesophyll tissue is encircled by sclerenchyma in the region of grooves. Small vascular bundles are situated in the ridges. Parenchyma forms ground tissue, leaving a central airspace. Numerous large vascular bundles are distributed throughout the ground tissue. Each vascular bundle is collateral, closed, and covered by sclerenchyma tissue.

Fruit anatomy

Rice fruit is called a caryopsis. Pericarp tightly covers the grain (seed) and is generally known as husk. The grain is covered by an aleurone layer consisting of proteins. Endosperm occupies the major portion in grain. The embryo (germ) is situated one

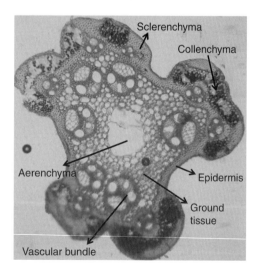

Fig 5.6. Transverse section of rachis (100×) showing sclerenchyma, collenchyma in the region of ridges, ground tissue, vascular bundles and central aerenchyma. Larger vascular bundles are distributed in ground tissue.

side of the bottom. The endosperm consists of protein starch-storing parenchyma cells called amyloplasts.

5.2 Maize

5.2.1 Introduction

Maize (*Zea mays*, 2n = 20) is the most important food and animal feed crop of the western hemisphere, and is grown throughout the world under a wide range of climates. Since pre-Hispanic time, it has been the basic food for the majority of people in Mexico and Latin American countries. From about AD 900 to AD 1300 there was a shift from intensive foraging with some agriculture to intensive farming with some foraging. This was accompanied by more densely settled population and greater cultural complexity among people in the mid-western hemisphere. The maize in the upper midwest around AD 600 was poorly adapted to the short growing season and photoperiod of the more northerly latitudes. After AD 1100, the strains known as Northern Flint Corns were developed with better photoperiodic response, yield, pest resistance and hardiness. In scientific and formal usage, 'maize' is normally used in a global context. Equally, in bulk-trading contexts, 'corn' is used most frequently.

5.2.2 Origin, evolution and domestication

Maize is considered to have originated on the Pacific slopes of the Mesoamerica 7000–10,000 years ago (Wilkes, 2004). Specimens of primitive cobs of maize have been recovered in caves in Oaxaca, Mexico, that date back to 8000 BC. Mexico is one of the centres of origin of maize, where several native landraces grow in abundance in diverse environments. From Mexico, it diffused to South America through the Andes during 6000–2000 BC. Studies on microsatellite variation of maize genomic DNA indicates that maize arose from a single domestication event around 9000 years ago (Matsuoka *et al.*, 2001).

Several factors were involved in the diversification of maize, including the existence of primitive landraces, the influence of exotic varieties from the northern part of the country, natural hybridization with other races as well as favourable geographical conditions for growth of short-season maize. Maize was considered as 'mother corn' by the Native American Indian communities. It was introduced in northern America around 700 years ago, where it was adapted within a very short time period. During the Spanish discovery of the New World, maize was brought to the soil of Europe, from where it spread rapidly to Asia and Africa, primarily through the establishment of Portuguese colonies in Africa and Asian countries. Today it is a major food crop of almost all the continents.

The origin of maize has been debated for a long time, but major research findings support that the most probable progenitor of maize is teosinte (*Zea mexicana*) or 'Balsas' teosinte (*Zea mays parviglumis*), a weedy member of genus *Zea* (Beadle, 1939). However, how the simple ear of teosinte evolved into the large maize cob within a short evolutionary period of about 10,000 years is a baffling question to maize researchers. Several theories have been put forward for explaining the origin of maize. The teosinte theory supported by major research groups is contradicted by other theories, which consider maize originated from *Tripsacum dactyloides* (Mangelsdorf and Reeves, 1938), *Zea diploperennis*, hybrids of *Tripsacum* and teosinte (tripartite theory), from transmutation of teosinte male ear to maize female cob (Iltis, 1983) or from ancestral maize species through duplication of chromosomes (Gaut and Doebly, 1997).

5.2.3 Genetic resource

Since a population of maize is a mixture of heterogeneous and heterozygous genotypes, it is best described as a race (a group of individuals having enough common traits for grouping). The studies on numerical taxonomy and molecular phylogeny of maize have helped to identify a number of races and their geographical relationships. Early classifica- tions of maize races were based on morpho- logical, physiological, genetic and agronomic features and cytogenetic traits. It has been observed that the Mesoamerican and South American regions show greater genetic diver- sity and race differences than the northern American regions. For example, more than 65 races of maize have been identified in Mexico, while in the USA, only 14 races are observed. The number of races of maize in South American countries such as Bolivia (77) and Peru (68) are higher than that of North American countries (Hernández, 2009).

However, based on kernel (seed) char- acteristics, the following subtypes of maize are commonly recognized by growers and consumers:

1. Dent corn (*Zea mays indentata*) – upon maturity, the kernels become indented in the crown of the seed. Dent corn is com- monly used as human food, for industrial purposes and livestock feeds all over the world. The denting is caused by different degrees of drying and shrinking of starch.
2. Flint corn (*Zea mays indurata*) – hard, smooth kernel of flint corn contains various proportions of hard and soft starch. It is popular in Central and South America.
3. Sweetcorn (*Zea mays* var. *saccharata*/*Z. mays* var. *rugosa*) – immature kernels are sweet (contains ~10% sugar) and consumed raw. In mature kernels, sugar is converted to starch, so the sweetness is reduced. The sweetness is caused by a recessive gene *sug- ary-1* (*su-1*), which retards conversion of sugar to starch in the endosperm.
4. Flour corn (*Zea mays amylacea*) – kernels are soft, smooth and filled, and mainly used for production of white flour. Its use is lim- ited in parts of the USA and South America because it is used mostly as feed grain.
5. Popcorn (*Zea mays everta*) – a type of flint corn with very hard exterior. The starchy endosperm explodes and becomes fluffy upon heating, giving the characteristic roundish shape of popcorn.
6. Waxy corn (*Zea mays*) – a mutant type, having waxy endosperm, where all the amy- lose of starch is replaced by amylopectin. It is caused by a recessive mutation 'waxy' (*wx*).

There are also other speciality types of maize such as blue corn, high lysine corn, high oil corn, baby corn or ornamental corn.

5.2.4   Utilization

Maize and cornmeal (ground dried maize) constitutes a staple food in many regions of the world. Maize meal is also used as a replacement for wheat flour, to make corn- bread and other baked products. Popcorn and corn flakes are a common breakfast cereal found in many countries all over the world. Sweetcorn, a genetic variety that is high in sugars and low in starch, is usually consumed in the unripe state. When maize is ground into flour, maize yields more flour, with much less bran, than wheat does. However, it lacks the protein gluten of wheat and, therefore, makes baked goods with poor rising capability and coherence. Certain varieties of maize have been bred to produce many additional developed ears. These are the source of the 'baby corn' used as a vege- table in Asian cuisine.

Grain alcohol from maize is tradition- ally the source of bourbon whiskey. Starch from maize can also be made into plastics, fabrics, adhesives and many other chemical products. The corn steep liquor, a plentiful watery by-product of maize wet milling process, is widely used in the biochemical industry and research as a culture medium to grow many kinds of microorganisms. Maize is increasingly used as a feedstock for the production of ethanol fuel. Maize leaves are used as forage in most countries and sometimes as mulching material.

5.2.5   Morphological description

Maize has been described in numerous botanical books and illustrations. The Aztec and Mayan civilizations had their own god named after maize, whose pictures revealed maize cobs. After the discovery of the New World, the first detailed description and illustration of maize can be found in 'Historia

Natural y General' by Gonzalo Fernandez de Oviedo (1535) of Spain, although a description of maize can be found in Columbus's chronicle as early as in 1511 (Finan, 1948). Since then, various descriptions of the maize plant have been documented in several books, monographs and articles. In this section, a brief general morphology of modern cultivated types of maize is outlined.

Vegetative characters

Maize is a mesophytic herb, less drought tolerant compared to sorghum and pearl millet; it grows up to 2–4 m in height. The root system is fibrous; stilt roots are formed on the lower nodes of the stem and a ring of crown roots arise from the stem just above the ground level. The stem is aerial, erect, greenish, hard and succulent, with nodes and internodes reaching 20–30 cm; the apex of the stem ends in the tassel (Fig. 5.7). Leaves are simple, exstipulate, alternate, sessile, lower leaves are 50–100 cm long and 5–10 cm wide, upper leaves somewhat short and narrow, arranged in two rows on stem, with an expanded leaf base with leaf sheath, wavy margin and parallel venation.

Floral characters

Maize has a diclinous inflorescence, with male and female inflorescences arranged on the same plant. The male inflorescence is known as tassel and is present at the apex (top) of the stem. The female inflorescence is known as a cob, and is present on the tips of lateral branches, which are developed from the axils.

The male inflorescence is modified into a panicle (inflorescence of male flowers), with spikelets in pairs. In each pair, one is sessile. The spikelets are covered by glumes.

The female inflorescences are the ears, which are tightly covered over by several layers of leaves, and so closed-in by them to the stem that they do not show themselves easily until the emergence of the pale yellow silks from the leaf whorl at the end of the ear. The silks are elongated stigmas that look like tufts of hair, at first green and later red or yellow. Spikelets are arranged in pairs on the cob and are covered by cupules; they are sessile, having a female flower.

Flowers are sessile, unisexual, zygomorphic, epigynous; in every flower there is lemma, palea and two lodicules. The androecium has three stamens, long anthers, dithecous and basally fixed. The gynoecium is unilocular, unicarpellary, single ovule in basal placentation, with a silky-like style and a feathery stigma.

When the tassel is mature and conditions are suitably warm and dry, anthers on the tassel dehisce and release pollen. Maize pollen is anemophilous (dispersed by wind) and because of its large settling velocity most pollen falls within a few metres of the tassel. Each silk may become pollinated to produce one kernel of maize.

Fruit is generally referred as grain or kernel and is botanically called a caryopsis. The kernel of maize has a pericarp of the fruit fused with the seed coat, typical of the grasses. The cob is close to a multiple fruit in structure, except that the individual fruits (the kernels) never fuse into a single mass.

Fig. 5.7. Maize plant morphology: tassel occurs at the apex of stem and silk at the centre of the stem from the ground level.

The grains are about the size of peas, and adhere in regular rows round a white pithy substance, which forms the ear. An ear contains from 200 to 400 kernels, and is from 12–25 cm in length. They are of various colours: blackish, bluish-grey, purple, green, red, white and yellow.

Seeds are endospermic and filled with carbohydrates; based on the nature of carbohydrates they are of different shapes and sizes.

5.2.6   Anatomical description

Anatomical features of maize have been well documented by different researchers. However, a comprehensive collection of anatomical features describing essential anatomical structures of root, stem, leaf, floral parts and seed along with their biological significance is lacking.

Root anatomy

There are two distinct types of roots in the maize root system, the embryonic roots that develop from young embryo and the postembryonic (adult) roots that develop during different stages of crop growth. The general anatomical features of both types of root are similar and are typical of monocots. The maize root is composed of epidermis, ground tissue (cortex), endodermis surrounding the vascular bundles that contain xylem and phloem in alternate bands (Fig. 5.8).

The epidermis consists of elliptical cells subtended by two layers of hypodermis. In the early stage of root development, the cortex is made up of ovoid parenchymatous cells with considerable intercellular space.

As the root ages, the cortical cells elongate to assume a plate-like appearance. A single layer of ovoid to cubical cells of cortical cells of cortical parenchyma (multilayered) encircles the endodermis. The endodermal cells are barrel- to boat-shaped and are thickened with suberin on the inner tangential wall. Crystals are present in the endodermal cells; their shape and size vary with cultivar and with age of the root. The pericycle below the endodermis is composed of one to several layers of thick-walled sclerenchymatous cells, of which the outermost layers are highly lignified.

Xylem and phloem tissues are present on a diffident radius and are called radial vascular bundles. Xylem parenchyma surrounding the metaxylem may be thickwalled or lignified. The xylem and phloem show a typical closed radial arrangement. Protoxylem bundles are present on the exterior side of the metaxylem. The size of the metaxylem bundle varies according to cultivar. The pith is solid with round intermediate to compactly arranged parenchyma cells.

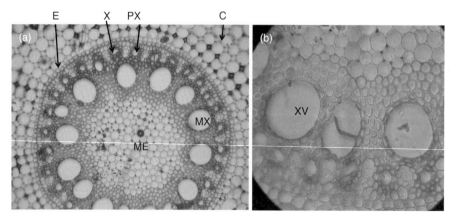

Fig. 5.8. (a) Transverse section of root (100×) showing different parts: cortex (C), endodermis (E), protoxylem (PX), metaxylem (MX) and medulla (ME). (b). Sector enlarged (400×) showing xylem vessels (XV).

Significance of variations in root anatomy

Cultivar differences are expressed in a number of root characteristics in maize (Fig. 5.9a, b). These are:

1. Presence or absence of sclerenchymatous exodermis in the cortex imparting strength and protection to internal tissue.
2. Thickness of endodermal cell walls.
3. Size and shape of crystals in the endodermal cell cavity.
4. Intensity of lignification in the pericycle cell layers and in cells surrounding the vascular bundles.

The presence of exodermis, higher thickness and higher lignification provides better structural integrity to the root system. Thus genotypes exhibiting these characteristics are expected to have better crop stand and resistance to lodging. Heavy lignification of exodermis of maize root gives higher bending strength, thus providing better anchorage (Ennos *et al.*, 1993). Thicker roots are advantageous for higher root growth and help in better water transport. Higher root diameter is associated with higher xylem diameter in maize, which helps in better solute transport. A number of drought-tolerant maize lines developed by the International Maize and Wheat Improvement Centre (CIMMYT) have thicker root systems; thus they are able to penetrate deeper in the soil for extraction of available moisture compared to drought-susceptible lines (Hund *et al.*, 2009).

Stem anatomy

Stem anatomy of maize is typical of that of the grasses. The ontogeny and anatomy of vascular bundles in maize has been described earlier by Esau (1943).

YOUNG STEM   The epidermis consists of a single layer of cells with thickened walls on the outer side. The ground tissue consists of thin-walled parenchyma. The vascular bundles, conjoint collateral in type, are scattered in the ground tissue.

MATURE STEM   In a transverse section, the epidermis consists of cubical to boat-shaped epidermal cells containing elongated crystals. Just below the epidermis, and opposite the ridges of the stem, there are alternate bands of large and small thick-walled sclerenchymatous patches. Each sclerenchymatous band alternates with a broad band of chlorenchymatous tissue corresponding to the furrows of the stem outline. These hypodermal bands of sclerenchyma are in turn connected to the subtending broad cylinder of sclerenchyma, which contains the ring of vascular bundles. There are four to five rows

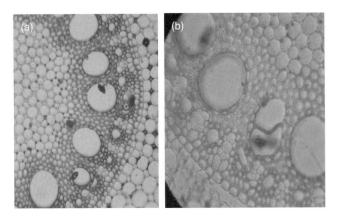

Fig. 5.9. Transverse section of root (400×) showing (a) Casparian thickening in endodermal cells and (b) Casparian thickening in endodermal cells and crystals in the pericycle.

of vascular bundles, the outermost in general being the smallest in size (Fig. 5.10a, b). In some sections, just below the outermost layers, large and small vascular bundles alternate with each other. The central vascular bundles are scattered in the ground tissue. Mechanical tissues in these are not as extensive as in the outer bundles, particularly in the peripheral region. They contain only a semi-lunar band of sclerenchyma adjacent to the protoxylem cavity towards the centre of the pith. The peripheral region of the central vascular bundles does not contain sclerenchyma. The ground tissue is composed of round parenchymatous cells. Pith is solid.

Significance of variations in stem anatomy

The internodes of maize vary in strength and supporting ability. Examination of transverse sections from seventh and eighth internodes, collected at physiological maturity, reveals considerable variation in distribution and amount of mechanical tissue (sterome) and vascular bundles, and in the size and shape of the protoxylem cavity (Fig. 5.11a, b). The vascular bundles are associated with protective fibre cells, which originate primarily from procambium and provide strength to the stem system (Esau, 1943).

Mechanical tissue, consisting of sclerenchyma and collenchyma, form the sterome system of the plant. On the basis of the amount of mechanical tissue present, the genotypes may be tentatively ranked in different classes. Maize genotypes show large variations in anatomical structure of stem with respect to the distribution and intensity of sclerenchyma, on the basis of which we can classify the genotypes as having a weak sterome system, intermediate system or a strong sterome system (Fig. 5.12a, b, c). Therefore when breeding for new genotypes, anatomical studies may help in better selection of genotypes according to the target environment and breeding objectives. Genotypes having strong sterome are expected to be resistant to lodging and exhibit better resistance to stem borer insects. High amounts of sclerenchyma give mechanical support (Murdy, 1960) and reduce the loss of water from internal parenchyma tissue by evapo-transpiration. Medium to large size and a high number of xylem vessels are required for efficient translocation of water under drought conditions. Thickening of hypodermis and higher amount of sclerociation in vascular bundles provide better tolerance to water stress (Murdy, 1960).

Anatomical variations in stem can also be used for prediction of growth behaviour

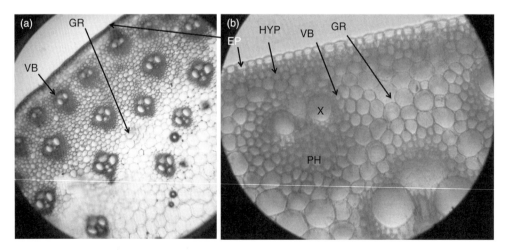

Fig. 5.10. (a) Transverse section of stem (100×) showing atactostele. (b) Sector enlarged (400×) showing epidermis (EP), hypodermis (HYP), scattered vascular bundles (VB) in ground tissue (GR), xylem (X) and phloem (PH).

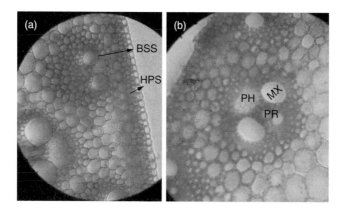

Fig. 5.11. (a) Transverse section of stem (400×) showing lignified patches of hypodermal sclerenchyma (HPS) and bundle sheath cells (BSS). (b) Transverse section of stem (400×) showing vascular bundles with protoxylem cavity (PR).

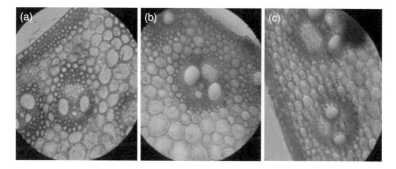

Fig. 5.12. (a) Strong stereome (400×) showing large vascular bundles. (b) Intermediate stereome (400×) showing medium size vascular bundles. (c) Weak stereome (400×) showing small bundles and smaller number of sclerenchyma patches below the epidermis.

of plants. Genotypes having higher thickening of hypodermis usually show a slower rate of internode growth, thus increasing the crop growth duration. Modern short-duration maize varieties in the USA have a high rate of internode growth and less thickening of the hypodermis compared to the slow-growing races of Mexico and Peru.

A number of stem borers infest maize by boring and feeding on the internode tissues. Stems having a greater thickness of parenchymatous tissues and a long internode have been correlated with resistance to stem borers (Santiago *et al.*, 2003), indicating these can be used as reliable predictors in developing insect-resistant

genotypes. A large variability is found among maize cultivars in the thickness of sclerenchyma present in the hypodermal regions, which may be utilized in breeding for lodging and stem borer resistance (Maiti, unpublished).

Stem anatomy studies of mutants have also allowed identification of important genes involved in cell wall biosynthesis in maize. A good example is the brittle stalk mutant of maize, which has reduced mechanical strength in the stem. Anatomical investigations revealed that the brittleness is due to reduced deposition of cellulose and uneven deposition of secondary cell wall material in the sub-epidermal and perivascular sclerenchyma (Ching *et al.*, 2006). By using the transposon tagging approach, a gene *Bk2* controlling cellulose deposition has been identified, which shows high expression in the vascular bundle in the wild type (*Bk2/bk2*) and reduced expression in the mutant (*bk2/bk2*).

Leaf anatomy

The anatomy of the maize leaf is characteristic of that of a mesophytic grass. There is some variation among the genotypes studied, but in general, leaf anatomy does not vary significantly from genotype to genotype. The epidermis is cuticularized, and composed of rectangular or oval epidermal cells. The lower epidermis is entire in outline. Silica crystals, prominent in some genotypes and not so in others, protrude from the adjoining cell walls of two adjacent epidermal cells towards the cuticle. Both the upper and lower epidermis contain unicellular long trichomes (Fig. 5.13a). Upper epidermal cells are wavy in outline and interspersed with zones of bulliform cells, the size of which differ in different genotypes. Two dumb-bell shaped guard cells cover the stoma, known as a graminaceous type of stomata (Fig. 5.13b). The chlorophyllous tissue surrounding the vascular bundles may be loose or compact. The surface structure of leaf shows variation in stomatal frequency, its size, and trichome density. The shape of the midrib in transverse section is almost semi-lunar. Bulliform cells are absent in the upper epidermis of the midrib and vascular bundles of different sizes are present below the upper epidermis. In some genotypes the midrib has three large vascular bundles, of which one is in the centre and one is to each side at the junction of leaf lamina midrib. Chlorophyllous tissue is present between the vascular bundles, but confined to the lower portion of the midrib. The sclerenchyma forms a band below the upper epidermis. The vascular bundles are of three types in the lamina and midrib: (i) the large central vascular bundle corresponding to the main vein; (ii) medium size vascular bundles; and (iii) very small vascular bundles. The first two types are generally fibro-vascular, containing patches of sclerenchyma on both sides (connecting the lower or upper epidermis), or at one side only (connected to the lower epidermis). The third type – the small vascular bundles – are generally present in the lamina, and are without any fibrous bands. These laminar bundles are again of two sizes, alternating with each other. Each vascular bundle remains surrounded by a bundle sheath consisting of thin-walled parenchyma cells (Fig. 5.14a,b). The cells of the bundle sheath generally contain large chloroplasts and form starch; grana are absent. This special structure was first identified by Haberlandt in 1882. The structure appears like a garland

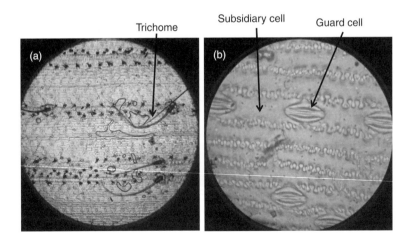

Fig. 5.13. (a) Leaf upper epidermal impression (40×) showing unicellular long trichomes. (b) Leaf upper epidermal impression (400×) showing epidermal cells, stomata (graminaceous type) consisting of dumb-bell shaped guard cells.

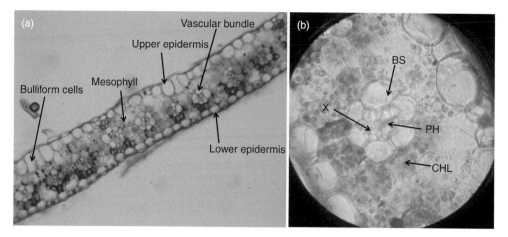

Fig. 5.14. (a) Transverse section of leaf (40×) showing different features. (b) Sector enlarged (400×) with fibro-vascular bundle showing chloroplasts (CHL), bundle sheath (BS), xylem (X) and phloem (PH).

(*Kranz* = wreath, garland), so it is known as Kranz anatomy. Kranz anatomy is invariably associated with C_4 photosynthesis. A typical feature of mesophyll cells is that they are found toward the exterior of the leaf so that they can be in contact with the intercellular airspace. The bundle sheath cells are arranged in the interior of the mesophyll cells.

Significance of variations in leaf anatomy

Maize leaf structure is not only important for understanding Kranz anatomy and the C_4 carbon cycle. There are various modifications in leaf anatomy with respect to leaf cuticularization, deposition of silica crystals in the upper epidermal cells (offering rigidness to the leaf as well as resistance to insects), size and shape of bulliform cells (help in rolling of the leaf thereby avoiding loss of transpiration under drought situations) (Fig. 5.15a, b), type and arrangement of vascular bundles both in the lamina and midrib and also for size and shape of bundle sheath chloroplasts (Fig. 5.16a, b).

Few studies have been directed on leaf anatomical traits in maize and its applications.

Adaption to drought

Kranz anatomy, which is essential for C_4 photosynthesis, provides distinct photosynthetic advantages to the maize plants under higher CO_2 concentration by overcoming photorespiration (Sage and Monson, 1999). The C_4 photosynthesis system is particularly advantageous under high temperature and high light intensity conditions prevailing in arid regions.

The maize leaves have developed a number of mechanisms for reducing evapotranspirational loss. The presence of thick cuticle on the epidermis of the leaf reduces transpirational loss in many xerophytic species (Fig. 5.17a, b). The thickness of cuticle is inversely proportional to leaf water loss in maize, particularly when the stomata are closed (Ristic and Jenks, 2002). Maize genotypes exhibit high variation in leaf glossiness related to variation in transpiration loss. Non-glossy leaves show higher cuticular wax deposition and a higher amount of trichome hairs. The glossy leaves show much higher cuticular transpiration than the non-glossy leaves (Traore *et al.*, 1989). Glossy sorghum's drought resistance is due to greater reflexion of radiation (Maiti *et al.*, 1994) and is also tolerant to shootfly because of the high density of trichomes. The high density of trichomes covering the surface reduces the direct sunlight effect and minimizes leaf temperature, which in effect reduces transpirational loss. Reduction in stomatal density on lower and upper epidermis minimizes the loss of water from the leaf, and the presence of large bulliform cells contributes to leaf rolling thereby

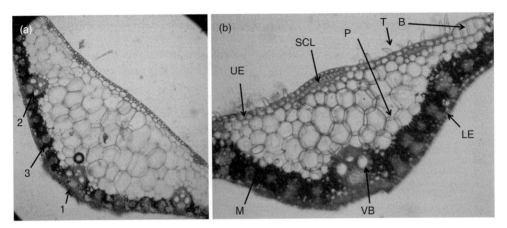

Fig. 5.15. (a) Transverse section of leaf at middle vein (100×) showing big (1), medium (2) and small (3) size vascular bundles. (b) Transverse section of leaf showing upper epidermis (UE) and lower epidermis (LE), bulliform cells (B) and trichomes (T) in UE. Sclerenchyma (SCL) patches are present below the UE, parenchyma (P) present between the UE and mesophyll (M).

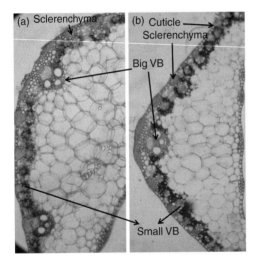

Fig. 5.16. Transverse section of leaf (100×) showing variations in thickness of sclerenchyma and the distribution of vascular bundles (VB): (a) large amount and (b) smaller amount of sclerenchyma.

preventing evapo-transpiration. Thus in breeding for drought-tolerant maize varieties and hybrids, these leaf anatomical traits are of considerable importance.

Panicle meristem

At the end of the vegetative stage (approximately 16 to 28 days after emergence) the shoot apex is transformed into a reproductive apex with the formation of panicle meristem (panicle initiation stage). The panicle meristem assumes a bulbous shape with a constriction at its base.

The panicle meristem shows a definite cellular organization. There are prominent tunica and corpus layers in the panicle axis, with two to three layers of seriately arranged tunica cells overlying the corpus. The cellular organization in the corpus layer is not regular. The protoplasm in the tunica cells and the nuclei with prominent chromatic material are distinct.

Spikelet primordia

Spikelet primordia develop first at the base and gradually progress towards the tip of the panicle meristem. The initiation of spikelet primordia is marked by the condensation of protoplasm and by an increase in volume of the flank meristem, i.e. the progenitor of spikelet primordia. In this region, two to three tunica cells with large nuclei bulge out and there are anticlinal divisions of the underlying cells which broaden the outline of the spikelet primordium. Subsequently, the cells beneath this region undergo both anticlinal and periclinal divisions, thus increasing the volume of the spikelet primordium. The spikelet primordium assumes a bulbous appearance

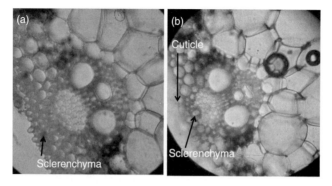

Fig. 5.17. Transverse section of leaf (400×) showing variation in thickness of sclerenchyma patches: (a) large amount and (b) smaller amount of sclerenchyma.

with a constriction on the sides, indicating the boundary between the adjacent spikelet primordia. This sequence of development of the spikelet primordia gradually progresses towards the tip of the panicle. The differentiating vascular traces in the panicle axis gradually establish vascular connection with the developing spikelets.

About 3 days after initiation, the panicle is much more elongated and the formation of the spikelet primordia has advanced towards the tip. Each primordium divides to give two spikelets (in pairs), which subsequently develop floral parts. Each spikelet is enclosed by a pair of glumes (one lower glume and one upper). Beginning at this stage, each individual spikelet primordium divides and extends laterally to give rise to two floret primordia. Each floret primordium then develops floral parts. Although spikelet development takes place acropetally, subsequent development displays a basipetal gradient.

Microsporogenesis

Microsporogenesis in maize is as in other angiosperms. The young anther (8 days after panicle initiation) appears as a mass of tissue in which the microsporangia are embedded. The microsporangia are present as four separate masses in the four lobes of the anther. These cells divide along periclinal walls, leading an inner mass of primary sporogenous tissue and outer primary parietal cells.

The ultimate cells of sporogenous tissue are pollen mother cells, which by reduction

and division ultimately lead to the formation of four pollen grains. In mature anthers approaching the time of dehiscence, some fringe-like outgrowths are observed all along the outer wall of the anther.

Structure of the ovule

The single ovule is present in the cavity of the ovary and is attached laterally to the ovarian wall by a short funicle. The ovarian wall is composed of a single layer of cubical epidermal cells, subtended by several layers of thin-walled cells.

Seed anatomy

The fruit is known as a caryopsis, and usually referred to as grain (seed). The seed consists of seed coat, endosperm, scutellum (cotyledon) and germ (embryo).

Seed coat

In a transverse section of the mature caryopsis the structure of the seed coat and endosperm is distinctly visible. Thick cuticle overlies the epidermis. The pericarp consists of epicarp, hypocarp and mesocarp. The epicarp is made up of elongated epidermal cells (Fig. 5.18a, b). The hypocarp cells are obliterated. Owing to compression, the mesocarp cells, which are two- to three-layered, have lost their cell outlines at some places. Elongated cross cells are only rarely visible and the tube cells which are

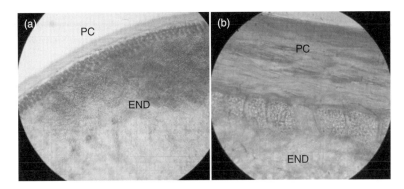

Fig. 5.18. (a) Longitudinal section of grain (caryopsis 100×) showing endosperm (END) and pericarp (PC). (b) Sector enlarged (1000×).

distinct at an early stage are mostly obliterated. Below the mesocarp there is a single layer of large regularly oriented cubical and rectangular cells called the aleurone layer. Staining bodies are present in the aleurone cell and starch grain variation among genotypes.

Endosperm

Below the aleurone layer is the peripheral corneous endosperm, a single layer of well demarcated cells with compactly arranged starch and protein granules. The endosperm occupies about two-thirds of the total seed and is located in the broader part of the maize grain. There is a continuous layer of the large cubical cells immediately beneath the hull called the aleurone layer. This aleurone layer contains protein granules. The rest of the endosperm consists of starch-laden cells, which also contain some lipid.

The cells of this layer are cubic, elongated, small and tightly pressed against the aleurone layer. Below this peripheral layer, the endosperm cells are loose, large-sized, and elongated and cubic to angular in shape. These constitute the floury endosperm. Starch and protein granules become increasingly less compact towards the centre of the endosperm.

Embryo

The embryo of the maize grain is located beneath the endosperm. It is demarcated from the latter by a single layer of epithelial cells. The embryo consists of a radicle and a plumule (Fig. 5.19). The radicle is partially covered and protected by coleorhiza. The width of the pericarp at the hilar region is much reduced. The radicle is partially covered and protected by coleorhiza. The plumule is partially covered and protected by the aleurone cells, which are small in size, compact, and rectangular in shape. The endosperm cells between the aleurone layer and the scutellum are compact. The scutellar cells are elongated and form a single layer surrounding the embryo. The embryo cells are highly compact, especially at the hilar region. The structure of the caryopsis at the black-layer region is similar, cells outside the black layer present at the hilar region are highly compact. The black layer shows a semilunar ring of vacuolated cells in a network pattern, which appears to cut off the vascular connection from the pedicel to the grain. Formation of this black layer coincides with the termination of grain development.

A longitudinal section of the maize grain soaked for 2 or 3 days should be taken along the endosperm–embryo axis and stained with dilute iodine solution in order to study its structure. The outermost coat enclosing the entire grains formed of inseparably fused fruit coat (pericarp) and seed coat is called hull. The hull is not stained by iodine. In a longitudinal section

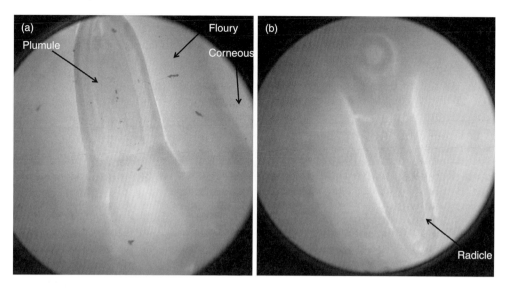

Fig. 5.19. (a) Longitudinal section of grain (caryopsis) showing plumule, floury endosperm, corneous endosperm and (b) radicle.

the seed shows two distinct regions, the endosperm and embryo.

Significance of variations in seed anatomy

Maize is considered as a model species for seed development and embryogenesis studies in grasses. Quite a large number of mutant types for kernel characters have been identified in maize. Anatomical studies of seed mutants in maize have helped to understand a number of basic and applied questions related to seed development, maturation and embryogenesis. Anatomies of embryogenic (*emb*) mutants, which are defective in different stages of embryogenesis, are commonly used for understanding the process of embryogenesis in maize and related species (Elster *et al.*, 2000). The kernel mutants are also used for programmed cell death (PCD), which leads to cellular differentiation, structuration of different organs such as scutellum and suspensor, mobilization of stored products, kernel denting and other processes involved with endosperm and embryo development.

Variation in distribution and amount of floury and corneous endosperm may determine the water uptake and nutritional quality of grains (Maiti, unpublished).

5.3 Wheat

5.3.1 Introduction

Wheat (*Triticum* spp.) is the most important cereal crop of the world in terms of production (World production 685.61 Mt in 2009), closely followed by rice (685.24 Mt in 2009) (FAO, 2011). China, India, the USA and the Russian Federation are the principal wheat growers of the world. The wheat crop refers to different species of genus *Triticum*, including diploid, tetraploid and hexaploid species. Bread wheat (*Triticum aestivum*, $2n = 6x = 42$) is the major species cultivated worldwide, covering more than 85% of wheat area. Other economically important species are macaroni or durum wheat (*Triticum durum*, which is used to make pasta, spaghetti and macaroni), poulard wheat (*Triticum turgidum*, $2n = 4x = 28$), club wheat (*T. aestivum* subsp. *compactum*, $2n = 6x = 42$), macha wheat (*T. aestivum* subsp. *macha*) etc. A few other wheat species were grown earlier, namely spelt wheat (*Triticum spelta*), emmer or wild emmer wheat (*Triticum diccoides*) and einkorn wheat (*T. monococcum*), but these have largely been replaced by *Triticum aestivum*, although they are locally cultivated in some pockets.

5.3.2 Origin and domestication

Wheat along with oat is considered to have originated in Persia. The archaeological record suggests that this first occurred in the regions known as the Fertile Crescent and the Nile Delta. These include southeastern parts of Turkey, Syria, the Levant, Israel and Egypt. Recent genetic and archaeological discoveries indicate that both emmer and durum (hard pasta wheat) wheats originated from the Karacadag region of south-eastern Turkey.

Wheat is one of the first cereals known to have been domesticated, and wheat's ability to self-pollinate greatly facilitated the selection of many distinct domesticated varieties. Cultivation and repeated harvesting and sowing of the grains of wild grasses led to the creation of domestic strains, as mutant forms ('sports') of wheat were preferentially chosen by farmers. It is widely grown in Afghanistan, America, India, China, Russia, France and Canada.

5.3.3 Utilization

Wheat is grown for its grain consumed as food and leaves as a forage crop for livestock, and its straw can be used as a construction material for roofing thatch. Wheat grain is a staple food used to make flour for leavened, flat and steamed breads, biscuits, cakes, breakfast cereal, pasta, noodles, couscous and for fermentation to make beer, other alcoholic beverages, or biofuel.

5.3.4 Morphological description

Several authors have provided good botanical and anatomical descriptions of different parts of wheat (for a comprehensive review see Kirby, 2002). A general description of the morphology and anatomy of different plant parts of wheat is presented below.

Vegetative characters

Wheat is an annual herb that grows up to 90–140 cm (Fig. 5.20). The root system is adventitious or fibrous and consists of two sets of roots: (i) the seminal or seedling roots, that belong to the embryo, are produced by the germinating seedling (these are five in number); and (ii) clonal roots that arise from the basal nodes of the plant and form the compact vegetative mass or 'crown'. The clonal roots form the permanent root system while the seminal roots dry after about 30 days of seedling emergence. Wheat culms or stems are erect, elastic, cylindrical, jointed and smooth. The bold joints, termed as nodes, separate the plant into sections called internodes, and these nodes and internodes are differentiated when the plant starts elongating. There are five to seven nodes and the lateral branch develops from the axil of the lower leaves. The lower internodes are shorter while the upper ones are progressively longer and they are six in number at the maturity of the plant. The main culm produces branches at the base,

Fig. 5.20. Morphology of wheat plants showing stem, leaves and spike inflorescence.

close to the ground, called tillers (primary), and the tillers produce additional tillers known as secondary and tertiary tillers having their own root system.

Wheat leaves are alternate, simple, sessile, exstipulate, with parallel venation arranged in two rows on the stem. Leaves consist of two parts, the leaf sheath, which encircles the stem, and the blade that bends away from the stem. The thickened junction is called a collar and the membranous outgrowth is called a ligule. The claw-like appendages attached to the base of the leaf and closely clasping the sheath constitutes the auricles. The foliage arrangement on the culm is opposite.

Floral character

The panicle is called an ear or spike. This is a compound distichous spike, the axis bearing two opposite rows of lateral spikelets and a single terminal spikelet, but in case of *T. monococcum* there is no terminal spikelet. The spikelets are sessile, have two glumes at their base and are arranged alternately. Each spikelet is attached to a rachis node and is known as a floret. Flowers are called florets. Flowers are stipulate, ligulate, bisexual, zygomorphic, trimerous and epigynous. The floret is composed of three stamens and a pistil, which consists of ovary. The flower is enclosed by a lemma (outer bract) and a palea (inner bract). The lemma tip is extended to form awns. Sepals and petals are together called tepals, and each floret consists of two tepals. There are three stamens, which are dithecous, versatile and yellow in colour. The gynoecium is unilocular, unicarpellary, with a single ovule on basal placentation, silky-like style and bifid feathery stigma. Anthesis occurs over about 3–10 days, depending on the environment, after the ear emerges from the flag leaf sheath, when a number of closely correlated events occur in a very short time. As the filament grows, the anther dehisces, each chamber developing a longitudinal split, starting at the tip of the anther, through which pollen is released. The stigma lobes, which are pressed together before anthesis, move apart, and the receptive branches are spread widely giving a large area for pollen

interception. The whole process is complete within about 5 min. Pollination is anemophilous. The pollen grain, which has a lifespan of about 5 h, when settled on a stigma, germinates in about 1.5 h to produce a pollen tube. At normal temperatures, the pollen tube reaches the embryo sac in about 40 min. The wheat fruit (usually called a grain) is a caryopsis with a thin-walled pericarp enclosing a single seed coat and testa is fused with pericarp. The shape is oval with a smooth and rounded dorsal surface and grooved or furrowed ventral surface at the centre. The furrow or groove is called a crease. The colour is either red, white or amber. The grain consists of grain coat, nucellar epidermis, endosperm and embryo. The seed is endospermic, oblong in shape and constitutes about 82–86% of the grain and consists of starch and gluten.

5.3.5 Anatomical description

Root anatomy

The young seminal root consists of a root cap. In the transverse section, the root shows three distinct regions, epidermis, cortex and vascular tissue (Fig. 5.21).

Epidermal cells have long unicellular root hairs in the root hair zone. These cells are compactly arranged and single layered.

The cortex with multilayered parenchymatous cells surrounds an inner cylinder of tissue called the stele. The inner layer of cortex that surrounds the stele shows Casparian stripes of endodermis.

The stele has alternate bands of xylem and phloem arranged around a central metaxylem vessel. In the root hair zone, lateral branch roots arise from the stele, adjacent to the phloem. They grow through the cortex and ramify into the soil, their structure resembling that of the main root. In the older regions of the root, the cortex dies leaving only the stele so that the root appears to become thinner as it gets older. In the deeper regions of the soil, the anatomy of the nodal roots is similar to that of a seminal root.

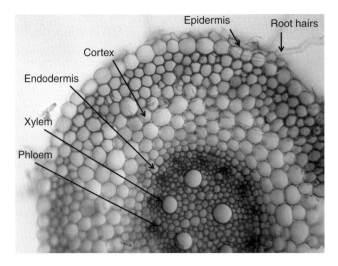

Fig. 5.21. Transverse section of root (100×) showing different parts.

Stem anatomy

At early stages the shoot apex is conical in shape and initiates leaves. As development progresses the apex becomes more cylindrical in shape, representing that the initiation of spikelet primordia has begun. The next identifiable stage is the double ridge stage. The lower, smaller ridge is a leaf primordium, the further growth of which is more or less completely stopped, but remnants may be seen at maturity beneath the lowest spikelet on the ear as a ridge of tissue sometimes referred to as a collar. The upper, larger ridge finally differentiates to become the spikelet. The double ridge stage is sometimes considered to be the commencement of floral differentiation, but it occurs from when 40–80% of the spikelets have been initiated.

A transverse section of wheat stem shows the typical structures present in other cereals, which include epidermis, ground tissue and vascular bundles (Fig. 5.22).

From the double ridge stage differentiation of the spikelets occurs, as the various floral structures (glumes, lemma and palea, lodicules, stamens and carpel) are initiated.

Ultimately, when about 20 spikelet primordia have been initiated, the final number of spikelets is determined by the formation of a terminal spikelet. This occurs when the

last initiated primordia, instead of becoming spikelet primordia, develop into glume and floret primordia. The terminal spikelet stage is regarded as a key stage in wheat phenology. At the terminal spikelet stage, the shoot apex is a fully formed embryo ear.

Studies of the anatomy of the shoot have revealed that the leaf and the florets originate in different tissues. Leaf primordia arise from the superficial layer of cells (dermatogen, tunica), while the spikelet primordia are initiated in the deeper layers of the apex, the corpus or core.

Significance of variation in stem anatomy

The monocot stems provide mechanical support for lodging resistance as well as ability to bend under wind pressure or when the top is heavier during the grain filling period. Wheat is an important lingo-cellulosic feedstock. Lodging resistance is also related to digestibility of the straw as animal feed. The high yielding semi-dwarf wheat varieties that are more resistance to lodging and produce stiffer straw are less palatable as animal feed. Lodging resistance is directly related to the diameter of culm, diameter of culm cavity and percentage of mechanical tissues (Li *et al.*, 2000). Higher deposition of lignin in the culm tissues also provides mechanical strength and thus lodging resistance in

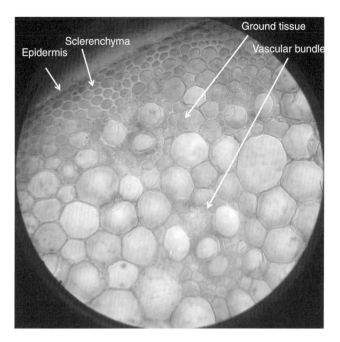

Fig. 5.22. Transverse section of stem (small portion 400×) showing epidermis, sclerenchyma, ground tissue and vascular bundle.

wheat (Zhu *et al.*, 2004). Thickness of sclerenchyma, thickness of epidermis and density of epidermis are important factors for digestibility of wheat straw as ruminant feed (Travis *et al.*, 1996). The cultivars less susceptible to lodging contain lower amount of neutral digestible fibre, and thus have less nutritive value as animal feed.

Leaf anatomy

There are three main features of the anatomy of the leaf, which includes epidermis, mesophyll and vascular bundles (Fig. 5.23).

The upper and lower epidermis of the mature leaf encloses the mesophyll, which is traversed at intervals by the vascular tissue. The upper epidermis is a complex tissue with several cell types. The bulliform cells are the largest cells lying between the veins at the bottom of the furrows. Next to the bulliform cells are long cylindrical cells with a smaller diameter than the bulliform cells, alternating in a regular manner with stomata. There is generally a single row of stomata between each rank of bulliform cells and the vascular tissue. Each stoma is made up of two characteristically shaped guard cells and has two associated accessory cells. There are more on the upper surface and these are more densely distributed towards the tip. On the other flank of the row of stomata, over the veins there are long cylindrical cells characterized by thickened wavy walls. The long cells are interspersed in a regular manner by short cells of two types, cork cells and silica cells. Short, unicellular hairs occur mainly over the veins and on either side of the row of stomata. The lower epidermis has fewer cell types, mainly the long cylindrical cells with wavy walls interspersed by short cells. Stomata occur in the same position relative to the veins as in the adaxial epidermis and, although hairs occur, they are less frequent than on the upper epidermis. The epidermis on both surfaces of the leaf has a cuticle with strongly developed epicuticular wax. The form of the wax depends upon the surface of the leaf and the position of the

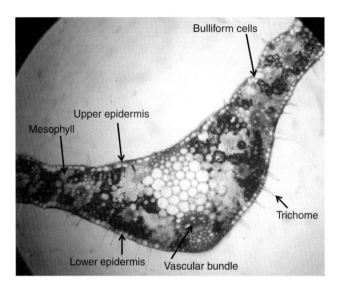

Fig. 5.23. Transverse section of leaf (40×) showing different parts.

leaf on the stem. This occurs as lobed plates, simple plates, flat ribbons and tubes, the amount and form of the wax depending on the position and surface of the leaf. The vascular tissue lies beneath the ridges of the lamina and the associated thickening capping the vascular bundle of the midrib, and the major veins extend from the ad- to the abaxial epidermis.

The mesophyll cells are of a complex lobed shape, resembling armed palisade cells. When viewed in transverse section, the sub-epidermal cells of the mesophyll are elongated similar to palisade cells. The cells in the middle layers of the leaf are not so elongated. Viewed in longitudinal section, the lobed nature of these cells is apparent. In the leaves at the base of the plant, the degree of lobing is low and the dimensions of the lobes are large. With ascending leaf position up the stem, the degree of lobing increases and the diameter of the lobes decreases. The effect of these changes is to increase the cell surface area per unit area of leaf with ascending leaf position up the stem. There is variation in the compactness and arrangement of the mesophyll cells. Some cultivars have a relatively loose arrangement of cells, while in others the cell arrangement is more compact and files

of cells radiate in a regular manner from the vascular bundles. Prominent sub-stomatal cavities occur, particularly beneath the stomata of the abaxial surface of the leaf. The vascular tissue and mesophyll are organized in alternate strips of tissue running parallel with each other along the long axis of the leaf.

The vascular bundle has the structure typical of a C_3 plant. The phloem is abaxial to the xylem and in the larger bundles consists of regularly arranged sieve tubes and companion cells. The xylem has two large, prominent xylem vessels between which are smaller metaxylem vessels and fibres. Adaxial to the metaxylem, there is an area of disrupted protoxylem. The conducting elements are surrounded by an inner (mestome) sheath and an outer (parenchyma) sheath, though these are not as clearly defined as in some other C_3 plants. The cells of the mestome sheath are small and thick-walled and are without chloroplasts. Those of the outer bundle sheath are large and thin-walled and contain chloroplasts. In longitudinal section, the cells of the bundle sheaths are elongated with blunt ends. The walls of the mestome sheath are lignified, and sometimes the wall adjacent to the conducting elements is thicker than the other walls of the cell. The complex fine structure of the mestome sheath is important in regulating the transport of water and solutes. The small veins that interconnect the main longitudinal veins consist only of a single sieve tube and xylem vessel and two files of parenchyma cells. They pass through the mestome and parenchyma sheaths and connect directly with the metaxylem and metaphloem of the main bundles. They do not have bundle sheaths, and the vessel walls have a complex fine structure depending on the adjacent mesophyll walls. The major bundles run parallel with each other the whole length of the leaf. The small transverse veins, which constitute about 7% of the total length, occur every 2.5–3 mm, towards the tip of the leaf, the smaller longitudinal bundles terminate in a Y-shape, the forks of the Y comprising small transverse veins that link to the longitudinal veins at either side. At the pointed tip of the leaf, the veins converge and connect with each other. The distance between the longitudinal veins varies from about 0.3 mm in the first leaf to about 0.15 mm in a culm leaf.

Significance of variation in leaf anatomy

Wheat is a crop having varied level of ploidy (diploid, tetraploid and hexaploid). Polyploidy has been observed to be associated with increased vegetative vigour in many plant species, but these variations are not prominent among the wheat species differing in ploidy. Both *T. monococcum* and *T. aestivum* exhibit similar growth habits. A common manifestation of polyploidy is increase in cell size to accommodate larger genome volume. *T. aestivum* cells are larger in volume than *T. monococcum* cells, but the number of mesophyll cells per unit area is more than double in diploid wheat compared to hexaploid wheat.

Thickness of leaf epidermis and hypodermis has been observed to be associated with resistance to biotic stresses and tolerance to drought in wheat. The glossy appearance of wheat leaf caused by suberization and wax deposition on the epidermis helps to reduce water loss, lower internal leaf temperature by reflecting sunlight and

act as a mechanical barrier to fungal penetration or insect-pest attack (Carver and Thomas, 1990). Leaf thickness is thus considered to be a selection criterion for development of drought-tolerant wheat cultivars. Wheat genotypes adapted to the very hot climates of Baluchistan region of Pakistan show increase in thickness of both epidermis and hypodermis under xeric adaptation (Hameed *et al.*, 2002). These drought-tolerant genotypes also exhibited lower stomatal density on the adaxial surface than the abaxial surface, a common mechanism observed for reducing transpiration in plants under drought stress. A lower ratio of xylem to mesophyll tissues, increase in lignin deposition and thickening of sclerenchyma have been suggested to be indicative of drought tolerance in wheat (Ridley and Todd, 1966).

The photosynthetic activity of wheat leaves is higher than other C_3 crops except rice, which is due to high number of mesophyll cells per unit area, arrangement of chloroplasts in cells to maximize solar radiation and the high concentration of rubisco protein in the leaves.

The seed

The embryo or germ is situated at the point of attachment of the spikelet axis, and the distal end has a brush of fine hairs. The embryo is made up of the scutellum, the plumule (shoot) and the radicle (primary root). The scutellum is the region that secretes some of the enzymes involved in germination and absorbs the soluble sugars from the breakdown of starch in the endosperm. The coleoptile is well developed in the embryo, forming a thimble-shaped structure covering the leaf primordia and the shoot meristem. Two vascular bundles are found laterally placed with reference to the scutellum–coleoptile plane. One or two lines of stomata extending from the base to the tip are found in the outer epidermis associated with each vascular bundle. They tend to be more numerous towards the tip. Stomata also occur in the inner epidermis though they are less frequent. The guard cells of the stomata do not have the characteristic dumb-bell shape, such as those found in the leaf, but are kidney-shaped, similar to dicotyledon guard cells.

Embryo development

Division of the fertilized egg nucleus commences later than that of the endosperm. The early divisions produce a five-celled embryo with a basal cell, although variation in the pattern of development has been observed. Continuing cell division produces at first a club-shaped structure, which ultimately differentiates to form a mature embryo in the ripe seed.

The growth of the embryo is supported by nutrients derived from the antipodal cells and from the hydrolysis of parenchyma cells of the nucellus and neighbouring endosperm cells. Later in development, transfer cells appear in the nucellar endosperm epidermis, near the base of the embryo and the provascular tissue.

Endosperm development

Following the fusion of the sperm nucleus and the polar nuclei, cell division is, for a time, synchronous, the number of endosperm cells doubling every 4–5 h. At first, the endosperm is coenocytic, but after about 3 days cell walls are formed. Cell wall growth commences at the edge of the embryo sac and furrows inwards to the central vacuole. The cells at the periphery of the endosperm divide, and eventually the entire embryo sac is cellular. After cell formation is complete, the sub-cellular structures, which will synthesize the protein bodies and the other cell components, are formed. Amyloplast division ceases before cell division, and starch grains differ in growth rate in different cell layers of the endosperm.

5.4   Sorghum

5.4.1   Introduction

The genus *Sorghum* belongs to the family Gramineae, tribe Andropogoneae, and includes annual species. In India it is commonly called Jwaarie or Jowar or Jondhahlaa. Sorghum

(*Sorghum bicolor*) is the fifth most important cereal crop and is the dietary staple of more than 500 million people in more than 30 countries. It is grown on 42 Mha in 98 countries of Africa, Asia and the Americas. Nigeria, India, the USA, Mexico, Sudan, China and Argentina are the major producers. Other sorghum-producing countries are Mauritania, Gambia, Mali, Burkina Faso, Ghana, Niger, Somalia and Yemen, Chad, Sudan, Tanzania and Mozambique. Sorghum is a self pollinating diploid ($2n=2x=20$) with a genome ($1C=735$ Mbp) about 25% the size of maize or sugarcane.

5.4.2   Origin and domestication

Nothing is known about when *S. bicolor* was first brought into cultivation, but Murdock (1959) stated that, along with several West African crops, it was domesticated some 7000 years ago.

The sorghum in India has been known since ancient times, but the wild species, more or less adapted, is found throughout the world. Engravings have been found in ruins in Asia that represent sorghum in 700 BC. It was probably introduced into the interior of Asia at the beginning of the Christian era, when sorghum was also known in the Mediterranean region and tropical Africa.

The domestication of sorghum was initiated by allelic changes in only two loci, resulting in different sections, followed by the innovation of the cropping techniques. Wild sorghum disperses its seeds by the rupture of inflorescence nodes with subsequent spacing of the seeds. The essential step adapted for the domestication was the harvest of the inflorescence, and the use of grain for sowing. The plants are cultivated in warmer climates worldwide. Species are native to tropical and subtropical regions of all continents in addition to the south-west Pacific and Australasia.

5.4.3   Utilization

Sorghum is a major source of food for millions of people in the semi-arid tropics. In tropical areas sorghum grain is important as food and as livestock feed. The stem and foliage are used as a green crop, hay silage and pasture. The stems are also used as fuel and building materials. In temperate areas it is a major source of cattle feed, except in China, where it is primarily used for food. Sorghum is used in the preparation of different types of food, and unleavened bread is the most common food made from sorghum flour. The dough is sometimes fermented before the bread is prepared and the grain is boiled to make a porridge or gruel. It is also used in the preparation of biscuits. Beer is prepared from sorghum grain in many parts of Africa. As well as these products, popped and sweet sorghum, which are parched, are also eaten in some countries (House, 1980). Sweet sorghum is emerging as a feedstock for ethanol production. Sorghum grain has high levels of iron (>70 ppm) and zinc (>50 ppm) and is hence being targeted as a means to reduce micronutrient malnutrition globally.

5.4.4 Morphological description

Sorghum seedlings can be morphologically glossy or nonglossy: seedlings with dark green leaves (normal) are nonglossy, and seedlings with light yellow-green and shining leaf surface are glossy. Epicuticular wax in varying intensity imparts variability in glossiness in sorghum seedlings. A systematic study of the world sorghum germplasm collection indicated a low frequency of accessions with the glossy trait. A large proportion (84%) of the glossy lines was of Indian origin but some were from elsewhere (Nigeria, Sudan, Cameroun, Ethiopia, Kenya, Uganda, South Africa and Mexico). Most of the glossy lines are in the *durra* group but some are from the taxonomic groups *guinea, caudatum* and *bicolor*. Glossy lines vary in morphological, anatomical and agronomic attributes, many being extremely late or photoperiod-sensitive and very tall. Some lines are early maturing, intermediate to dwarf in height and agronomically good. Studies have indicated that glossy lines contribute to shootfly resistance

and seedling drought resistance. Source material for the glossy trait is maintained by the Genetic Resources Unit at ICRISAT.

Vegetative characters

Sorghum is a genus of numerous species of grasses. It reaches to a height ranging from 2 to 4 m, having various stems with tillers, each bearing one inflorescence in a panicle. One fertile sessile spike is accompanied by two sterile pendant spikes, the characteristics of the genus. It is well adapted to semi-arid conditions; most varieties are drought tolerant and heat tolerant, and are especially important in arid regions where the grain is the staple or one of the staples for poor and rural people. It has a fibrous root system, and forms stilt roots. The stem is aerial, erect, greenish and hard, succulent with nodes and internodes. Leaves are simple, exstipulate, alternate, sessile and long, arranged in two rows on stem, with an expanded leaf base with leaf sheath, wavy margin and parallel venation. Deposition of silica crystals in the leaves is a characteristic feature of grasses.

Floral characters

Bisexual flowers are formed on sessile spikelets. Flowers are sessile, unisexual, zygomorphic, epigynous; in every flower there is lemma, palea and two lodicules. There are three stamens with long anthers, dithecous, base fix. The gynoecium is unilocular, unicarpellary, with a single ovule on basal placentation, a silky-like style and bifid feathery stigma. The fruit is a caryopsis, generally called grain or seed. Seeds are endospermic (Fig. 5.24).

5.4.5 Anatomical description

In cereals, the formation of safety zones in the root–shoot junction could protect the vessels of roots from embolism originating in the shoot. The root–shoot junction was examined both anatomically, with a light microscope, and experimentally, using a pressurized-air method, in the base of seminal

Fig. 5.24. Sorghum plants with spike inflorescence.

and adventitious roots of maize (*Z. mays* L. cv.), a corngrass mutation of maize (*Cg* mutant), sorghum (*S. bicolor* L. cv.), winter oats (*Avena sativa* L. cv.), spring wheat (*T. aestivum* L. cv.), winter wheat (*T. aestivum* cv.), winter barley (*Hordeum vulgare* L. cv.), spring rye (*Secale cereale* L. cv.) and winter rye (*S. cereale* cv.). Two types of hydraulic architecture were found in the cereal roots: (i) a very safe root vessel system, as in winter rye, in which the vessels of the roots are separated from those of the shoots by tracheids; versus (ii) a completely unsafe system, as in corngrass, where the vessels in the root are continuous with the vessels in the shoot. The xylem anatomy of the seminal roots is generally correlated with the species-specific overall root morphology. Rye, wheat and barley, which develop four to six seminal roots, show a high degree of vascular segmentation resulting in the formation of safe root vessels; maize, sorghum and oats, which typically develop a primary seminal root, contain unsafe vessels that are continuous through the mesocotyl and through the first node. As significant differences in vascular segmentation of the root–shoot junction occur not only between species, but also between cultivars, we suggest that selection based on the occurrence of safety zones might be used in breeding programmes designed to improve adaptation of cereals to drought and cold temperatures (Aloni and Griffith, 1991).

Root anatomy

The root anatomy of *S. bicolor* consists of an outer piliferous layer surrounding a cortex composed of outer solid parenchymatous layer without air cavities and the innermost layer, the endodermis, which envelopes the central stele (Fig. 5.25). Endodermal cells contain lignosuberin thickening called the Casparian band. Xylem and phloem are arranged in the form of separate strands on different radii. They alternate with each other; hence the vascular strands are described as separate and radial. Xylem is exarch having protoxylem towards the pericycle and metaxylem towards the centre or medulla (centripetal development). Xylem is polyarch and protoxylem points range from 25 to 30. Cambium is absent and hence vascular bundles are described as closed type. Pith is scanty. The non-vascular sclerenchymatous tissue left between xylem and phloem strands is called conjunctive tissue.

Little information is available on root anatomy of sorghum in relation to adaptation. The anatomy of nodal roots is similar to that of seminal roots, but extensive sclerenchyma regions develop in both the hypodermis and stele. In sorghum roots, silicon is accumulated in specialized idioblasts termed 'silica cells'. These cells are dumb-bell shaped. Silicification in root endodermis and leaf epidermis is observed to be higher in drought-tolerant genotypes. Lux *et al.* (2002) suggested

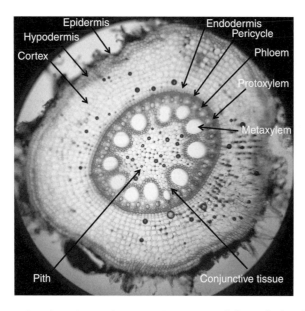

Fig. 5.25. Transverse section of root (100×) showing various parts. Radial vascular bundles are visible in the stele with prominent pith.

that high root endodermal silicification might be related to drought tolerance.

Significance of variations in root anatomy

Cultivar differences are expressed in a number of root characteristics in sorghum. These are:

1. Presence or absence of sclerenchymatous exodermis in the cortex.
2. Thickness of endodermal cell walls (Fig. 5.26).
3. Density, size and shape of crystals in the endodermal cell cavity.
4. Intensity of lignification in the pericycle cell layers and in cells surrounding the vascular bundles.

The presence of exodermis, higher thickness and higher lignification provides better structural integrity to the root system. Thus genotypes exhibiting these characteristics are expected to have better crop stand and resistance to lodging.

Stem anatomy

A transverse section of a stem shows cuticularized epidermis followed by a sub-epidermal fibro-vascular bundle with thick-walled sclerenchyma, which is of smaller sizes, followed by rings of bigger vascular bundles in the cortex (Fig. 5.27).

In transverse section, the epidermis consists of cubical to boat-shaped epidermal cells with thickened outer wall with cutin. The ground tissue consists of thin-walled parenchyma. The vascular bundles are pointed collaterally and scattered in the ground tissue. There is a continuous layer of sclerenchyma just below the epidermis. There are alternate bands of large and small vascular bundles below the hypodermal sclerenchyma band. The vascular bundles towards the centre are larger. The mechanical tissues in the outer vascular bundles are extensive, particularly in the peripheral region.

In the stem, different types of vascular bundles are found:

1. Large bundle is sheathed by sclerenchymatous mass on lateral and xylary polar ends and a voluminous phloic cap of sclerenchyma just beneath the epidermal layer.
2. Similar to above, but without epidermal sclerenchyma sheath.
3. Comparatively small with smaller xylary fibrous strands and the phloic cap equal in

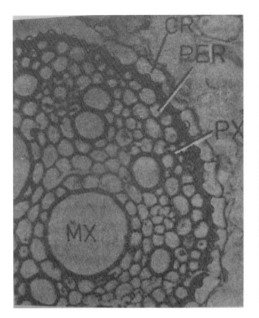

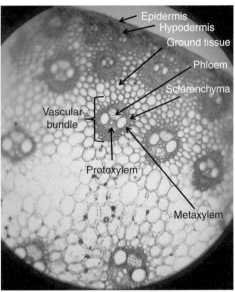

Fig. 5.26. Transverse root section (400×) of sorghum showing highly thickened endodermal cell wall and silica crystals in the endodermal cell cavity: CR, silica crystal; MX, metaxylem; PER, pericycle (Maiti, 1996).

Fig. 5.27. Transverse section of stem (100×) illustrating epidermis, hypodermis, and scattered vascular bundles in ground tissue. Each vascular bundle is surrounded by a high number of sclerenchyma. Smaller vascular bundles are situated at the periphery, with larger bundles at the centre of the stem.

volume to that of the xylic sclerenchyma strand.

4. Smallest vascular bundles intermingled between the above three types with the absence of phloic elements.

Significance of variations in stem anatomy

Stem anatomy has been studied in a number of genotypes to determine variability, if any, in internal structure. Examination of transverse sections, collected at physiological maturity reveals considerable variation in distribution and amount of mechanical tissue (sterome) and vascular bundles, and in the size and shape of the protoxylem cavity.

Sorghum stem is a common animal feed. Digestibility of sorghum as animal feed is linked to anatomical features. As observed in wheat, lignification reduces digestibility of sorghum as fodder.

Mechanical tissue, consisting of sclerenchyma and collenchyma, form the sterome

system of the plant. On the basis of the amount of mechanical tissue present, the genotypes may be tentatively ranked in the following classes: weak sterome system, intermediate system and strong sterome system. While breeding for new genotypes, anatomical studies may help in better selection of genotypes according to the target environment and breeding objectives.

Leaf anatomy

The upper epidermis may be easily identified due to the presence of xylem and bulliform cells towards it. The epidermis is uniseriate and composed of more or less oval cells (Fig. 5.28). The outer walls of the epidermal cells are highly cutinized, which reduces the loss of water due to transpiration. Epicuticular wax is present on the cuticle of the epidermis (Fig. 5.29b, c). The graminaceous type of stomata (stomata are covered by two dumb-bell shaped guard

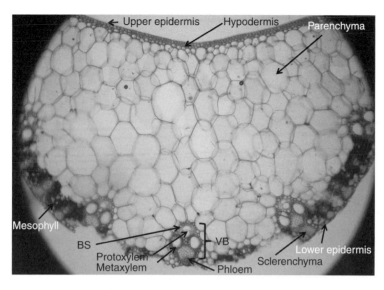

Fig. 5.28. Transverse section of leaf (100×) showing various parts, including upper epidermis, sclerenchyma, mesophyll and fibro-vascular bundle (VB), bundle sheath (BS), protoxylem and phloem.

cells) are present in both of the epidermal layers, but the stomata are usually more frequent in the lower epidermis. All the epidermal cells except the guard cells are colourless. Epidermis is rough in nature due to the presence of silica crystals (Fig. 5.29a). Two types of hairs can be observed in sorghum, non-glandular and glandular trichomes. Leaf trichome density (non-glandular pointed trichomes) and glossiness intensity were considered reliable and very simple indicators in the selection of sorghum genotypes tolerant to shootfly. The presence of bilobed glandular trichomes (Fig. 5.29d) induces susceptibility to this insect in sorghum. As the leaf is isobilateral, the mesophyll is not differentiated into palisade and spongy tissues, but is composed of compactly arranged thin walled, isodiametric chlorophyllous cells having well developed intercellular spaces among them. The leaf receives the vascular supply from the stem. The vascular tissue occurs in the form of discrete bundles called veins. The vascular bundles are collateral and closed. Most of the bundles are small in size but fairly large bundles also occur at regular intervals; the xylem is found towards the upper side and phloem towards the lower side of the bundles. As

sorghum is a C_4 plants it shows Kranz anatomy (Fig. 5.30). Usually each bundle remains surrounded by a bundle sheath consisting of thin-walled parenchyma cells. The cells of the bundle sheath generally contain large chloroplasts devoid of photosystem I (without grana) and starch grains in them.

Leaf wax – cuticle

Analysis of *S. bicolor* bloomless (*bm*) mutants with altered epicuticular wax (EW) structure uncovered a mutation affecting both EW and cuticle deposition. The cuticle of the mutant line was about 60% thinner and approximately one-fifth the weight of the wild-type parent cuticles. Reduced cuticle deposition was associated with increased epidermal conductance to water vapour. The reduction in EW and cuticle deposition increased susceptibility to the fungal pathogen *Exserohilum turcicum*. Evidence suggests that this recessive mutation occurs at a single locus with pleiotropic effects. Chemically induced mutants had essentially identical EW structure, water loss and cuticle deposition (Jenks *et al.*, 1994).

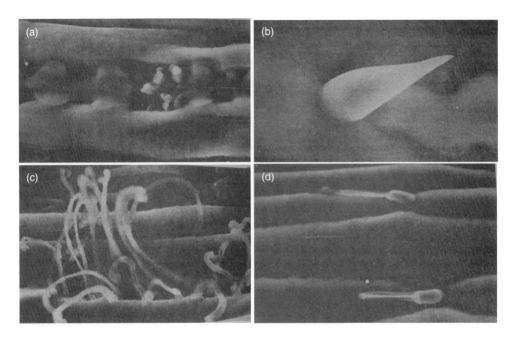

Fig. 5.29. Leaf surface structures (SEM): (a) bilobed and trilobed silica crystals; (b) glossy line showing a smooth wax layer and a trichome (SEM); (c) Genotype IS 4776 showing long wax strands (SEM); (d) Genotype IS 844 showing glandular bicellular trichomes (Maiti, 1996).

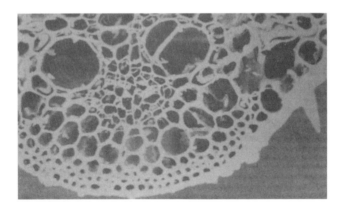

Fig. 5.30. Scanning electron micrograph of the midrib showing thick-walled epidermis, trichome, sclerenchyma sheath and Kranz anatomy (Maiti, 1996).

Trichome

Sorghum leaf surfaces show the presence of non-glandular pointed trichomes oriented at acute angles.

Two types of hairs can be observed in sorghum: are non-glandular and glandular trichomes. Leaf trichome density (non-glandular

pointed trichomes) and glossiness intensity were considered reliable and very simple indicators in the selection of sorghum genotypes resistant to shootfly. The presence of bilobed glandular trichomes induces susceptibility to this insect in sorghum. The presence of non-glandular trichomes acts as barriers to the laying of eggs and migration of larva downwards through the collar. In addition, the glossy leaf surface offers difficulty egg-laying, which may cause the eggs to slide off. There is a large variation in the intensity of trichomes and it has been found that shootfly tolerance increases with intensity of trichomes (Maiti *et al.*, 1984; Maiti, 1996).

Genetic study has been undertaken on the inheritance of glossy leaves in sorghum. In a study it was reported that the glossy leaf is a genetic trait that showed pleiotropic effects on disease and pest resistances. True-glossy (*tg*), glossy (*g*) and non-glossy (*G*) were multiple alleles located in a glossy locus (Tarumoto, 2006).

Significance of variations in leaf anatomy

The different sorghum genotypes show differences in the following leaf anatomical features:

1. Amount of cuticularization.
2. Intensity of silica crystals in the upper epidermal cells offering rigidity to the leaf as well as resistance to leaf insects.
3. Size and shape of bulliform cells helps in leaf rolling, thereby avoiding loss of transpiration under drought situations.
4. Type and arrangement of vascular bundles both in the lamina and midrib.
5. Characteristic size and shape of bundle sheath chloroplasts.

From observing these variations, the following inferences can be drawn on the significance of leaf anatomical features:

1. Presence of thick cuticle on the epidermis of the leaf reduces transpirational loss of water.
2. High density of trichomes covers the surface and reduces the direct sunlight effect and minimizes leaf temperature, thereby reducing transpirational loss of water.
3. High density of trichomes offers insect resistance, especially sucking pest tolerance.
4. More numerous and larger size vascular bundles help in efficient translocation of water.

5. Reduction in stomatal density on lower and upper epidermis minimizes the loss of water from leaf.
6. Presence of long size bulliform cells prevents evapo-transpiration.

The seed

Two types of endosperm are found in the seed, corneous and floury (Fig. 5.31). The corneous endosperm is also called vitreous endosperm and is present at the periphery, while the floury endosperm is found in the inner core of the main bulk of the endosperm. Cells in the corneous endosperm are oval or slightly elongated with compactly arranged starch and protein granules, while the cells in the floury part are much more elongated and broad with loose arrangements of starch and protein granules in the lateral and dorsiventral sections of the endosperm. In lateral section, the scutellar tissue, along with the embryo can be seen clearly near the broad hilar region. The pointed end of the embryo is inclined towards the lateral wall of the seed. The corneous endosperm encloses the inner mass of floury endosperm.

The seeds can be classified according to the intensity of corneous over floury endosperm (Fig. 5.32a–d):

1. Almost the entire endosperm is corneous – only a very small floury portion may be present in the centre; completely corneous (IBPGR).
2. Over 50% of the endosperm is corneous – the endosperm is broad; almost corneous (IBPGR).
3. The endosperm is almost floury – a thin layer of corneous endosperm may be present that should not exceed 10% of the entire mass that surrounds the broad floury endosperm mass; almost starchy (IBPGR).
4. More than 75% of the endosperm is floury – the endosperm layer is medium broad; partly corneous (IBPGR).
5. The whole endosperm is floury – a thin corneous field of the endosperm may be present, full of starch (IBPGR).

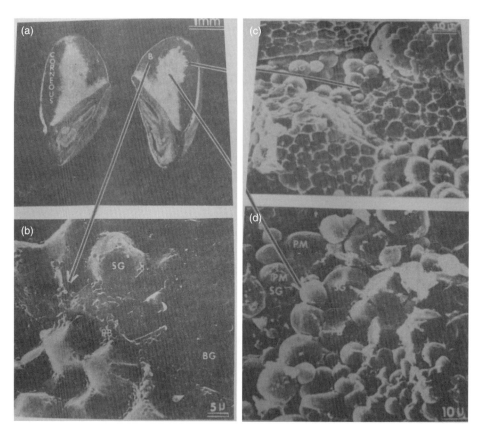

Fig. 5.31. Scanning electron microscopic analysis of the corneous and floury endosperm of the sorghum seed. (a) Longitudinal section of the corneous and floury endosperm. (b) Typical structure of the corneous endosperm with a continuous face between the flour and the protein, resulting in a space without air between the starch granules; the corneous endosperm appears extremely dense and vitreous. (c) Transition section with corneous and floury endosperm cells. (d) The floury endosperm contains air spaces between the starch granules with a clustered and loose appearance. SG, starch granules; PM, protein matrix; PB, protein bodies; BG, broken granules (Courtesy L. Rooney, Texas A&M University).

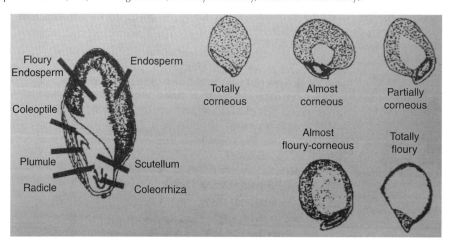

Fig. 5.32. Parts of the seed and texture of the endosperm (Maiti, 1996).

Anatomy of mature kernel

The structure and chemical components of a mature seed are of great interest to the food chemist. In mature seed, the seed wall is fused with the fruit wall, which is a characteristic of monocot seed. The caryopsis consists of three main parts: the pericarp or outer covering, the endosperm or storage tissue and the germ or embryo (Fig. 5.33). Details of these structures have been discussed by many authors (Rooney and Clark, 1968; Rooney and Sullins, 1977). Glueck and Rooney (1978) have given a detailed account of the chemistry and structure of the grain in relation to mould resistance. The following section describes the structure of different parts of a caryopsis, such as pericarp, testa, embryo and hilum.

Description of the parts of a caryopsis

The pericarp, which forms the peripheral boundary of the seed, consists of epicarp, mesocarp and endocarp. The outermost layer or epicarp is made up of two to three layers of long and rectangular cells, which contain wax and pigments. The hypocarp consists of two to three layers of cells. This is followed by seven to eight dense layers of mesocarp cells containing small starch granules embedded in a fluid proteinaceous material within the cell. Mesocarp cells vary in thickness. Sorghum cultivars with a thin mesocarp withstand weathering better than those with thick mesocarp and have a better milling quality (Glueck and Rooney, 1976). The innermost layer of the pericarp is the endocarp, which is composed of cross cells and tube cells. The main function of cross cells is to transport water. This the breakage point of the inner mass of the endosperm. Mesocarp pigments influence grain colour during milling. The highly pigmented layer just beneath the pericarp is known as the testa. Testa pigmentation is associated with polyphenols or tannins (Glueck *et al.*, 1978). The testa and pericarp thickness varies greatly among sorghum varieties and genotypes.

Endosperm

The endosperm consists of an aleurone layer, the peripheral corneous endosperm and the central floury portion. The aleurone layer is located beneath the pericarp, which consists of a single layer of narrow rectangular cells, which do not contain starch grains. The continuous portion of the endosperm possesses a continuous interface between starch and protein.

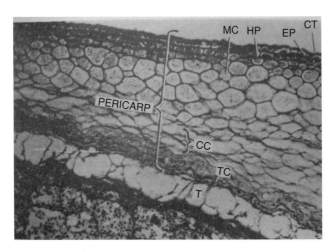

Fig. 5.33. Transverse section of a mature caryopsis (400×) showing cuticle (CT), epicarp (EP), hypocarp (HP), mesocarp (MC), cross cells (CC) and tube cells (TC) (Courtesy L. Rooney, Texas A&M University).

Just beneath the corneous portion of the endosperm are located several layers of large elongated and vacuolated cells forming the floury endosperm portion. The starch granules are spherical and are not held together by a protein matrix. The protein bodies and matrix are present in the floury endosperm.

Embryo

The embryo contributes 10% of the total dry weight of the seed. The scutellum, consisting of vacuolated parenchyma cells, has a well-developed vascular system and helps in the translocation of nutrient from the endosperm into the developing roots and leaf tissues of the embryonic axis during germination. The hilum helps in the translocation of nutrients from the vegetative plant parts into the ovule during caryopsis development. Translocation of the nutrients into developing endosperm takes place through specialized transfer cells in the scutellum.

Research trends on sorghum anatomy

Remarkable research advances have been undertaken with respect to anatomy of different parts of the plants with respect to insect and drought tolerance. With respect to general anatomy there exists a large variation in leaf, stem and root anatomical structures among sorghum genotypes. There also exists a large variation in leaf surface structures with respect to glossiness, non-glossy trait, types of trichomes and intensity and silica content, bulliform cells and Krantz tissue. Similarly, variation exists in stem anatomy with special reference to the intensity of mechanical tissue-stereome, which offers resistance to stem borers as well as drought. The presence and intensity of non-glandular pointed trichomes and glossiness are inversely correlated with shootfly resistance. The intensity of trichomes and glossiness on the other hand may create a microclimate thereby reducing radiation load and transpiration, emphasizing the great importance of these traits in sorghum crop adaptation to both insect

and drought stress. Therefore, there is a great necessity to screen sorghum cultivars and germplasm for these traits and to utilize these traits in sorghum crop improvement. There is also a necessity to select sorghum genotypes with minimum sclerenchyma in the stem for increasing the acceptability as fodder for cattle.

5.5 Pearl Millet

5.5.1 Introduction

Pearl millet (*Pennisetum glaucum* (L.) R. Br.) is an important grain crop as well as food for semi-arid and arid regions of world. In India it is known as *bajra*, and is primarily consumed in the states of Haryana, Rajasthan, Gujarat and Madhya Pradesh. It is used as hay and for pasture. It is one of the most important food and forage cereals in the semi-arid tropics of many African countries, in India, Pakistan, Bangladesh, Burma, Sri Lanka, Argentina and the USA. India is the largest producer of pearl millet. Pearl millet is the staple food and fodder crop of farmers in the semi-arid areas of north-west India.

5.5.2 Origin, domestication and adaptation

Pearl millet has been present in the African semi-arid tropics since 1100 BC. The origin of pearl millet has been traced to tropical Africa, with its centre of diversity for the crop in the Sahel zone of West Africa. Cultivation subsequently spread to east and southern Africa, and southern Asia. Records exist for cultivation of pearl millet in the USA in the 1850s, and the crop was introduced into Brazil in the 1960s. The greatest morphological diversity of pearl millet can be found today in the Sahel zone, located in western Africa, south of the Saharan desert and north of the forest zones. It has been concluded that pearl millet is the product of multiple domestications due to the high diversity present.

Pearl millet is well adapted to production systems characterized by drought, low soil fertility and high temperature. It performs well in soils with high salinity or low pH. Because of its tolerance to difficult growing conditions, it can be grown in areas where other cereal crops, such as maize or wheat, would not survive.

5.5.3 Utilization

Pearl millet grains are consumed as food in most of the African countries and some states of India (Haryana, Rajasthan, Gujarat and Madhya Pradesh). It is also used as pasture. Pearl millet is an important food across the Sahel; it is often ground into flour, rolled into large balls, parboiled, and then consumed as porridge with milk. Sometimes it is prepared as a beverage.

5.5.4 Morphological description

Pearl millet is a genus of numerous species of grasses. *Pennisetum americanum* is an annual erect plant with profuse tillering habit that grows up to 2–4 m.

Vegetative characters

Pearl millet is a herb, height 2–4 m, and is a mesophyte, well adapted to semi-arid conditions. The root system is somewhat large and fibrous and therefore has a great capacity for efficient use of the soil. It can also utilize soil nutrients better than other crops due to the deep penetration of its roots into the interior layers of the soil. The plant is aerial, erect, greenish and hard, with succulent nodes and internodes. The nodes are slightly swollen, softly haired (pubescent), while the internodes are cylindrical and glabrous. The length of the internodes decreases gradually from the base of the stem upwards. One light groove can be found above each node containing the axillar bud. The bud is slightly swollen and produces the ring of adventitious root primordium in

its base, usually fuzzy and sometimes hairy. The fuzz is rough and rigid. Leaves are simple, exstipulate, alternate, sessile, long, arranged in two rows on stem, with an expanded leaf base with leaf sheath, wavy margin, and parallel venation. The leaves are straight, 20–100 cm long and 0.5–5 cm wide, usually glabrous with spaced fuzz. The base of the leaf is slightly auricular. The ligule is narrow, membranous, with a strip of fuzz that almost surrounds the stem. The leaf sheath surrounds the stem completely and is thicker than the blade; the external surface of the sheath can be glabrous or covered with rough hairs in some fuzzy varieties.

Floral characters

Bisexual flowers are formed on sessile spikelets and staminate or male flowers are formed on pedicillate spikelets. This is a cross-pollinated crop (80%). The terminal inflorescence is a cylindrical, bulbous panicle. The peduncle is thin, cylindrical and covered with soft fuzz under the spikelets. The rachis is straight, cylindrical, solid and covered with soft and short fuzz. The small rachis produces a series of florets placed in a spiral form on the rachia. The rachillas are small and covered with short hairs, shorter in the apex of the panicle and denser approximately in the middle portion. Flowers are sessile, unisexual, zygomorphic and epignous; in every flower there is lemma, palea and two lodicules. Each spikelet consists of two sterile glumes and two florets, the lower floret is generally male and the one above hermaphrodite; glume 1 is small or 0, orbicular; glume 2 is small or 0, variable. Lemma 1 is oblong to ovoid, palliated; the palea is membranous and staminated; lodicules are absent. There are three stamens: typical pencilated, long anthers, dithecous, basally fixed. The carpel is unilocular, unicarpellary, with a single ovule on basal placentation, silky-like style and bifid feathery stigma. The gynoecium is protogynous. The grain is grey, pearl-white, and rarely yellow and endospermic. It is a caryopsis (Fig. 5.34).

Fig. 5.34. Pearl millet plant at the grain filling stage.

5.5.5 Anatomical description

Root anatomy

The anatomical structure of the root of pearl millet is characteristic of monocots. It is composed of an outer epidermis, ground tissue, cortex below the epidermis, an endodermis surrounding the vascular bundles that consist of xylem and phloem in alternating bands, and pith (Fig. 5.35). The epidermis consists of elliptical cells subtended by two layers of hypodermis. In the early stages of growth, the cortex of the root epidermis consists of ovoid parenchymatous cells with considerable intercellular space. As the root ages, the cortical cells elongate and take a plate-like shape. A single layer of ovoid or cubical parenchyma cells surrounds the endodermis. The endodermal cells are barrel- to boat-shaped, and are thickened with suberin on the inner tangential wall. Crystals are present in the endodermal cells, their shape and size varying with the cultivar and age of root (Fig. 5.36). The pericycle below the endodermis is composed of one to several layers of thick-walled sclerenchyma cells, of which the outer layer is highly lignified; the parenchyma that surrounds the metaxylem may be thick-walled or lignified. The size of the metaxylem varies among cultivars.

Significance of variations in root anatomy

Cultivar differences are expressed in a number of root characteristics in millet. These are:

1. Presence or absence of sclerenchymatous exodermis in the cortex imparting resistance to the tissue.
2. Thickness of endodermal cell walls.
3. Density, size and shape of crystals in the endodermal cell cavity.
4. Intensity of lignification in the pericycle cell layers and in cells surrounding the vascular bundles.

The presence of exodermis, higher thickness and higher lignification provides better structural integrity to the root system. Thus genotypes exhibiting these characteristics are expected to have better crop stand and resistance to lodging.

Stem anatomy

The anatomy of the young pearl millet stem is characteristic of the grasses. The epidermis consists of a single cell layer with thickened walls on the outer side. Thick cuticle covers the epidermal cells and reduces the transpiration. The ground tissue consists of thin-walled parenchyma (Fig. 5.37). Vascular bundles of the collateral type are scattered in the ground tissue. A transverse section through a mature stem shows an epidermis consisting of cubical to boat-shaped epidermal cells, containing elongated crystals. Just below the stem epidermis, there are alternating bands of large and small thick-walled sclerenchyma patches. Each sclerenchyma band alternates with a broad band of sclerenchymatous tissue that corresponds to the furrows in the stem outline. These hypodermal bands of the sclerenchyma are in turn connected to the subtending broad sclerenchyma cylinder, which contains the ring of vascular bundles. In a young stage plant, hypodermis consists of collenchyma tissue offering great photosynthetic capacity (Fig. 5.38). There are four to five rows of vascular bundles, the smaller ones found toward the outside. Alternating small and

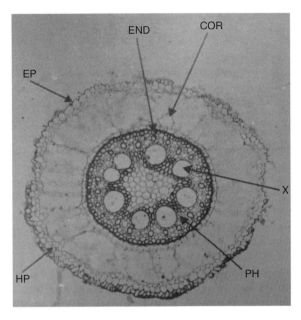

Fig. 5.35. Transverse root section showing the pattern of cortical cells and arrangement of vascular bundles (75×): EP, epidermis; HP, hypodermis; COR, cortex; END, endodermis; PH, phloem; and X, xylem (Maiti and Wesche-Ebeling, 1997).

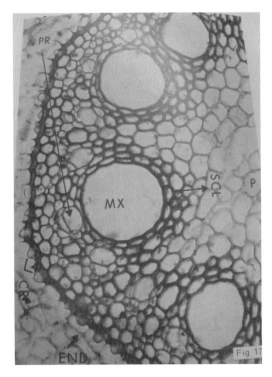

Fig. 5.36. Transverse section of stem (400×) showing endodermal cell wall thickening, crystals, the pericycle and the vascular bundles. Note: crystals are very large: inner tangential wall highly suberized; pericyclic cells less lignified. END, endodermis; CR, crystals; SCL, sclerenchyma; MX, metaxylem; P, pith; PR, protoxylem (Maiti and Wesche-Ebeling, 1997).

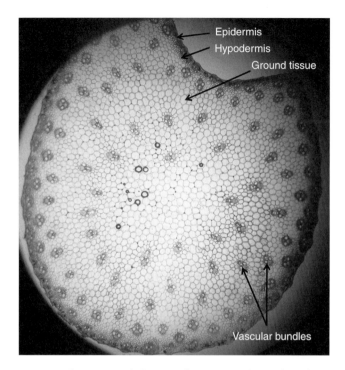

Fig. 5.37. Transverse section of stem (ground plan 40×) illustrating epidermis, hypodermis and scattered vascular bundles in ground tissue. Note: the large amount of sclerenchymatic hypodermis offers great strength to the stem.

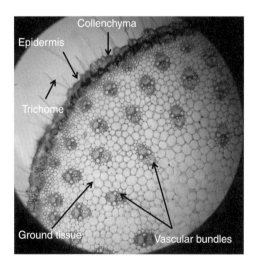

Fig. 5.38. Transverse section of young stem enlarged (100×) showing collenchyma tissue in the region of the hypodermis offering photosynthetic capacity and a high density of trichomes covering the epidermis.

large bundles may be seen just below the external layers. The central vascular bundles are scattered in the ground tissue. The mechanical tissue consists of the sclerenchyma and collenchyma that form the stereome system of the plant.

Significance of variations in stem anatomy

Stem anatomy has been studied in a number of genotypes to determine variability, if any, in internal structure. Examination of transverse sections from seventh and eighth internodes, collected at physiological maturity reveals considerable variation in distribution and amount of mechanical tissue (stereome) and vascular bundles, and in the size and shape of the protoxylem cavity.

Mechanical tissue, consisting of sclerenchyma and collenchyma, form the stereome system of the plant. Millet genotypes show large variations in anatomical structure of stem with respect to the distribution and intensity of sclerenchyma, on the basis of which we can classify the genotypes into three groups: with a weak, intermediate or strong stereome system. While breeding for new genotypes, anatomical studies thus may help in better selection of genotypes according to the target environment and breeding objectives. Some points to be noted are:

1. It is expected that the genotypes having a strong stereome will be resistant to lodging. A large amount of sclerenchyma gives mechanical support and avoids the loss of water from internal parenchyma tissue by evapo-transpiration.

2. It is expected that medium to large size and a high number of xylem vessels are required for efficient translocation of water under drought conditions.

Leaf anatomy

The anatomy of pearl millet leaves is characteristic of a mesophytic grass. The anatomy of the leaves does not vary significantly from genotype to genotype. The cuticularized epidermis is composed of oval or rectangular epidermal cells. The lower epidermis is entire in its outline. Silica crystals are prominent in some genotypes, but not in others. The upper epidermal cells are wavy in outline and interspersed with zones of bulliform (motor) cells, that show high variation in size among genotypes. The sclerenchyma surrounding the vascular bundles may be loose or compact. In transverse section bulliform cells are present in the upper epidermis of the midrib and vascular bundles of different sizes can be found. The midrib has three vascular bundles in some genotypes, one in the centre and the other two at the junction of the leaf lamina and midrib. The vascular bundles in the lamina and midrib can be of three types: (i) a large central vascular bundle present in the main vein; (ii) medium sized vascular bundle; and (iii) very small vascular bundle. The first two types are fibro-vascular and contain patches of sclerenchyma in both sides connecting the upper or lower epidermis. Each vascular bundle remains surrounded by a bundle sheath consisting of thin-walled parenchyma cells (Fig. 5.39). The cells of the bundle sheath generally contain large chloroplasts, grana are absent and large starch grains occur in the chloroplasts. Mesophyll cells contain normal chloroplast. This special structure was first identified by Haberland; it appears like a garland (*Kranz*=wreath,

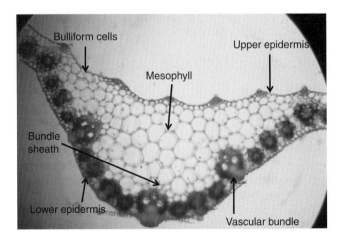

Fig. 5.39. Transverse section of leaf (100×) showing upper and lower epidermis, bulliform cells, mesophyll and vascular bundles. Each vascular bundle is covered by bundle sheath cells. The large size vascular bundle is at the central (mid-vein) region.

garland), so it is known as Kranz anatomy. Kranz anatomy is the special characteristic of C$_4$ plants. Kranz anatomy is one of the adaptations for drought resistance in plants of semi-arid zones.

Significance of variations in leaf anatomy

The different millet genotypes show differences in the following leaf anatomical features:

1. Amount of cuticularization.
2. Intensity of silica crystals in the upper epidermal cells offering rigidness to the leaf, thereby offering resistance to the leaf insects.
3. Size and shape of bulliform cells helps in leaf rolling, thereby avoiding loss of transpiration under drought situations.
4. Type and arrangement of vascular bundles, both in the lamina and midrib.
5. Characteristic size and shape of bundle sheath chloroplasts.

From observing these variations, the following inferences can be drawn on the significance of leaf anatomical features:

1. Presence of thick cuticle on epidermis of leaf reduces transpirational loss of water.

2. High density of trichomes cover the surface and reduces the direct sunlight effect and minimizes leaf temperature, thereby reducing transpirational loss of water.
3. High density of trichomes offers insect resistance, especially sucking pest tolerance.
4. A greater number and large size vascular bundles help in efficient translocation of water.
5. Reduction in stomatal density on lower and upper epidermis minimizes the loss of water from the leaf.
6. Presence of long size bulliform cells prevents evapo-transpiration by rolling the leaves.

Reproductive growth – panicle development

At the end of the vegetative stage approximately 16–28 days after emergence, the shoot apex is transformed into a reproductive apex with the formation of a panicle meristem (stage of panicle initiation) (Fig. 5.40). The panicle meristem assumes a bulbous shape with a constriction in its base with definite cellular organization. There are prominent tunica and corpus layers in the panicle axis, with two to three layers of serially arranged tunica cells overlying the corpus. The cellular organization

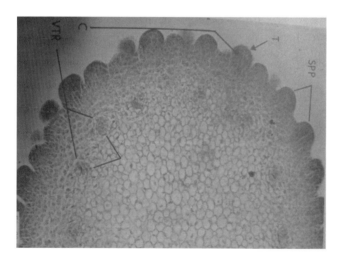

Fig. 5.40. Transverse section of panicle meristem, showing elongation of spikelet primordia with distinct tunica and corpus (75×): SPP, spikelet primordia; T, tunica; C, corpus; VTR, vascular trace (Maiti and Wesche-Ebeling, 1997).

in the corpus layer is irregular. The protoxylem in the tunica cells and the nuclei with prominent chromatic material are distinct.

The spikelet primordial develops initially at the base and gradually progresses towards the tip of the panicle meristem. Initiation of the spikelet primordial is marked by the condensation of protoplasm and by an increase in the volume of the flank meristem. The spikelet primordium assumes a bulbous appearance (Fig. 5.41) with constriction at the sides, indicating the boundary between the adjacent spikelet primordia (Maiti and Bidinger, 1981).

About 3 days after initiation, the panicle is much more elongated and the formation of the spikelet primordial has advanced toward the tip. Each primordium divides to give two spikelets, which subsequently develop flower parts. Each spikelet is enclosed by a pair of glumes, the upper and lower (Fig. 5.42). Subsequently each individual spikelet primordium divides and extends laterally to give rise to two floret primordia which develop into flower parts.

Two types of florets exist in each spikelet in pearl millet, the upper is perfect and the lower is male. Each floret is enclosed by two glumes. In a perfect flower (bisexual), the flower primordium develops in the axil of the lateral primordial, which form two glumes, upper and lower at the base (Fig. 5.43). The floret primordium gives rise to the primordial forming palea and lemma and the central primordium, which functions as the carpel primordium.

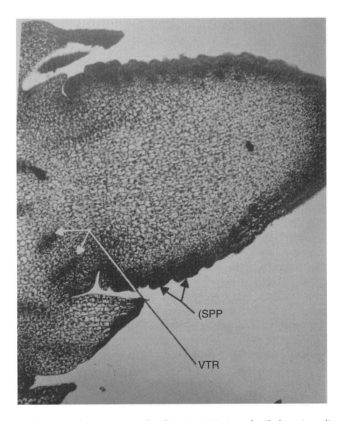

Fig. 5.41. Longitudinal section of a young panicle, showing initiation of spikelet primordia at the base. Note: spikelet primordia are developing at the base of the panicle; towards the tip, spikelet primordia are not yet developed. SPP, spikelet primordia; VTR, vascular trace (Maiti and Wesche-Ebeling, 1997).

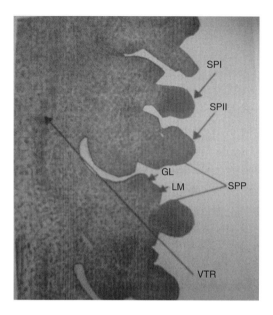

Fig. 5.42. Longitudinal section of a young panicle, showing more developed spikelet primordia at the base, compared to less developed primordia at the top (75×). SP, spikelet primordia; GL, glume; LM, lemma; VTR, vascular trace (Maiti and Wesche-Ebeling, 1997).

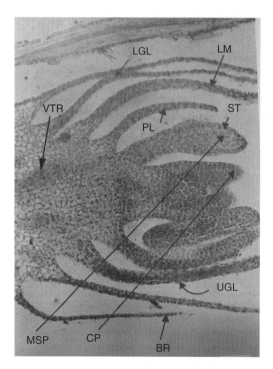

Fig. 5.43. Longitudinal section through a spikelet, showing the anther at the archesporial stage and the developing carpel (45×). VTR, vascular trace; ST, stamen; CP, carpel; MSP, microsporangia; PL, palea; LM, lemma; LGL, lower glume; UGL, upper glume; BR, bristle (Maiti and Wesche-Ebeling, 1997).

Three primordial stamens develop from the central primordium afterwards. The developmental pattern of a male flower is similar to that of a bisexual flower, except for the presence of the carpel. It has a lemma and lacks a palea. The central primordium gives rise to the three stamens in the male flower.

Microsporogenesis

Eight days after panicle initiation a tissue mass in which the microsporangia are embedded appears in the anther. Microsporangia are present as four separate masses in the four lobes of the anther. These cells divide along periclinal walls, leading towards an inner mass of primary sporangial tissue and outer primary parietal cells. The ultimate cells of sporangial tissue are the pollen mother cell, which by reduction division ultimately leads to the formation of four pollen grains (Fig. 5.44). In mature anthers approaching the time of dehiscence, some fringe-like outgrowths are observed all along the outer wall of the anther.

Structure of the ovule

A single ovule is present in the cavity of the ovary and is laterally attached to its wall by a short funicle. The ovary wall is composed of a single layer of cubical epidermal cells, subtended by several layers of thin-walled cells (Fig. 5.45).

Fruit (caryopsis)

In a transverse section of a mature caryopsis the structure of the seed testa and endosperm become visible. A thick cuticle covers the epidermis; the pericarp consists of an epicarp, hypocarp and endocarp. The epicarp is composed

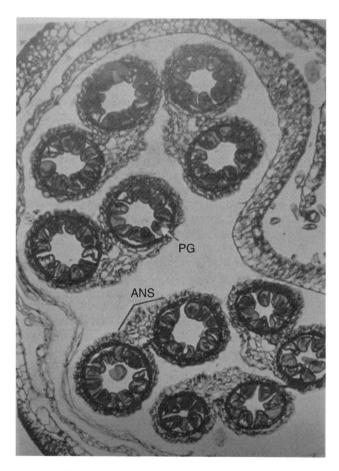

Fig. 5.44. Transverse section through the anther, showing the pollen grains (75×). PG, pollen grains; ANS, anther sac (Maiti and Wesche-Ebeling, 1997).

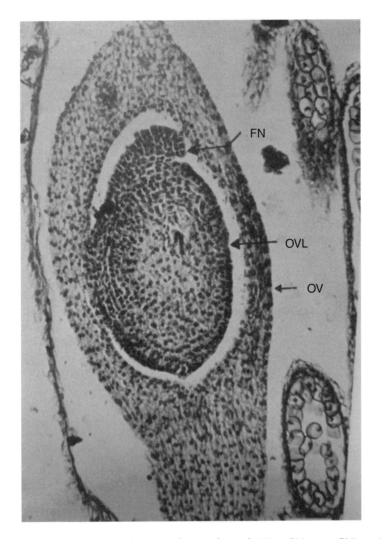

Fig. 5.45. Longitudinal section through the ovary, showing the ovule (75×). OV, ovary; OVL, ovule; FN, funicle (Maiti and Wesche-Ebeling, 1997).

of long epidermal cells, the hypocarp cells are thickened. The two to three layers of mesocarp cells have lost their outline due to the compression of outer cell layers. Elongated cross cells are only rarely visible and tube cells, which are distinct at an early stage, are mostly obliterated. Below the mesocarp exists a single layer of large cubic and rectangular cells, regularly oriented, called the aleu-

rone layer; they contain staining bodies (Fig. 5.46).

Endosperm

Below the aleurone layer is the peripheral corneous endosperm, made up of a single, well-demarcated layer of cells, with closely arranged starch (Fig. 5.47) and protein granules. The large, long, cubic to rectangular

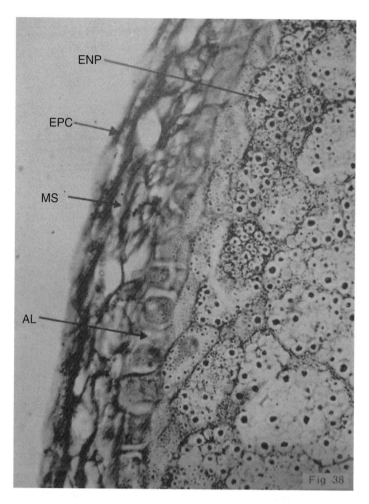

Fig. 5.46. Transverse section through a developing seed, showing epicarp, mesocarp, aleurone layer and endosperm (75×). EPC, epicarp; ENP, endosperm; MS, mesocarp; AL, aleurone layer (Maiti and Wesche-Ebeling, 1997).

endosperm cells are loose below this peripheral layer and contain the floury endosperm. The starch and protein granules become progressively less compact towards the midpoint of the grain endosperms, from the milky stage to maturity.

Embryo

The aleurone cells in the embryo region are small, compact and rectangular. The endosperm cells between the aleurone layer and the scutellum are compact. The scutellar cells are elongated and form a single layer that surrounds the embryo. The embryo cells are compact, especially in the hilar region. The cells outside the black layer in the hilar region are compact. The black layer shows a semi-lunar ring, composed of vacuolated cells. The formation of this layer coincides with the termination of grain development.

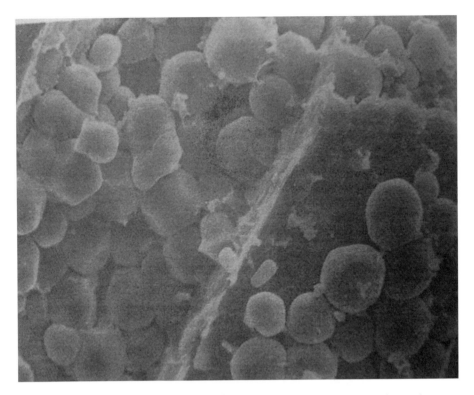

Fig. 5.47. Scanning electron microscope micrograph of grain endosperm showing starch granules (Maiti and Wesche-Ebeling, 1997).

References

Aloni, R. and Griffith, M. (1991) Functional xylem anatomy in root-shoot junctions of six cereal species. *Planta* 184(1), 123–129.

Beadle, G.W. (1939) Teosinte and the origin of maize. *Journal of Heredity* 30, 245–247.

Carver, T.L.W. and Thomas, B.J. (1990) Normal germling development by *Erysiphe graminis* on cereal leaves freed of epicuticular wax. *Plant Pathology* 39(3), 367–375.

Ching, A., Dhugga, K.S., Appenzeller, L., Meeley, R., Bourett, T.M., Howard, R.J. and Rafalski, A. (2006) Brittle stalk 2 encodes a putative glycosylphosphatidylinositol anchored protein that affects mechanical strength of maize tissues by altering the composition and structure of secondary cell walls. *Planta* 224, 1174–1184.

Clark, L.H. and Harris, W.H. (1981) Observations on the root anatomy of rice (*Oryza sativa* L.). *American Journal of Botany* 68(2), 154–161.

Elster, R., Bommert, P., Sheridan, W.F. and Werr, W. (2000) Analysis of four embryo-specific mutants in *Zea mays* reveals that incomplete radial organization of the proembryo interferes with subsequent development. *Development Genes and Evolution* 210, 300–310.

Ennos, A.R., Crook, M.J. and Grimshaw, C. (1993) The anchorage mechanics of maize, *Zea mays*. *Journal of Experimental Botany* 44, 147–153.

Esau, K. (1943) Ontogeny of the vascular bundle in *Zea mays*. *Hilgardia* 15, 327–368.

Finan, J.J. (1948) Maize in the great herbals. *Annals of the Missouri Botanical Garden* 35, 149–191.

Food and Agriculture Organization (FAO) (2011) World Wheat, Corn and Rice. Oklahoma State University, FAOSTAT.

Gaut, B.S. and Doebley, J.F. (1997) DNA sequence evidence for the segmental allotetraploid origin of maize. *Proceedings of the National Academy of Sciences* 94, 6809–6814.

Glueck, J.A. and Rooney, L.W. (1976) Physical and chemical characterization of sorghum lines with resistance to grain deterioration. *Cereals Foods World* 21, 436–437.

Glueck, J.A. and Rooney, L.W. (1978) Chemistry and structure of grain in relation to mold resistance in sorghum disease. A world review. In: Williams, R.J., Frederiksen, R.A., Mughogho, L.K. and Bengiston, G.D. (eds) *Proceedings of the International Working Group on Sorghum Disease*. ICRISAT, Patancheru, India, pp. 119–140.

Glueck, J.A., Rooney, L.W., Rosenow, B.T., Miller, F.R. and Lichtendalner, R.E. (1978) Physical and structural properties of weathered sorghum grain. *Miscellaneous Publications of the Texas Agriculture Experimental Station* 1375, 12–30.

Hameed, M., Mansoor, U., Ashraf, M. and Rao, A.R. (2002) Variation in leaf anatomy in wheat germplasm from varying drought-hit habitats. *International Journal of Agriculture & Biology* 4, 12–16.

Hernández, J.A.S. (2009) *The Origin and Diversity of Maize in the American Continent*. Greenpeace, Mexico, 33 pp.

House, L.R. (1980) *A Guide to Sorghum Breeding*. ICRISAT, Patancheru, A.P., India, 238 pp.

Hund, A., Ruta, N. and Liedgens, M. (2009) Rooting depth and water use efficiency of tropical maize inbred lines, differing in drought tolerance. *Plant and Soil* 318, 311–325.

Iltis, H.H. (1983) From teosinte to maize: the catastrophic sexual transmutation. *Science* 222, 886–894.

Jenks, M.A., Joly, R.J., Peters, P.J., Rich, P.J., Axtell, J.D. and Ashworth, E.N. (1994) Chemically induced cuticle mutation affecting epidermal conductance to water vapor and disease susceptibility in *Sorghum bicolor* (L.) Moench. *Plant Physiology* 105(4), 1239–1245.

Kirby, E.J.M. (2002) Botany of the wheat plant. In: Curtis, B.C., Rajaram, S. and Macpherson, H.G. (eds) *Bread Wheat, Improvement and Production*. FAO Plant Production Series No. 30, Rome.

Kondo, M., Aguillar, A., Abe, J. and Moreta, S. (2000) Anatomy of nodal roots in tropical upland and low land rice varieties. *Plant Soil Science* 3(4), 437–445.

Li, H.B., Bai, K.Z., Kuang, T.Y., Hu, Y.X., Jia, X. and Lin, J.X. (2000) Structural characteristics of thicker-culms in the high yield wheat cultivars. *Acta Botanica Sinica* 42, 1258–1262.

Lux, A., Luxová, M., Hattori, T., Inanaga, S. and Sugimoto, Y. (2002) Silicification in sorghum (*Sorghum bicolor*) cultivars with different drought tolerance. *Physiologia Plantarum* 115(1), 87–92.

Maiti, R.K. (1996) *Sorghum Science*. Science Publishers, New Delhi, India, 352 pp.

Maiti, R.K. and Bidinger, F.R. (1981) Growth and development of the pear millet plant. *Research Bulletin No. 6*. International Crops Research Institute for the Semi-Arid Tropics, Patancheru, Andhra Pradesh, India, pp. 1–9.

Maiti, R.K. and Wesche-Ebeling, P. (1997) *Pearl Millet Science*. Science Publishers, New Delhi, India, 232 pp.

Maiti, R.K., Prasada Rao, K.E., Raju, P.S. and House, L.R. (1984) The glossy trait in sorghum: its characteristics and significance in crop improvement. *Field Crops Research* (9), 279–289.

Maiti, R.K., De La Rosa, M.I. and Sandoval, N.D. (1994) Genotypic variability in glossy sorghum lines for resistance to drought, salinity and temperature stress at the seedling stage. *Journal of Plant Physiology* 143, 241–244.

Makino, A. (2011) Photosynthesis, grain yield, and nitrogen utilization in rice and wheat. *Plant Physiology* 155, 125–129.

Mangelsdorf, P.C. and Reeves, R.G. (1938) The origin of maize. *Proceedings of National Academy of Sciences* 24, 303–312.

Matsuoka, Y., Vigouroux, Y., Goodman, M.M., Sanchez-Gonzalez, J., Buckler, E. and Doebley, J. (2001) A single domestication for maize shown by multilocus microsatellite genotyping. *Proceedings of National Academy of Sciences* 99, 6080–6084.

Murdock, G.P. (1959) Staple subsistence crops of Africa. *Geographic Review* 50, 521–540.

Murdy, W.H. (1960) The strengthening system in the stem of maize. *Annals of the Missouri Botanical Garden* 47, 205–226.

Ridley, E.J. and Todd, G.W. (1966) Anatomical variations in the wheat leaf following internal water stress. *Botanical Gazette* 127, 235–238.

Ristic, Z. and Jenks, M.A. (2002) Leaf cuticle and water loss in maize lines differing in dehydration avoidance. *Journal of Plant Physiology* 159(6), 645–651.

Rooney, L.W. and Clark, L.E. (1968) The chemistry and processing of sorghum grain. *Cereal Science Today* 13, 259–264, 285.

Rooney, L.W. and Sullins, R.D. (1977) The structure and sorghum and its relation to processing and nutritional value. In: Dendy, D.A.V. (ed.) *Proceedings of the Symposium on Sorghum and Millets as Human Food*, Vienna, Austria. Tropical Products Institute, London, pp. 1–109.

Sage, R.F. and Monson, R.K. (1999) *C4 Plant Biology*. Academic Press, USA, 596 pp.

Sage, T.L. and Sage, R.F. (2009) The functional anatomy of rice leaves: implications for refixation of photorespiratory CO_2 and efforts to engineer C4 photosynthesis into rice. *Plant and Cell Physiology* 50(4), 756–772.

Santiago, R., Souto, X.C., Sotelo, J., Butrón, A. and Malvar, R.A. (2003) Relationship between maize stem structural characteristics and resistance to pink stem borer (Lepidoptera: Noctuidae) attack. *Journal of Economic Entomology* 96, 1563–1570.

Tarumoto, I. (2006) Genetic studies on glossy leaves in sorghum (*Sorghum bicolor* L. Moench). *Japan Agricultural Research Quarterly* 40(1), 13–20.

Traore, M., Sullivan, C.Y., Rosowski, J.R. and Lee, K.W. (1989) Comparative leaf surface morphology and the glossy characteristic of sorghum, maize, and pearl millet. *Annals of Botany* 64(4), 447–453.

Travis, A.J., Murison, S.D., Hirst, D.J., Walker, K.C. and Chesson, A. (1996) Comparison of the anatomy and degradability of straw from varieties of wheat and barley that differ in susceptibility to lodging. *Journal of Agricultural Science* 127, 1–10.

Uga, Y., Ebana, K., Abe, J., Morita, S., Okuno, K. and Yano, M. (2009) Variation in root morphology and anatomy among accessions of cultivated rice (*Oryza sativa* L.) with different genetic backgrounds. *Breeding Science* 59(1), 87–93.

Wilkes, G. (2004) Corn, strange and marvelous: but is a definitive origin known? In: *Corn: Origin, History, Technology, and Production*. C. Wayne Smith (Publishing House), Wiley & Sons, pp. 3–63.

Wilson, J.R., Mertens, D.R. and Hatfield, D. (1993) Isolates of cell types from sorghum stems: digestion, cell wall and anatomical characteristics. *Journal of the Science of Food and Agriculture* 63(4), 407–417.

Zhu, L., Shi, G.X., Li, Z.S., Kuang, T.Y., Li, B., Wei, Q.K., Bai, K.Z., Hu, Y.X. and Lin, J.X. (2004) Anatomical and chemical features of high-yield wheat cultivar with reference to its parents. *Acta Botanica Sinica* 46, 565–572.

6

Pulses

Pulses are important sources of proteins and other nutrients. This chapter concentrates only on a few important pulses, such as chickpea, common bean, green gram, red gram, cowpea and pea.

6.1 Chickpea

6.1.1 Introduction

Chickpea (*Cicer arietinum*), the world's third most important food legume, is currently grown on about 11 Mha, with 96% of cultivation in the developing countries. It is also called Bengal gram, Garbanzo bean, Indian pea, ceci bean. Chickpea is a self-pollinating diploid ($2n=2x=16$) with genome size $1C=740$ Mbp. It is the second most important dry grain crop.

Chickpea falls into two types: desi and kabuli. The desi type plants are shorter with smaller leaves, white, pink or blue flowers and pigmented seeds, and are grown in the semi-arid tropics. The kabuli types are taller plants with larger leaves, white flowers and longer, more rounded, cream-coloured seeds and are grown in temperate regions. The desi types predominate in the Indian subcontinent and kabuli types in other countries. Kabuli types dominate American production.

Chickpea plants mainly adapt to drought environments either through escape, or tolerance mechanisms. In a previous study early maturity chickpea lines were found, which can escape terminal drought stress. The results showed that tolerant lines had a significant difference at 1% level of probability over susceptible lines for drought tolerance. These lines produced a significantly higher yield than control drought tolerant lines. To produce the highest yield, lines require drought tolerance with high adaptability, early maturity and large seed size (Sabaghpour *et al.*, 2006).

6.1.2 Origin and distribution

Chickpea is an ancient crop, first grown in Turkey about 7450 BC and in India about 4000 BC. Chickpea originated in Mediterranean countries and is cultivated in Asia, south Europe and Mexico. The probable place of origin is the region between Cuacaso and north-eastern Persia (van der Maesen, 1972), however, Ladizinsky (1975) proposes that the centre of origin was south-eastern Turkey. The most important importing countries are Spain, Algeria, Iran, Libya and the USA.

6.1.3 Utilization

Chickpea is used as food in the following ways. Chickpea is in great demand for its nutritive seeds, which have a high protein content of 25.3–28.9% after dehulling. Sprouted seeds are eaten as a vegetable or salads. Young plants and green pods are eaten like spinach. Tinned chickpea is sold in Turkey and Latin America. Mature chickpeas can be cooked and eaten cold in salads, cooked in stews, ground into a flour called gram flour (also known as *besan* and used primarily in Indian cuisine), cooked and ground into a paste or roasted, spiced and eaten as a snack. Chickpeas are low in fat and most of this is polyunsaturated. The nutrient profile of desichana (the smaller variety) is different, especially the fibre content, which is much higher than the light coloured variety. Chickpea is also used as animal feed.

Chickpea is also used for medicinal purposes. Chickpea can use as a hypocholesteraemic agent. Germinated chickpea was effective in controlling cholesterol levels in rats. Glandular secretions of the leaves, stems and pods contain malic and oxalic acids, giving a sour taste. In India, these acids are collected from the crop canopy by spreading thin muslin over the crop during the night. This solution is a popular medicine used in curing bronchitis, catarrh, cutamenia, cholera, constipation, diarrhoea, dyspepsia, flatulence, snake bite, sunstroke and warts. Raw chickpeas contain protease (enzyme that breaks down proteins) inhibitors, which counteract the digestion enzymes in our body.

6.1.4 Morphological description

Vegetative characters

Chickpea is an herbaceous annual crop. The growth habit varies from prostate to erect. It is grown in tropical, subtropical and temperate regions. It possesses a strong taproot system with three or four rows of lateral roots. The parenchymatous tissues of the root are rich in starch. All the peripheral tissues disappear at plant maturity, and are substituted by a layer of cork. The roots grow 1.5–2.0 m deep and bear *Rhizobium* nodules. They are of the carotenoid type, branched with laterally flattened ramifications, sometimes forming a fanlike lobe.

The stem is erect, branched, viscous, hairy, terete, herbaceous, green and solid. The branches are usually quadrangular, ribbed and green. There are primary, secondary and tertiary branches. Thick, strong and woody primary branches (ranging from one to eight) arise from ground level as they develop from the pumular shoot as well as lateral branches of the seedling. Secondary branches (from 2 to 12) develop at buds located on the primary branches. They are less vigorous than the primary branches. The number of secondary branches determines the total number of leaves, and hence the total photosynthetic area. Tertiary branches arise from the secondary branches. The primary branches form an angle with a vertical axis, ranging from almost a right angle to an acute angle. Usually stems are incurved at the top, forming a spreading canopy. Chickpea leaves are petiolate, compound (Fig. 6.1) and uniparipinnate (pseudoimparipinnate). Some lines have simple leaves. The rachis is 3–7 cm long with grooves on its upper surface. Each rachis supports from 10 to 15 leaflets, each with a small pedicel. The leaflets do not end at the true terminal position, but at the subterminal position. The leaflets are 8–17 mm long and 5–14 mm wide, opposite or alternate with the terminal leaflet. Leaflets are serrated, the teeth covering about two-thirds of the foliar blade. Leaflets are obovate to elliptical with the basal and top portions cuneate and rounded. Leaves are pubescent. The stipules are ovate to triangular in shape and serrated with from two to six teeth, 3–5 mm long and 2–4 mm wide. The longest margin is toothed and the smallest one entire. The external surface of the chickpea plant, except the corolla, is densely covered with glandular or non-glandular hairs. The hairs vary in dimensions: short stalked, multicellular stalked (both glandular and non-glandular) and unicellular, some genotypes, however, do not possess any hair.

Floral characters

The solitary flowers are born in an axillary raceme. Sometimes there are two to three flowers on the same nodes; such flowers

Fig. 6.1. Chickpea plant morphology showing compound leaves and pods.

possess both a peduncle and pedicel. The racemose peduncle is 60–30 mm in length. At flowering, the floral and racemal portions of the peduncle form a straight line, giving the appearance that the flowers are placed on the leafy axil by a single peduncle. Chickpea flowers are bisexual and have a papilionaceous corolla. They are white, pink, purple or blue in colour. In coloured flowers, the peduncles may be of different colours, the floral part purplish and the raceme green. The axillary inflorescence is shorter than the subtending leaf. The calyx is dorsally gibbous at the base. There are five sepals with deep lanceolate teeth. The teeth are longer (5–6 mm) than the tube (3–4 mm) and have prominent midribs. The five sepals are subequal. The two dorsal (vexillar) sepals are closer to each other then they are to the two lateral ones in the ventral position. The fifth calyx tooth is separate from the others. The peduncles and the calyx are glabrous. The calyx tube is oblique. Chickpea flowers have five sepals, which are generally celeste and purplish red or light pink in colour. The petals are polypetalous, i.e. consisting of vexillum, wings and keel. The vexillum is obovate, 8–11 mm long, 7–10 mm wide, and either glabrous, or pubescent with no glandular hair on its external surface. The wings are also obovate with short pedicles. They are 6–9

mm long and about 4 mm wide with an auriculate base. The auricules are over the pedicel and form a packet in the basal upper part, which is covered by the vexillum. The keel is 6–8 mm long, rhomboid, with a pedicel 2–3 mm long. Two-thirds of the frontal side of its ventral face is adnate. The wings do not show concrescence with the keel. There are ten stamens in diadelphous (9)+1 condition. The filaments of nine of the stamens are fused, forming an androecial sheath; the tenth stamen is free. The stamina column is persistent. The fused part of the filament is 4–5 mm long and the free part 2–3 mm, upturned and dilated at the top. The apex of the stem is oblique. The stamens facing the petals are a little longer than the others. The anthers of these stamens are bicelled, basifixed and round. The other anthers are dorsifixed, ovate, and longer than the basifixed ones at flowering. The anthers burst longitudinally. The pollen grains are orange. The gynoecium is monocarpellary, unilocular and superior, with marginal placentation. It is ovate with a pubescent surface. The ovary is 2–3 mm long and 1–15 mm wide. There are 1–3 ovules, rarely four. The style is 3–4 mm long, linear, upturned, and glabrous except at the bottom. The stigma is globose and capitates. Sometimes it may be as small as the style. Anther dehiscence takes place inside the bud 1 day before the opening of the flower. When pollen is first liberated, the stigma is still above and quite free from the base of the anthers. The filament gradually elongates to carry the anthers above the stigma. This process is completed before the flower opens, thus facilitating self-pollination. Anthesis in chickpea is occurs through the day.

Pod and seed development

Pod formation begins 5–6 days after fertilization. The number of pods per plant varies between 30 and 150 depending on the environmental conditions and the genotype. The pod wall is 0.3 mm thick with three layers: exocarp, mesocarp and endocarp. The exocarp is hairy and glandular, the mesocarp has six to eight layers of parenchyma and the endocarp consists of three to four cell layers with fibres in its outermost region and five to six layers of parenchyma. Pod size ranges from 15 to 30 mm in length, 7 to 14 mm in thickness and 2 to 15 mm in width. The pod shape varies from rhomboid, oblong to ovate. The number of seeds per pod ranges from one to two, with the maximum being three.

The number of trichomes and length of trichomes plays an important role in pod borer resistance. Pod trichomes' length and density showed negative correlation to pod borer damage in chickpea, i.e. genotypes having longer pod trichomes and or more pubescent pods received less pod borer damage. The chickpea genotypes having a thicker pod wall received lower pod borer damage. Pod length, breadth and area of respective genotypes showed a significant effect in resistance mechanism against pod borer damage (Hossain *et al.*, 2008). Therefore, there is a great necessity in screening peanut cultivars and selecting cultivars for trichome density on pods, which may then be incorporated into the breeding programme for pod borer resistance.

Seed

The seeds are owl's head shaped, and the surface may be smooth or wrinkled. The two cotyledons are separated by a groove in highly wrinkled seeds. The beak above the micropyle is produced by the tip of the radicle. The shape of cotyledons varies from semi-spherical to oviform. The length of the seed ranges from 4 to 12 mm and its width from 4 to 8 mm. The seed mass varies from 0.10 to 0.75 g per seed. The seed colour ranges from whitish (even chalky) and cream to deep black, although many other colours such as red, orange, brown, green and yellow may be found. The cotyledons are cream, green, or orange coloured.

Kabuli seeds contained fewer pectic polysaccharides and less protein, and had a thinner seed coat due to thinner palisade and parenchyma layers. The outer palisade layer thickness varied from one to two cells, leading to a textured and sometimes wrinkled appearance of the seed surface. The desi palisade layers were rigid and extensively thickened. Hourglass cells were homogeneous for both seed types, but not in an interspecific desi line (containing *Cicer echinospermum* parentage), which had heterogeneous cells. The inner surface of the seed coat contained both pectic and proteinaceous materials (Wood *et al.*, 2011).

6.1.5 Anatomy

Root anatomy

A transverse section of the root shows three parts, epidermis, cortex and stele (Fig. 6.2).

The epidermis consists of uniseriate compactly arranged barrel-shaped cells without intercellular spaces and has unicellular root hairs.

The cortex is the middle region, lying between the epidermis and the stele, and is multiseriate, parenchymatous and relatively homogeneous. Outer layers of the cortex consist of loosely arranged thin-walled parenchyma cells with prominent intercellular spaces and contain food deposits. The cells are usually round or oval, colourless and store starch. This region helps in lateral conduction of water and salts. The inner layer of the cortex consists of compactly arranged barrel-shaped cells with Casparian strips or Casparian bands, called endodermis.

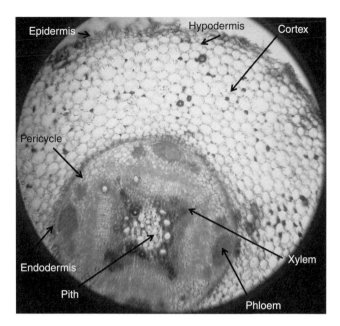

Fig. 6.2. Transverse section of root (100×). Note: secondary growth in the stele has started and the endodermis and cortex are undergoing compression due to enlargement of the stele.

In transverse section, Casparian bands appear as a lens-shaped thickening on the radial walls of the endodermis.

Pericycle is the outer most region of the stele. It is uniseriate and parenchymatous. Cells of the pericycle retain meristematic activity. Lateral roots arise endogenously from the pericycle. Stele consists of xylem and phloem and is arranged in the form of separate strands on different radii, alternating with each other. Xylem is exarch having protoxylem towards the pericycle and metaxylem towards the centre. Generally four (tetrarch) protoxylem regions are present and form xylem tissue. Phloem occurs below the pericycle and consists of a large amount of sclerenchyma. The centre of the stem is occupied by a small amount of pith (Fig. 6.2).

Root nodules

Members of the Leguminosae family usually are found in symbiosis with a bacterium of the family Rhizobiaceae, which associates with the root cells of the plant forming specialized structures termed root nodules.

Chickpea, being a leguminous plant, has the capacity to fix atmospheric nitrogen with the help of rhizobium present in such root nodules. The symbiotic response of chickpea and rhizobium is dependent on genetic and physiological factors of both species. *Rhizobium* spp. (fast-growing species) and *Bradyrhizobium* spp. (slow-growing species) are predominantly aerobic chemoorganotrophic soil genera that form root nodules by symbiotic association with legumes and non-legumes, but specially legumes.

ROOT INVASION AND NODULATION   Rhizobia are rhizosphere organisms multiplying on the root surfaces and in the surroundings. They move to the rhizosphere in response to the nutrients and other substances containing the root excretions. In this active soil volume, population growth is stimulated until the rhizobia occupy the whole chickpea root surface and attach to epidermal cells, particularly to the root hairs (Fig. 6.3). Some bacteria specifically interact with newly emerging root hairs, however, and initiate a pronounced curling of these growing hair cells. To initiate this reaction rhizobia have evolved flavonoids.

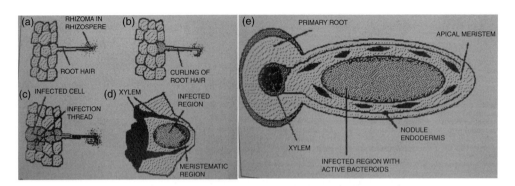

Fig. 6.3. Sequential events in root invasion and nodulation up to differentiation of an active nodule.
(a) Rhizobia moves to the rhizosphere in response to the nutrients and other substances present in the root
secretions. (b) After detection of specific flavonoids released by root cells some rhizobia interact with newly
emerging root hairs and initiate a pronounced curling of hair cells. (c) Invasion of plant cells via induction
of an infection thread that penetrates the plant tissue and continues to grow and ramify in the root cortex.
(d) Infection thread eventually invades a focus of dividing plant cells and rhizobia are released into these
cells. The bacteria continue to grow and ultimately differentiate into bacteroids capable of fixing nitrogen.
(e) An active root nodule is formed (Maiti and Wesche-Ebeling, 2001).

These compounds are released by the plant into the root rhizosphere and act as positive regulators of infection genes (*Nod* genes). Following the initiation of an infection process, rhizobia entrapped within curled root hair cells began the invasion of these plant cells. Invasion occurs via the induction of an infection thread that penetrates the plant tissue and continues to grow and ramify in the root cortex. The infection thread releases rhizobia and these are embedded within a plant membrane called the peribacteroidal membrane. The bacteria continue to grow and ultimately differentiate into bacteroids capable of fixing nitrogen.

NODULE GROWTH AND MORPHOLOGY   Chickpeas have a nodule type termed intermediate (Fig. 6.4), which is characterized by the presence of a defined meristem during nodular growth, an open vascular system connecting the root vascular system with the nodule meristem, vacuolated infected cells, aspargine/glutamine as the major translocation products of nitrogen fixing activity, and enlarged simple bacteroids.

The features of developing and senescing bacteroids include the presence of persistent meristematic tissue at the distal end of the multi-lobed nodules, and a gradient of cells at different stages of development toward the proximal point of attachment of the nodule to the parent root. The cytoplasm of infected cells in the nitrogen-fixing region of the nodules is densely packed with symbiosomes, most of which contain a single bacteroid. In later stages of development infected cells become enlarged and highly vacuolated, and eventually lose their contents. Uninfected cells in the central region are smaller than infected cells and are also highly vacuolated.

Stem anatomy

The outline of the stem in transverse section is quadrangular in shape. It shows an outer epidermis, central cortex and inner stele (Fig. 6.5).

Epidermis contains uniseriate compactly arranged cells with unicellular long trichomes. The epidermis is cover by thick cuticle. Below the epidermis, four to six layers of collenchyma with some chlorenchyma patches are present. Collenchyma gives considerable strength, flexibility and elasticity to young stems. Having chloroplasts, it may carry out photosynthesis.

Multilayered parenchymatous cortex is present between the collenchyma and stele.

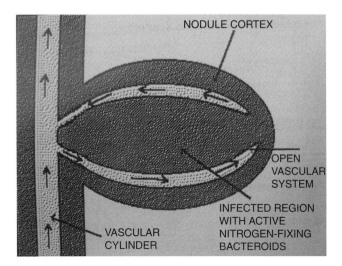

Fig. 6.4. Schematic structure of an indeterminate root nodule. The open vascular system distinguishes it from a determinative root nodule (Maiti and Wesche-Ebeling, 2001).

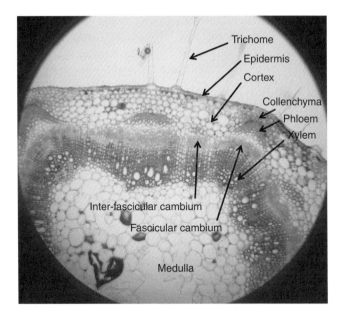

Fig. 6.5. Transverse section of stem (100×) showing unicellular long trichomes, collenchyma, single layer of epidermis and the medulla. Secondary growth is just initiated, so inter-fascicular cambium and fascicular cambium are forming secondary vascular tissues.

Cells of the inner layer of the cortex, the endodermis, are thickened by Casparian strips. It is uniseriate with stored food material and covers the stele.

Stele is the central region of the stem and larger than the cortex. It consists of vascular bundles, medulla and medullary rays. Xylem and phloem organize in the form of vascular bundles; 10–12 wedge-shaped vascular bundles are arranged in one ring, called eustele. It contains xylem towards the centre of the axis and phloem towards the periphery.

In the vascular bundle, xylem and phloem are present together (conjoint) on the same radius (collateral), with a strip of cambium (fascicular cambium) between xylem and phloem. Hence the vascular bundle is described as conjoint, collateral and open type. Xylem is endarch with protoxylem present towards the centre. Medulla and medullary rays are distinct, present at the centre of the stem and usually parenchymatous. The cells are round or oval with intercellular spaces. Medulla occupies a large area compared to the other parts of the stem, and stores food materials. The extensions of parenchymatous pith between vascular bundles are called primary medullary rays, which help in lateral conduction and may give rise to secondary meristem (inter-fascicular cambium).

Leaf anatomy

The leaf shows a dorsi-ventral nature, i.e. the leaf blade consists of distinct dorsal and ventral surfaces. Leaf transverse section shows three distinct regions, the epidermis, mesophyll and vascular system.

The epidermis is single layered with compactly arranged barrel-shaped (tabular) cells. The outer surface of the epidermis is covered by a thin continuous layer of cutin called cuticle. Anisocytic type of stomata are present in both epidermal layers. The stomatal index of upper epidermis is 28 to 33 and lower epidermis is 35 to 37 under 40× magnification. Stomata facilitate the exchange of gases between the leaf and environment. All the epidermal cells except guard cells are colourless.

The ground tissue of the leaf, present between the upper and the lower epidermal layers, is called mesophyll. The mesophyll is chlorenchymatic and concerned mainly with photosynthesis. The mesophyll is differentiated into an upper palisade tissue and lower spongy tissue. The palisade tissue is present below the upper epidermis. The palisade cells are thin walled, cylindrical and contain numerous chloroplasts. The cells are compactly arranged in one layer and perpendicular to the upper epidermis. Narrow, intercellular spaces are present in the palisade tissue and allow the loss of water by transpiration. Palisade tissue is the most highly

specialized type of photosynthetic tissue. The upper surface of the leaf is dark green in colour due to palisade tissue. The lower part of the mesophyll present towards the lower epidermis is the spongy tissue. Cells of the spongy tissue are thin walled, irregular in shape, and arranged loosely with large continuous intercellular spaces. Stomata open into large intercellular spaces called substomatal chambers. The cells of spongy parenchyma contain a smaller number of chloroplasts, hence the lower surface of the leaf is light green in colour.

The vascular tissue occurs in the form of discrete bundles called veins. Veins are interconnected to form reticulate venation. Vascular bundles (veins) occur in between the palisade and spongy tissue (median in position). They are collateral and closed. In the vascular bundle, xylem is present towards the upper epidermis and phloem towards the lower epidermis. The vascular bundle is surrounded by parenchymatous bundle sheath. The cells of the bundle sheath are called border parenchyma.

A large number of trichomes are present on the leaf surface of chickpea (Fig. 6.6). During late vegetative growth, chickpea leaves and stem are covered with aqueous glandular droplets. If these droplets persist at low humidity for a long time, there may be substantial loss of water via glandular trichomes. It is concluded that water loss via trichomes could lower leaf temperature by several degrees within a wide range of atmospheric conditions. Seasonal variations were observed in hydrochloric acid, malic acid and Ca ions secreted by chickpea trichomes. It is suggested that malate, chloride ions and Ca ions are linked to the amount of daily sunlight and to the day of the year, while the pH of secretions is not directly related to sunlight level (Lazzaro *et al.*, 1995).

Role of glandular trichomes
in insect resistance

Chickpea leaves bear glandular trichomes on the leaf surface. Little research has been undertaken into the role of these trichomes in insect resistance.

Late vegetative growth chickpea leaves and stems can be covered with aqueous

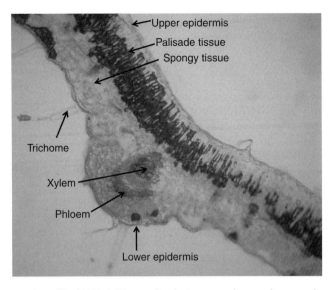

Fig. 6.6. Transverse section of leaf (100×). Note: palisade tissues are long and compactly arranged, and epidermis consists of bilobed trichomes. Xylem is towards the upper epidermis and phloem towards the lower epidermis.

glandular droplets. In a study it was concluded that water loss via the glandular trichomes can be enough to lower leaf temperature by several degrees centigrade within a wide range of atmospheric conditions. The exudate solutes were primarily malic, hydrochloric and oxalic acids. Without the strong acids a chickpea leaf, wet even on dry days, would be ripe for parasitic attack (Lauter and Munns, 1986).

The role of exudates from trichomes were studied as follows: the plant characters responsible for the absence of egg parasitoids and the feasibility of increasing parasitism levels on chickpea by mass-releasing *Trichogramma chilonis* Ishii were investigated. The residence time of female *T. chilonis* on chickpea leaves was affected by trichomes and the acidic trichome exudates secreted on all green parts of the plant. The parasitoids spent a longer time on chickpea leaves where the acidic trichome exudates washed off than on unwashed leaves. When placed on unwashed chickpea leaves, 6.8% of the parasitoids were trapped and killed by the exudates (Romeis *et al.*, 1999).

Effects of malic acid and oxalic acid on oviposition of *Helicoverpa armigera*

were investigated. However, there was a significant negative correlation between pod damage and oxalic acid levels. Oxalic acid, which had been reported to have an antibiotic effect on *H. armigera* larvae, has an important role in resistance to this pest in chickpea (Yoshida *et al.*, 1997).

Chickpea trichome exudates contain malic acid, oxalic acid and succinic acid. Oxalic acid has an antibiotic effect on the larvae of the pod borer, *H. armigera*, which results in reduced pod damage. A dense mat of non-glandular trichomes in chickpea species prevents the small larvae from feeding on the pods. Glandular trichomes and their chemical exudates may also be harmful to insect predators and parasites (Peter and Shanower, 1998).

The comparative assessment of density of glandular hairs, population and size of stomatal apertures in chickpea cultivars were studied. In the resistant reaction group, there were higher number of glandular hairs on the dorsi-ventral sides and higher number of leaf stomata as compared to the susceptible group. The resistant group had the smallest stomatal apertures (Randhawa *et al.*, 2009).

6.2   Common Bean

6.2.1   Introduction

Common bean (*Phaseolus vulgaris*) is a staple food product of great importance and high protein diet, mainly in Latin America and Africa as well as in other countries. It is an herbaceous annual plant domesticated independently in ancient Mesoamerica and the Andes, and is now grown worldwide for its edible bean, popular both dry and as a green bean. Unlike cereal crops, a leguminous crop like common bean has a high capacity for nitrogen fixation by *Rhizobium* bacteria in the root nodule. Brazil and India are the largest producers of dry beans while China produces, by far, the largest amount of green beans, almost as much as the rest of the top ten growers altogether http://en.wikipedia. org/wiki/Common_bean - cite_note-FAO-0

6.2.2   Origin, evolution and domestication

Different authors proposed different centres of origin of the common bean, though the south of Mexico and the regions of Guatemala and Honduras are generally considered the centre of origin. Though the number of species belonging to the genus *Phaseolus* is not known, it is considered to reach to 150.

The origin, evolution and domestication of *Phaseolus* have been postulated on the basis of data on the molecular biology. Wild species of *Phaseolus* showed variability in the banding pattern of storage proteins by electrophoresis. The wild accessions studied showed the CH and B phaseolin electrophoresis banding patterns. The 77 cultivated genotypes showed in order of decreasing frequency, the S, T and C patterns and the novel B pattern. It is suggested that Colombia is a meeting place for Andean (mainly T phaseolin) and Central American (mainly S phaseolin) germplasms of *P. vulgaris* and also as a domestication centre for the species. The Central American wild forms exhibited phaseolin electrophoresis banding patterns similar to the S pattern found in cultivated forms and novel banding patterns

designated M types were identified. It is concluded that there existed two primary areas of domestication, with that in Central America giving rise to small-seeded cultivars with T (and possibly C, H and A) phaseolin patterns. Lines were selected for the desirable white to beige colour, 1000-seed weight and higher yield.

6.2.3   Utilization

Similar to other beans, the common bean is high in starch, protein and dietary fibre and is an excellent source of iron, potassium, selenium, molybdenum, thiamine, vitamin B6 and folic acid. The dry beans are a food product of great importance and form a high protein diet. The leaf is occasionally used as a leaf vegetable and the straw is used for fodder. Beans are a legume and thus acquire their nitrogen through an association with rhizobia, a species of nitrogen-fixing bacteria. The commercial production of beans is well distributed worldwide with countries in Asia, Africa, Europe, Oceania, South and North America all among the top bean growers.

6.2.4   Morphological description

Vegetative characters

The common bean is a highly variable species with a long history. Bush varieties form erect bushes 20–60 cm tall, while pole or running varieties form vines 2–3 m long. All varieties bear alternate, green or purple leaves, divided into three oval (trifoliate), smooth-edged leaflets, each 6–15 cm long and 3–11 cm wide.

Floral characters

The white, pink, or purple flowers are about 1 cm long, and give way to pods 8–20 cm long, 1–1.5 cm wide, green, yellow, black or purple in colour, each containing from four to six beans. The beans are smooth, plump, kidney-shaped, up to 1.5 cm long, range widely in colour, and are often mottled in two or more colours. Seeds are large and reniform. The hilum is in the form of a half-moon shape. Raphe formed by two elevated locules of clear shiny coffee colour. The micropyle is round. Cotyledons are large, hard and a cream colour. The seed surface is reticulate, lobular in appearance and of variable colour, often coffee with dark spots.

6.2.5 Anatomical description

Root anatomy

The epidermis is uniseriate, having a single layer of compactly arranged thin-walled living cells. Both cuticle and stomata are absent. The epidermis gives protection and absorption of water and minerals. Epidermal cells produce unicellular root hairs, which are useful in increasing the surface area for absorption of water and helps in obtaining water from soil. The root hairs are generally short lived, and the epidermis ruptures after secondary growth.

Cortex is the middle region, lying between the epidermis and the stele, and is well developed and larger than the stele. It is multiseriate, relatively homogenous and consists of general cortex and endodermis. Cortex is extensively developed in the root and consists of loosely arranged thin-walled parenchyma cells with prominent intercellular spaces and flavonoid deposits. The cells are colourless and store starch and are usually round or oval. This region helps in the lateral conduction of water and salts. *Rhizobium* bacteria lives as symbionts in cortical cells and fix biological nitrogen. After secondary growth the general cortex ruptures.

Endodermis is very distinctive in roots. It is single layered with compactly arranged barrel-shaped cells. Endodermal cells are characterized by the presence of Casparian strips or Casparian bands made up of suberin. In transverse section, Casparian bands appear as a lens-shaped thickening on the radial walls of the endodermis. The endodermis is also called the starch sheet layer, as starch grains are present in the endodermal cells. Endodermis of the root acts as a barrier between cortex and stele. The endodermis ruptures due to secondary growth.

Pericycle is the outermost region of the stele, and is uniseriate and parenchymatous. Cells of the pericycle retain meristematic activity. Lateral roots arise endogenously from the pericycle and help in absorption of water. During secondary growth the pericycle forms phellogen or cork cambium and a small amount of vascular cambium.

In the vascular tissue, xylem and phloem are arranged in the form of separate strands on different radii, alternating with each other. Hence the vascular strands in root are described as separate and radial. Xylem is exarch having protoxylem towards the pericycle and metaxylem towards the centre and scattered secondary xylem vessels. Cambium is absent and hence vascular bundles are described as closed type.

Stem anatomy

A transverse section of primary stem shows three regions, the epidermis, cortex and stele.

Epidermis is the outermost region surrounding the stem. It is uniseriate, having a single layer of compactly arranged barrel-shaped living cells with trichomes. The outer walls of the epidermal cells are highly cutinized, which reduces transpiration. Few stomata are present in the epidermis for exchange of gases and transpiration. All the epidermal cells are colourless except the guard cells. Multicellular hairs are present on the epidermis.

The cortex is the middle region present between the epidermis and stele, and is smaller than the stele. Hypodermis is multilayered (from two to six layers) and collenchymatic; the collenchyma is living mechanical tissue. The hypodermis provides considerable strength, flexibility and elasticity to young stems. Having chloroplasts, it may carry on photosynthesis. The endodermis is the innermost layer of the cortex, and is usually distinct in stems without Casparian bands. The pericycle is multiseriate and sclerenchymatic. Sclerenchyma is mechanical tissue and gives strength and rigidity to the stem.

Xylem and phloem form the vascular tissues. They organize into wedge-shaped vascular bundles, with 15–20 vascular bundles arranged in one ring called a eustele. The vascular bundle consists of xylem towards the centre of the axis and phloem towards the periphery. Both xylem are phloem are present together (conjoint) on the same radius (collateral), having a strip of cambium (fascicular cambium) between xylem and phloem (open), hence the vascular bundle is described as conjoint, collateral and open type. Xylem is endarch with protoxylem present towards the centre. The xylem consists of many vessels arranged in rows. The extensions of parenchymatous pith between vascular bundles are called primary medullary rays, which help in lateral conduction and may give rise to secondary meristem (inter-fascicular cambium).

Leaf anatomy

The leaf epidermal cell surface of a common bean shows hook-shaped trichomes on the abaxial and adaxial leaf surfaces. In a transverse section, the leaf epidermal cells of various sizes are covered by a thin cuticle. This is followed by one layer of non-compact palisade parenchyma with abundant chloroplasts.

Leaflets of *P. vulgaris* contain calcium oxalates in the adaxial bundle sheath extensions. A comparative study was made on the leaf epidermal cell structure of four bean cultivars. Glandular trichomes were generally present on the veins. It was found that there is an inverse relation between leaf growth and the number of stomata, epidermal cells and trichomes at the lower and upper surface of the leaves. The decrease in number of these anatomical variables was associated with the elongation and expansion of the cells with growth.

Leaf pubescence in *P. vulgaris* has been associated with resistance to *Uromyces appendicularis*. The density of trichomes varied among cultivars and among leaves from different nodes. There was an inverse relationship between density of hairs and hooked trichomes. In a study, wrinkled leaves had more trichomes and stomata and smaller, more irregularly shaped epidermal cells than the smooth leaves.

Leaf anatomy and adaptation to drought and insects

Leaf anatomical traits play important roles in adaptation to biotic and abiotic stress factors.

A study was conducted to find out genetic variability for trichome distribution and density among three diverse dry bean (*P. vulgaris* L.) cultivars, and to characterize the types of trichomes present among the cultivars. Different types of trichomes were identified, on both the abaxial and adaxial leaf surfaces of the cultivars 'Bill Z', 'Pompadour Checa' and 'Diacol Calima'. More straight trichomes are present on the abaxial leaf surface (Dahlin *et al.*, 1992).

In common bean (*P. vulgaris* L.) leaf epidermal characteristics may play a role in plant defence against pathogens and/or drought. Crystalloids were present on both surfaces of the leaves which were devoid of wax. Trichome density and distribution differed among the common bean types. The abaxial leaf surface always presented more trichomes than the adaxial surface. Common bean leaves presented paracytic, anomocytic and anisocytic stomatal types. All common beans presented the adaxial epidermis of the leaves with a lower density of bigger stomata than on the abaxial epidermis (Sebastian Stenglein *et al.*, 2004).

Cuticle thickness plays a more important role than calcification of cell walls in the resistance of older plants to *R. solani* (Stockwell and Hanchey, 1983).

Seed anatomy

In a transverse section the cuticle visible and macrosclereids are thin and elongated in the form of a cylinder. There is a large number of starch granules distributed uniformly in the cotyledon cells. Protein granules are amorphous and distributed around starch granules. The species showed large variations in form, size, position of hilum, micropyle, and seed surface and colour.

6.3 Green Gram

6.3.1 Introduction

Green gram (*Vigna radiata*) is a pulse crop and belongs to the Fabaceae or bean or pea family. Grams are annual legume crops grown for their seed. The green grams are the most commonly grown pulses. Usually no fertilizers are applied to green gram. Over the centuries, green gram's adaptation to stable performance in marginal environments has resulted in a low yield potential, which limits responsiveness to better environments and improved cultural practices.

6.3.2 Origin, domestication and adaptation

Grams are native crops of India. Often called green gram or golden, it is cultivated in several countries of Asia, Africa and the Americas. Green gram usually matures in 60 to 90 days. The early maturing varieties can often produce before drought destroys many bean species.

Green gram grows best at an altitude of 0–1600 m above sea level and under warm climatic conditions (28–30°C). They are well adapted to red sandy loam soils, but also do reasonably well on not too exhausted sandy soils. Green gram is not tolerant of wet, poorly drained soils, but is drought tolerant and will give reasonable yields with as little as 650 mm of yearly rainfall. Heavy rainfall results in increased vegetative growth with reduced pod setting and development.

6.3.3 Utilization

The dried beans are prepared by cooking or milling and are eaten whole or split. The seeds or the flour may be utilized in a variety of dishes such as soups, porridge, snacks, bread, noodles and even ice cream. Green gram also produces great sprouts, which can be sold in health food shops or eaten at home.

Crop residues of *V. radiata* are a useful fodder. Green gram is sometimes specifically grown for hay, green manure or as a cover crop. Green gram is grown mainly on small-holdings, often as mixed crops or intercrops. This practice increases the nitrogen level in the soil by biological nitrogen fixation with symbiotic association of *Rhizobium* bacteria in root nodules (fixing atmospheric nitrogen in the form of nitrates). Associated crops are usually of longer duration than green gram (sugarcane, cotton, sorghum). To make use of a short cropping period, short-duration green gram is often relay-cropped.

6.3.4 Morphological description

Vegetative characters

Green gram is an annual herb, growing up to 50–75 cm, is mesophytic and well adapted for hot and drought conditions. It grows in tropical and temperate regions. It has a tap root system with high branching, and root nodules are present.

The stem is weak, aerial, erect, greenish, rounded, pubescent, with compound leaves, exstipulate, alternate, with two to three pairs of leaflets present on a rachis, ovate shape, acute apex and reticulate venation (Fig. 6.7).

Floral characters

The flowers are of the terminally racemose type, and are yellow in colour, pedicellate, bracteate, complete, bisexual, zygomorphic, pentamerous, and perigynous. There are five yellow gamosepalous sepals. The flower has five petals, yellow in colour, free, in a papilionaceous arrangement posterior petal is large, called a vexillum. Wing petals are present both side of the vexillum, and on the anterior side keel petals are present. There are ten stamens, diadelphous, dimorphic, dithecous, introse, unicarpellary and unilocular. From one to three ovules are present on marginal placentation, and a simple stigma with hairs. The fruit is leguminous, with five to ten rounded green seeds per fruit.

Fig. 6.7. Green gram plant morphology illustrating simple leaves with ovate shape and acute apex, yellow flowers and slender pods.

6.3.5 Anatomical description

Root anatomy

The root is covered by uniseriate compactly arranged barrel-shaped cells called epidermis. The epidermis ruptures due to the secondary growth.

Cortex is present below the epidermis, consisting of parenchymatous multilayered general cortex and uniseriate compactly arranged endodermis. The cells are usually round or oval in shape. This region helps in lateral conduction of water and salts and stores starch. The inner layer of the cortex is called endodermis and is uniseriate, with compactly arranged cells with stored food materials.

Stele consists of pericycle and vascular strands. Pericycle is the outermost region of the stele. Vascular strands contain xylem and phloem and are separate and radial. Xylem is exarch and triarch in condition. The centre part of the stem contains a small amount of pith.

ROOT NODULE   Experiments have been carried out on green gram (*V. radiata* cv. PS 16) to study the interaction between symbiosis and root rot in terms of plant growth and nitrogen fixation. The microtomy of the infected roots showed distortion of the outer layers as a possible cause for antagonistic interaction of these two bio-processes (Kush, 1982).

Stem anatomy

The outline of the stem in transverse section shows ridges and grooves (generally six ridges and six grooves).

The outermost region of the stem is surrounded by uniseriate epidermis, which is covered by thick cuticle. Unicellular trichomes are present on the epidermis.

The arrangement of the cortical layer varies from ridges to groove, showing three to six layers below the ridges (Fig. 6.8) and three layers below the grooves. There are one to two layers of collenchyma below the epidermis. Chlorenchyma patches are also distributed in the parenchyma tissue. The endodermis is not very distinct.

Vascular tissue consists of two to three layers of multiseriate sclerenchymatic pericycle covering the stele. There are 12–15 wedge-shaped vascular bundles arranged in one ring. Vascular bundles are conjoint, collateral and open type. Large sized vascular bundles are present

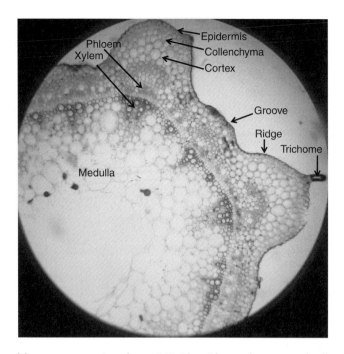

Fig. 6.8. Outline of the transverse section of stem (100×) has ridges and grooves and collenchyma is present below the ridges only. In the vascular bundle xylem is towards the periphery and phloem towards centre. The central portion is occupied by the medulla.

below the ridges and small vascular bundles below the grooves. Xylem is endarch with protoxylem present towards the centre. The xylem consists of many vessels, which are arranged in rows. Pith is prominent and occupies a larger amount of space.

Leaf anatomy

Leaf shows a dorsi-ventral nature, with uniseriate compactly arranged epidermis present on both surfaces of the leaf. Epidermis is covered by a thin layer of cuticle. The anisocytic type of stomata is present in both epidermal layers. The stomatal index of the upper epidermis is 25 and of the lower epidermis is 28.

The mesophyll is chlorenchymatic and differentiated into palisade and spongy tissue. The palisade cells are thick walled, cylindrical and loose and contain numerous chloroplasts (Fig. 6.9). Mesophyll consists of parenchyma with large intercellular spaces. Stomata open into large intercellular spaces called sub-stomatal chambers.

Vascular bundles (veins) occur between the palisade and spongy tissue (median in position). They are collateral and closed. In the vascular bundle, xylem is present towards the upper epidermis and phloem towards the lower epidermis.

The presence of loose palisade cells leads to a large amount of transpiration loss and makes the plant drought susceptible.

Petiole anatomy

A transverse section of the petiole shows three regions, the epidermis, ground tissue and vascular bundles (Fig. 6.10).

The epidermis is uniseriate, having a single layer of compactly arranged barrel-shaped living cells. Unicellular hairs are present on the epidermis.

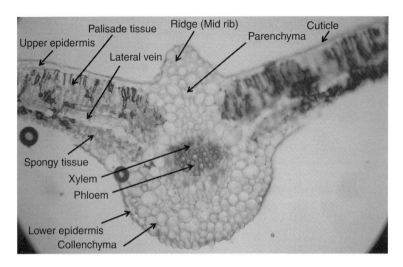

Fig. 6.9. Transverse section of leaf (100×) showing arrangement of tissues. Palisade cells are not very compactly arranged. At the centre of the leaf (midrib) two to three layers of collenchyma are present above the lower epidermis.

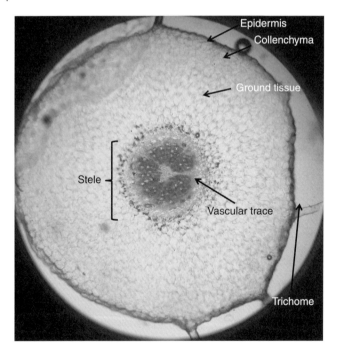

Fig. 6.10. Transverse section of petiole (100×) showing epidermis with trichomes, collenchyma, a large amount of ground tissue made up of parenchyma and central (stele) vascular tissue. Note: Vascular trace is clearly visible in the stele making the stele appear like a 'C' shape.

Ground tissue consists of thin-walled parenchymatous cells with well defined intercellular spaces among them.

Vascular bundles (stele) are arranged in the form of a 'C' shape in parenchymatic ground tissue. Vascular trace is present between the vascular bundles, which are wedge-shaped. Xylem is always found towards the outer side, whereas phloem is always towards the inner side.

6.4   Red Gram (Pigeonpea)

6.4.1   Introduction

Red gram (*Cajanus cajan* (L.) Millsp) is an important pulse crop. It is also known as pigeonpea, arhar or tur. Red gram belongs to the family Leguminosae. Red gram is mainly cultivated and consumed in developing countries of the world, and is widely grown in India. India is the largest producer and consumer of red gram in the world.

The life span of red gram is up to 5 years. It is extremely variable from the genetic viewpoint, hence the very many cultivars. The reproductive system is ca. 60% autogamous, with a chromosome number $2n=22$. There are 16,000–18,000 seeds/kg and some 10% hard seeds. The most favourable temperature range is 18–30°C. The crop is grown at a wide range of elevation. Red gram can be grown in almost all soil types that are not very deficient in lime and are not subjected to waterlogging. Optimum growth and yield are obtained in deep loam soils. Red gram can be grown as a mixed crop with groundnut, paddy or tapioca or as a pure crop.

Red gram is a protein-rich staple food. It contains about 22% protein, which is almost three times that of cereals. Red gram supplies a major proportion of the protein requirement for the vegetarian population in India. Red gram is mainly consumed in the form of split pulse as dal, which is an essential supplement of a cereal-based diet. It is particularly rich in lysine, riboflavin, thiamine, niacin and iron.

6.4.2   Origin and distribution

Red gram has an unknown origin, although it is probably Indian and African. It has been cultivated in ancient Egypt, Africa and Asia since prehistoric times, and was later introduced to America. Red gram is now acclimatized in several tropical countries. The major producer is India with over 100 cultivars, 2.4 Mha cultivated and 90% of world production. In India there are numerous varieties. Commercially there are the late, tall, long-podded, many-grained 'arhar' type of north-west and eastern India, and the early, small, few-seeded 'tur' type of Madhya Pradesh, western and peninsular India.

6.4.3   Utilization

Seeds are used as pulses. As a human food, under the Sahel conditions pigeonpea produces some 650 kg of beans/ha/year, having approximately 25% crude protein, with a well balanced composition in amino acids, except for methionine and cysteine, which are too low for an ideal human food. The pods may be consumed green, used as green peas, or as ripe beans, like cowpea. The plant has a high feeding value for beef and dairy cattle, swine, sheep and goats. Nitrogen fixing ability for a density of 7,000–10,000 plants/ha is of the order of 100–120 kg N_2/ha/year. The crop has long been used as a windbreak and as shade for young coffee trees, forest seedling nurseries and vegetable beds, and is an important honey-producing plant. The canning of green pigeonpeas is a major industry in Puerto Rico and Trinidad. Pigeonpea constitutes an important domestic fuel over large areas, albeit its caloric power is low. Leaves remain on the plant throughout the dry season. Yields of fodder vary widely with the ecological conditions and the care given to the crop. Tall perennial varieties are amenable to trimming as fodder, but also as green manure (2.6% nitrogen).

Due to its deep rooting system pigeonpea offers little competition to associated crops and is therefore much used in intercropping systems with cereals such as millet, sorghum and even maize; it also provides a good means to improve fertility in fallows.

6.4.4   Morphological description

Vegetative characters

Red gram is mesophytic and also grow in the semi-arid tropics. *Cajanus cajan* is very heat-tolerant, and prefers hot, moist conditions. It tolerates a wide range of soils, from sands to heavy black clays, and also tolerates a wide pH range, but the most favourable range is pH 5.0 to 7.0. It is sensitive to salt spray, high salinity and to waterlogging.

Cajanus cajan is one of the most drought-tolerant legume crops, with a wide range of rainfall tolerance, but prefers more than 625 mm and in elevated areas exceeding 2000 m cold nights and cloudy weather interfere with fertilization of flowers. It flowers well in regions where rainfall is 1500–2000 mm. However, it will grow on deep, well-structured soils where the rainfall is only 250–375 mm.

Red gram is an annual, or more usually short-term perennial shrub, that may reach 4–5 m in height, but usually only 1–2 m, woody at the base, with a variable habit but usually erect. It has a taproot system, which is deep and quick growing. Root nodules are present on the lateral roots, with *Rhizobium* bacteria living as symbionts in the root nodules and performing the biological fixation of atmospheric nitrogen. Generally root nodules appear on the roots 3 weeks after seeding, and the young nodules are the most active in fixing nitrogen. The stem is green to brown in colour and thick. The angular stem results from three ribs starting from the base of each petiole.

Leaves are trifoliate and alternate, set in a spiral along the stem (Fig. 6.11). Leaflets are oblong, lanceolate, 5–10 cm long × 2–4

Fig. 6.11. Red gram plant morphology showing spirally arranged trifoliate leaves, yellow flowers and red coloured leguminous fruit.

cm wide and pubescent, as is the stem. Lateral petioles are 2–3 mm long, with the terminal petiole reaching 10–20 mm. Stipules are linear and 2–3 mm long, whereas stipulets are filiform and 1–2 mm long.

Floral characters

Flowers form in racemes, having from five to ten flowers on top of an axillary, little-divided peduncle. Flowers are usually yellow but they may also be striated with purple streaks or plain red. Flowers are pedicellate, bracteate, complete, bisexual, zygomorphic, pentamerous and perigynous. Usually two flowers open at the same time on one inflorescence. The flowers are self-pollinated, pollination taking place before the flowers open. The calyx is 10–12 mm long with five linear teeth. Pods are flat, with an acuminate tip, pubescent and of variable colour, 5–9 cm long × 12–13 mm wide. There are five gamosepalous sepals. The corolla is 20–25 mm long, with the flag 18–20 mm wide. There are five petals, red in colour, free, in a papilionaceous arrangement, and the posterior petal is large, called a vexillum. Wing petals are present both sides of the vexillum; keel petals are present on the anterior side. There are ten stamens, diadelphous, dimorphic, dithecous, introse, unicarpellary, unilocular. From one to three ovules are present on marginal placentation, simple hairy stigma. The fruit is leguminous, with green pods containing two to nine seeds in shades of brown, red or black and rounded. Husks bearing deep oblique furrows underline the septa between the seeds.

6.4.5 Anatomical description

Root anatomy

Internal tissues of the root are covered by uniseriate compactly arranged thin-walled living cells called epidermis.

The cortex is the middle region, lying between the epidermis and the stele. Cortex is well developed and is larger than the stele. It is multiseriate and relatively homogeneous and consists of parenchyma.

The innermost layer of the cortex is called the endodermis. It is single layered, having compactly arranged barrel-shaped cells. The endodermis is also called the starch sheet layer, as starch grains are present in the endodermal cells.

The stele is covered by uniseriate parenchymatous pericycle. Stele (vascular strands) consists of xylem and phloem. Xylem is used for conduction of water and salts and phloem for the transport of food materials. Vascular strands in root are described as separate and radial. Xylem is exarch, having protoxylem towards the pericycle and metaxylem towards the centre and scattered secondary xylem vessels.

Stem anatomy

The primary structure of the stem consists of large and small vascular bundles forming a cycle. Vascular bundles are collateral type. The vascular tissue of the xylem in the secondary structure of stem contains large quantities of fibres and medullary rays.

Epidermis is the outermost region surrounding the stem. It is uniseriate with a single layer of compactly arranged barrel-shaped living cells. The outer walls of the epidermal cells are cutinized. Moreover, the epidermis is covered by a separate layer of cutin called cuticle on its outer surface (cuticularized). Few stomata are present in the epidermis for gaseous exchange and transpiration. All the epidermal cells are colourless except the guard cells. Unicellular trichomes are present on the epidermis.

The cortex is the middle region present between the epidermis and stele. Cortex is smaller in volume than the stele. The hypodermis is multilayered with four to six layers and is collenchymatic (Fig. 6.12). Collenchyma is living mechanical tissue. The hypodermis gives considerable strength, flexibility and elasticity to young stems. Having chloroplasts, it may carry out photosynthesis. The endodermis is innermost layer of the cortex. It is usually distinct in stems, without Casparian bands. In young stems the innermost layer of the cortex contains abundant starch, hence this layer is called the starch sheath. The starch sheath layer is homologous to endodermis, but morphologically unspecialized, hence this layer may be termed endodermoid. After secondary growth the endodermis ruptures.

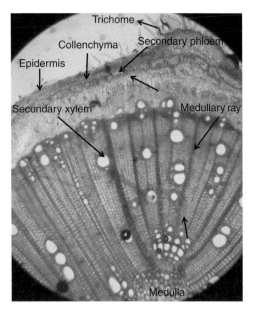

Fig. 6.12. Transverse section of stem (100×) showing arrangement of tissues. Note: the epidermis is fully covered by a high density of trichomes. Primary organization of the stem is disturbed by secondary growth and the cortex is compressed. Secondary xylem vessels are larger in diameter and medullary rays are clearly visible.

Stele is the central region of the stem and is larger than the cortex. Pericycle is multiseriate and sclerenchymatic. Sclerenchyma is mechanical tissue and gives strength and rigidity to the stem. There are 15–20 wedge-shaped vascular bundles arranged in one ring called the eustele, which consists of xylem towards the centre of the axis and phloem towards the periphery. In the vascular bundle xylem and phloem are present together (conjoint) on the same radius (collateral), having a strip of cambium (fascicular cambium) between xylem and phloem (open), hence the vascular bundle is described as conjoint, collateral and open type. Xylem is endarch with protoxylem present towards the centre. The xylem consists of many vessels arranged in rows.

Medulla and medullary rays are distinct, present at the centre of the stem and are usually parenchymatous. The cells are round or oval with intercellular spaces. The medulla stores food materials. The extensions of parenchymatous pith between vascular bundles are called primary medullary rays, which help in lateral conduction and may give rise to secondary meristem (interfascicular cambium).

Leaf anatomy

The leaf shows a dorsi-ventral nature, and uniseriate compactly arranged epidermis is present on both surfaces of the leaf. The epidermis is covered by a thin layer of cuticle. There is a high density of unicellular trichomes on the upper and lower epidermis (Fig. 6.13).

The mesophyll is chlorenchymatic and differentiated into palisade and spongy tissue. The palisade cells are thick walled,

cylindrical and contain numerous chloroplasts. Spongy tissue consists of parenchyma with large intercellular spaces. Vascular bundles (veins) occur in between the palisade and spongy tissue (median in position) and are collateral and closed. In the vascular bundle, xylem is present towards the upper epidermis and phloem towards the lower. The leaf is bifacial and covered with multicellular hairs. The stoma were more dense in the lower epidermis than the upper epidermis, and both sides of the leaf had a corneous layer (Yin *et al.*, 2004).

The presence of a thick cuticle, high density of trichomes and compactly arranged palisade tissues reduce drastically transpiration, thereby contributing to the drought resistance of red gram.

Role of trichomes in insect resistance

Three types of glandular (types A, B, and E) trichomes and two non-glandular (types C and D) types were identified in red gram. Types A, B, C and D were present on leaves, pods and calyxes of all three *Cajanus* spp., except for Type A, which was not found on pods and calyxes of most *C. scarabaeoides* accessions examined. Because of their small size and rarity, pods of *C. scarabaeoides* were the most densely pubescent, followed by pods of *C. cajan* and *C. platycarpus*.

Trichome density on pods varied significantly among red gram genotypes and different accessions of *C. scarabaeoides*. The resistance of *C. scarabaeoides* pods to *Helicoverpa armigera* (Hübner) larvae reported in an earlier study is due to the high density of non-glandular trichomes. This wild species may thus be an important source for developing insect-resistant red gram (Romeis *et al.*, 1999).

Trichome length and density, sugars, proteins and phenols were found to be associated with resistance to *M. vitrata* in short-duration red gram genotypes. Trichome density on upper and lower surfaces of the leaf, and length and trichome density and length on pods were found to be negatively correlated with the resistant genotype (Sunitha *et al.*, 2008).

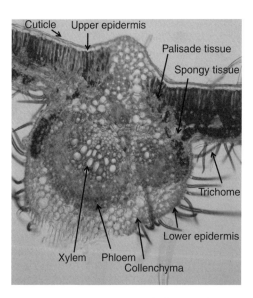

Fig. 6.13. Transverse section of leaf (100×) showing high density of trichomes on the lower epidermis. Note: palisade cells are long and extremely compactly arranged and the lower epidermis is fully covered by trichomes, there is a thick cuticle and large xylem vessels. All these characters contribute drought resistance to red gram compared with other legume crops.

Petiole anatomy

The epidermis is uniseriate, having a single layer of compactly arranged barrel-shaped living cells. The outer walls of the epidermal cells are cutinized. Unicellular hairs are present on the epidermis. Multilayered (two to six layers) hypodermis (Fig. 6.14) of angular collenchyma cells is found immediately beneath the epidermis. Collenchyma is living mechanical tissue. The hypodermis gives considerable strength, flexibility and elasticity to young petioles. Just beneath the hypodermis, ground tissue is found. It consists of thin-walled parenchymatous cells having well defined intercellular spaces among them. Vascular bundles are arranged in the form of a half-moon shape in ground tissue. The vascular bundles are of various sizes in the same petiole.

Conclusion

In conclusion, the anatomical features of red gram (such as thick cuticle, long compactly arranged palisade cells in the leaf; dense trichomes, thick cuticle and thick

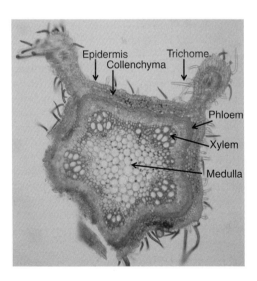

Fig. 6.14. Transverse section of petiole (100×) is pentagonal in outline and shows a high density of trichomes on the epidermis. Thick six-layered collenchyma tissue lies beneath the epidermis. There is a prominent medulla at the centre and five vascular bundles in the stelar region.

collenchyma in the petiole; and highly lignified hard wood) mentioned above contribute to drought resistance and its natural adaptation to drought situations. At the same time, the high density of long trichomes contributes tolerance to various insects.

6.5 Cowpea

6.5.1 Introduction

Cowpea (*Vigna unguiculata*) is one of the most important food legume crops in the semi-arid tropics, covering Asia, Africa, southern Europe and Central and South America. The growth habit is climbing, spreading or erect and cowpeas belong to the bean family (*Leguminosae*, subfamily *Papilionaceae*). A drought-tolerant and warm-weather crop, cowpeas are well-adapted to the drier regions of the tropics, where other food legumes do not perform well. Cowpeas are mainly important in the marginal rainfall areas because they are well adapted to the dry climate and suitable for a variety of intercropping systems. Cowpeas are basically annual crops grown for their leaves and seed. They are a common food item in the southern USA, where they are often called field peas, and an integral part of the cuisine in the southern region of India.

6.5.2 Origin, domestication and adaptation

Cowpea is native to Africa where it was domesticated over 4000 years ago. It has been well utilized in South-east Asia for more than 2000 years, and is an important crop in China and India.

Cowpea is generally tolerant of drought and low light conditions, but is very susceptible to a variety of insects and diseases and does not do well in poorly drained and cool areas. Local landraces of cowpea grown by farmers in West Africa are well adapted so that they start to flower at the end of the rains at a particular locality. The optimum temperature for their growth and development is 20–35°C. Cowpea can grow in a wide range of soils, is well adapted to light sandy soils where most other crops produce poorly, and they also do well on acid soils. On heavy fertile soils they show a vigorous vegetative growth, but not necessarily a good grain yield. Most varieties need a minimum rainfall of 200 mm during a growing season.

In a study, ten crop species were evaluated for their relative drought tolerance in the seedling stage. Soybean appeared to be the most drought susceptible and cowpea was the most drought tolerant (Singh *et al.*, 1999).

6.5.3 Utilization

Cowpea is cultivated for the seeds (shelled green or dried), the pods or leaves that are consumed as green vegetables or for pasture, hay, silage and green manure. Tender cowpea leaves and shoots contain 4% protein, 4% carbohydrates and are rich in calcium, phosphorus and vitamin B.

Dried seeds contain 22% protein and 61% carbohydrates. According to the USDA food database, cowpea has the highest percentage of calories from protein among vegetarian foods. The leaves may be dried and stored for later use. Cowpea has the useful ability to fix atmospheric nitrogen through its root nodules, and it grows well in poor soils with more than 85% sand and with less than 0.2% organic matter and low levels of phosphorus. In addition, it is shade tolerant and therefore compatible as an intercrop with maize, millet, sorghum, sugarcane and cotton. This makes cowpea an important component of traditional intercropping systems, especially in the complex and elegant subsistence farming systems of the dry savannahs in sub-Saharan Africa. In Africa, where cowpea is the preferred food legume, they are consumed in two basic forms: (i) cooked together with vegetables, spices and often palm oil, to produce a thick bean soup, which accompanies the staple food (cassava, yams, plantain); and (ii) decorticated and ground into a flour and mixed with chopped onions and spices and made into cakes.

6.5.4 Morphological description

Vegetative characters

The crop exhibits much variation in growth habit, leaf shape, flower colour, seed size and colour.

Cowpea is an annual mesophytic herb and grows well in the semi-arid tropics. It has a taproot system with bacterial root nodules. The stem is aerial, greenish, rounded and pubescent. The leaf base is pulvinus, and the leaves are simple, alternate, an ovate shape with an acute apex and reticulate venation (Fig. 6.15).

Floral characters

Flowers are in a terminal racemose inflorescence and are pedicellate, bracteate, complete, bisexual, zygomorphic, pentamerous and perigynous. There are five gamosepalous sepals and five petals, free, in a papilionaceous arrangement; the posterior petal is large, called the vaxillum. Wing petals are present on both sides of the vexillum; keel petals are present on the anterior side. The male flower contains ten stamens, which arise in two whorls of five each. The stamens of the outer whorl are opposite to the sepals and the inner whorl is opposite to the petals; hence it is called diplostemonous. Anthers are diadelphous, dithecous, basifixed and introrse. The female flower is unicarpellary and unilocular, with one to three ovules present on marginal placentation, with simple stigma and a hairy style. The fruit is a legume, long, slender and green in colour.

There are 12–20 seeds per fruit, rounded, and showing variation in colour among varieties.

6.5.5 Anatomical description

Root anatomy

A transverse section of the root shows three regions, the epidermis, cortex and vascular tissue.

Epidermal cells are compactly arranged as thin-walled living cells. The cortex is

Fig. 6.15. Cowpea plant morphology, showing simple leaves, yellow flowers and long slender pods.

the middle region, lying between the epidermis and the stele.

The cortex is well developed and is larger than the stele. It is multiseriate and relatively homogeneous and consists of parenchyma. The innermost layer of the cortex is called the endodermis. This is single-layered with compactly arranged barrel-shaped cells. The endodermis is also known as the starch sheet layer, as starch grains are present in the endodermal cells. The stele is covered by uniseriate parenchymatous pericycle.

The stele (vascular strands) consists of xylem and phloem. Xylem conducts water and salts and phloem transports food materials. Vascular strands in the root are separate and radial. Xylem is exarch, having protoxylem towards the pericycle and metaxylem towards the centre and scattered secondary xylem vessels.

ROOT NODULES   The role of nitrogen fixation is shown by the structure and organization of nodular tissues and bacteroids of cowpea induced by *Rhizobium* sp., 4 to 5 weeks after inoculation, when nitrogenase activity reaches a peak. All cell types in cowpea nodules were larger. The inner cortex of cowpea had an 'endodermis-like' layer of cells. Many cells of cowpea remained free of bacteroids. Cells in cowpea were mostly

elongated with a nucleus and more vacuoles. The bacteroids within cowpea cells were arranged without any particular order with more space for host cellular material. They were mostly present singly in peribacteroidal membrane sacs, which sometimes fused to enclose more than one bacteroid. In the differentiation of nodular tissue, the hosts seem to play the dominant role in the morphogenesis of bacteroids in symbiotic systems induced by the same strain of *Rhizobium* (Sen *et al.*, 1986).

Stem anatomy

The epidermis is uniseriate, having a single layer of compactly arranged barrel-shaped living cells. The outer walls of the epidermal cells are cutinized. The epidermis is covered by a separate layer of cutin called cuticle on its outer surface (cuticularized). Few stomata are present in the epidermis for the exchange of gases and transpiration. All the epidermal cells are colourless except the guard cells. Multicellular trichomes are present on the epidermis.

The cortex is the middle region present between the epidermis and stele, and is smaller than the stele. The hypodermis is multilayered (four to six layers) and collenchymatic. Collenchyma is a living mechanical tissue. The endodermis is the innermost layer of the cortex without any Casparian strips.

Xylem and phloem are organized into wedge-shaped vascular bundles, with 15–20 bundles arranged in one ring called a eustele. It consists of xylem towards the centre of the axis and phloem towards the periphery. In the vascular bundle xylem and phloem are present together (conjoint) on the same radius (collateral) and have a strip of cambium (fascicular cambium) between xylem and phloem (open); hence the vascular bundle is described as conjoint, collateral and open type. The xylem is endarch with protoxylem present towards the centre. The xylem consists of many vessels arranged in rows.

Leaf anatomy

Unicellular trichomes are present on both the epidermal layers and thick cuticle covers the epidermis. Well developed mesophyll is distributed between the upper and lower epidermis.

The mesophyll is highly differentiated into palisade and spongy tissue. The palisade cells are cylindrical in shape, thick walled and consisting of numerous chloroplasts. Sponge tissue consists of parenchyma with intercellular spaces and is for storage of food materials.

Vascular bundles are closed, collateral and occur between the palisade and spongy tissue (median in position).

Role of trichomes to insect resistance

Cowpea possesses leaf trichomes. Few studies have been undertaken the role of trichomes to various biotic stresses.

The role of pubescence was investigated in the resistance of two wild cowpea varieties. Resistance to *Maruca testulalis* (Geyer) was based on trichomes in the first instance and phytochemicals, with previous reports indicating that phytochemicals may be the principal factor in the resistance of TVNu 72 and TVNu 73 to this species. Glandular and non-glandular trichomes were found to be present on both the cultivated and wild cowpea. Two types of cowpea differed

significantly only in trichome number (susceptible cultivated cowpeas have more) and non-glandular trichome length. Trichome length and angle to pod surface seemed to be more important than density (Jackai and Oghiakhe, 1989).

Damage to cowpea (*Vigna unguiculata*) by the legume pod borer *M. testulalis* (Geyer) showed that trichome cover on individual cultivars varied in trichome length and density. Trichome density on different parts decreased with increasing plant age. Significant ($P < 0.05$) negative correlations were obtained between total trichome density on pods, pod infestation and damage severity. Results suggest that trichome length may be less important than density in reducing pod damage by *M. testulalis* in cultivated cowpeas. It is concluded that breeding for a higher level of trichomes in high yielding and agronomically desirable cowpea cultivars will serve as an important component in the integrated management of *M. testulalis* (Oghiakhe *et al.*, 1992).

Studies revealed that there was a significant positive correlation between larval penetration time on pods and length of non-glandular trichomes. Weight loss, number of feeding punctures and number of larvae on whole pods of wild varieties were lower and significantly different from cultivated and semi-cultivated. This study showed that it would be advantageous to use wild varieties in a breeding programme to incorporate pubescence into high-yielding commercial cultivars for resistance against *M. testulalis* and possibly other major pests (Oghiakhe, 1995).

Disease resistance

The infection process of *Colletotrichum destructivum*, a hemibiotrophic anthracnose fungus, was studied in two cowpea (*V. unguiculata*) cultivars that differ in disease reaction type. In the susceptible cultivar there was formation of large, multilobed, intracellular infection vesicles, followed by necrotrophic, radiating, secondary hyphae produced in tissues. But in the resistant cultivar, the production of appressoria and their melanization resulted in reduced penetration. Where penetration occurred, the initially infected epidermal cells underwent a hypersensitive response, restricting the growth of multilobed vesicles and thereby blocking the destructive necrotrophic phase of disease development (Latunde-Dada *et al.*, 1999).

Petiole anatomy

There is a close similarity between petiole and stem with regard to the structure of epidermis. The ground parenchyma of the petiole is like the stem cortex in arrangement of cells and in number of chloroplasts. The supporting tissue is collenchyma in relation to the arrangement of vascular tissues in the stem; the vascular bundles of the petiole are collateral.

The epidermis is uniseriate, having a single layer of compactly arranged barrel-shaped living cells. The outer walls of the epidermal cells are cutinized, which reduces transpiration. Unicellular hairs are present on the epidermis.

Multilayered (three to six layers) hypodermis of angular collenchyma cells is found immediately beneath the epidermis. Collenchyma is living mechanical tissue. The hypodermis gives considerable strength, flexibility and elasticity to young petioles. Just beneath the hypodermis ground tissue is found. This consists of thin-walled parenchymatous cells with well-defined intercellular spaces among them.

Vascular bundles are arranged in the form of a half-moon shape in ground tissue. The wedge-shaped vascular bundles are of various sizes in the same petiole.

Research needs

In view of the literature discussed above, it may be concluded sufficient research inputs have been directed on the roles of glandular trichomes in insect resistance in cowpea. The cuticle also has some role but studies are not available on the roles of these important traits on drought resistance. Several field crops discussed in this volume show the important traits for drought resistance; the presence of profuse trichomes and cuticle reduces transpiration loss.

Therefore, there is a great necessity to screen and select genotypes for combination of some of these traits that may be utilized in breeding for insect and drought resistance.

6.6 Pea

6.6.1 Introduction

A 'pea' is most commonly the small spherical seed or the seed-pod of the legume *Pisum sativum* Linn. Each pod contains several peas. Peapods are botanically a fruit, since they contain seeds developed from the ovary of a (pea) flower. However, peas are considered to be a vegetable in cooking. The name is also used to describe other edible seeds from the Fabaceae such as the pigeonpea (*C. cajan*), the cowpea (*V. unguiculata*) and the seeds from several species of *Lathyrus*.

6.6.2 Origin, distribution and adaptation

The wild pea is restricted to the Mediterranean basin and the Near East. The earliest archaeological finds of peas come from Neolithic Syria, Turkey and Jordan. In Egypt, early finds date from ca. 4800–4400 BC in the Nile delta area, and from ca. 3800–3600 BC in Upper Egypt. The pea was also present in Georgia in the 5th millennium BC. Farther east, the finds are more recent. Peas were present in Afghanistan ca. 2000 BC, in Harappa, Pakistan, and in northwest India in 2250–1750 BC. In the second half of the 2nd millennium BC this pulse crop appeared in the Gangetic basin and southern India.

The pea is a green, pod-shaped vegetable, widely grown as a cool-season vegetable crop. The seeds may be planted as soon as the soil temperature reaches 10°C (50°F), with the plants growing best at temperatures of 13–18°C (55–64°F). They do not thrive in the summer heat of warmer temperate and lowland tropical climates but do grow well in cooler high-altitude tropical areas. Many cultivars reach maturity about 60 days after planting. Generally, peas are grown outdoors during the winter, not in greenhouses. Peas grow best in slightly acidic, well-drained soils.

6.6.3 Utilization

Pisum are used as vegetables. The peapod is eaten flat, while in sugar/snap peas, the pod becomes cylindrical but is eaten while still crisp, before the seeds inside develop. The seeds have a high protein content.

6.6.4 Morphological description

Pisum sativum is an annual plant, with a life cycle of 1 year. It is a cool-season crop grown in many parts of the world; planting can take place from winter through to early summer depending on location. The average pea weighs between 0.1 and 0.36 g. The species is used as a vegetable, fresh, frozen or canned, and is also grown to produce dry peas such as the split pea. These varieties are typically called field peas.

The veining cultivars grow thin tendrils from leaves that coil around any available support and can climb to 1–2 m high. A traditional approach to supporting climbing peas is to thrust branches pruned from trees or other woody plants upright into the soil, providing a lattice for the peas to climb. Branches used in this fashion are sometimes called 'pea brush'. Metal fences, twine, or netting supported by a frame are used for the same purpose. In dense plantings, peas give each other some measure of mutual support. Pea plants can self-pollinate.

Vegetative characters

Pisum is an annual herb, mesophytic, and has a taproot system with bacterial root nodules. The stem is weak and a tendril climber (Fig. 6.16), aerial, greenish, rounded and pubescent. The leaf base is pulvinate and the leaves are pinnately compound, alternate, ovate, with an acute apex and reticulate venation. Terminal leaflets become tendrils; stipules are foliaceous.

Floral characters

Flowers are a terminal racemose type and are pedicellate, bracteate, complete, bisexual, zygomorphic, pentamerous and perigynous. There are five gamosepalous sepals and five free petals in a papilionaceous arrangement;

Fig. 6.16. Pea plant morphology. The pea has a weak stem and is a tendril climber, with pinnately compound leaves, and a terminal racemose inflorescence.

the posterior petal is large, called a 'vexillum'. Wing petals are present on both sides of the vexillum, with keel petals present on the anterior side. There are ten stamens, which actually arise in two whorls of five each. The stamens of the outer whorl are opposite to the sepals and inner whorl of stamens is opposite to the petals, hence it is called diplostemonous. The gynoecium is diadelphous, dithecous, basifixed, introse, unicarpellary and unilocular. There are from one to three ovules present on marginal placentation, simple hairy stigma. The fruit is leguminous (pod), slender and green coloured, with from five to ten seeds per fruit, rounded, and green in colour.

6.6.5 Anatomical description

Root anatomy

Internal tissues of the root are covered by uniseriate compactly arranged thin-walled living cells called epidermis.

The cortex is the middle region, lying between the epidermis and the stele. Cortex is well developed and is larger than the stele (Fig. 6.17). It is multiseriate and relatively homogeneous and consists of parenchyma. The innermost layer of the cortex is called the endodermis, which is single layered with compactly arranged barrel-shaped cells. The endodermis is also called the starch sheet layer, as starch grains are present in the endodermal cells.

The stele is covered by uniseriate parenchymatous pericycle. Stele (vascular strands) consists of xylem and phloem. Xylem conducts water and salts and phloem transports food materials. The xylem is exarch, with protoxylem towards the pericycle and metaxylem towards the centre and scattered secondary xylem vessels.

Stem anatomy

The stem is triangular in transverse section. A transverse section of the stem primarily shows epidermis, cortex and vascular bundles.

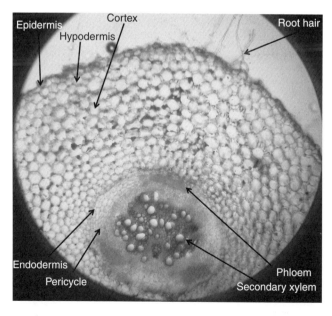

Fig. 6.17. Transverse section of root (100×) showing organization of various tissues. The epidermis has root hairs, and multilayered parenchymatous cortex occupies the major portion of the root. Note: secondary growth in stele has started with secondary xylem vessels formed towards the centre and phloem towards the outer side.

The epidermis is uniseriate with a single layer of compactly arranged barrel-shaped living cells. The epidermis is covered by a thin cuticle.

The cortex is multilayered and parenchymatic. Below the epidermis there are four to five layers of collenchyma restricted to the ridges (Fig. 6.18). The cells are round with prominent intercellular spaces. The endodermis is innermost layer of the cortex. Due to the secondary growth, vascular bundles are spread into the cortex. Pericycle is the outermost non-vascular region of the stele. It is multiseriate and sclerenchymatic. Sclerenchyma is mechanical tissue and gives strength and rigidity to the stem.

Xylem and phloem form the vascular tissues and are organized into wedge-shaped vascular bundles. From 10–15 vascular bundles are arranged in one ring called a eustele. A weak stereome system can be observed. In the vascular bundle xylem and phloem are present together (conjoint) on the same radius (collateral), having a strip of cambium (fascicular cambium) between xylem and phloem (open), hence the vascular bundle is described as conjoint, collateral and open type. The xylem is endarch with protoxylem present towards the centre. The xylem consists of many vessels arranged in rows.

Leaf anatomy

Epidermis covering the upper surface of the leaf is called upper epidermis or ventral epidermis or adaxial epidermis and on the lower side of the leaf is called lower epidermis or dorsal epidermis or abaxial epidermis. The epidermis is single layered with compactly arranged barrel-shaped (tabular) cells. The outer walls of the epidermal cells are cutinized. Anisocytic stomata are present in both epidermal layers, but the stomata are usually more frequent in lower epidermis.

The ground tissue of the leaf, present between the upper and the lower epidermal layers, is called mesophyll. The mesophyll is not clearly differentiated between an upper palisade tissue and lower spongy tissue (Fig. 6.19). Mesophyll

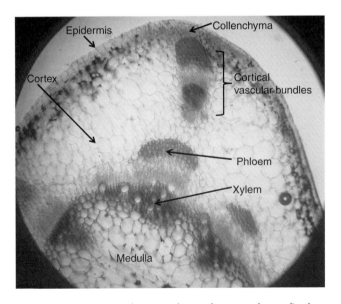

Fig. 6.18. Transverse section of stem (100×) showing ridges and grooves; four to five layers of collenchyma are present specially below the ridges. Cortical vascular bundles can be seen (special feature in some legumes). Secondary growth has just initiated.

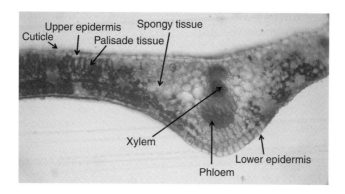

Fig. 6.19. Transverse section of leaf (100×) illustrating the arrangement of various tissues.

is compact below the upper epidermis and contains intercellular spaces above the lower epidermis.

Vascular bundles occur between the palisade and spongy tissue. The central part of vascular bundle consists of xylem, and phloem is on both sides of xylem.

Conclusion

The anatomy of the stem, leaf and petiole in pea, with a lack of trichomes, a thin cuticle and short loosely arranged palisade layer, contributes to the susceptibility of the crop to drought.

References

Dahlin, R.M., Brick, M.A. and Ogg, B.J. (1992) Characterization and density of trichomes on three common bean cultivars. *Economic Botany* 46(3), 299–304.

Hossain, A., Haque, A. and Prodhan, M.Z.H. (2008) Effect of pods characteristics on pod borer, *Helicoverpa armigera* (Hubner), investigation in Chickpea. *SAARC Journal of Agriculture* 6(1).

Jackai, L.E.N. and Oghiakhe, S. (1989) Pod wall trichomes and resistance of two wild cowpea, *Vigna vexillata*, accessions to *Maruca testulalis* (Geyer) (Lepidoptera: Pyralidae) and *Clavigralla tomentosicollis* Stål (Hemiptera: Coreidae). *Bulletin of Entomological Research* 79(4), 595–605.

Kush, A.K. (1982) Interaction between symbiosis and root pathogenesis in green gram (*Vigna radiata* L. Welczek). *Plant and Soil* 65(1), 133–135.

Ladizinsky, G. (1975) A new *Cicer* from Turkey. Notes Royal Botanical Garden, Edinburgh 34, 201–202.

Latunde-Dada, A.O., O'Connell, R.J., Bowyer, P. and Lucas, J.A. (1999) Cultivar resistance to anthracnose disease of cowpea (*Vigna unguiculata* (L.) Walp.) caused by *Colletotrichum destructivum* O'Gara. *European Journal of Plant Pathology* 105(5), 445–451.

Lauter, D.J. and Munns, D.N. (1986) Water loss via the glandular trichomes of chickpea (*Cicer arietinum* L.). *Journal of Experimental Botany* 37(5), 640–649.

Lazzaro, M.D., Thomson, W. and William, W. (1995) Seasonal variation in hydrochloric acid, malic acid, calcium ions secreted by the trichomes of chickpea (*Cicer arietinum*). *Physiologia Plantarum* 94, 291–297.

Maiti, R.K. and Wesche-Ebeling, P. (2001) Biological nitrogen fixation. In: Maiti, R.K. (ed.) *Advances in Chickpea Science*. Science Publishers, Enfield, USA, pp. 134–170.

Oghiakhe, S. (1995) Effect of pubescence in cowpea resistance to the legume pod borer *Maruca testulalis* (Lepidoptera: Pyralidae). *Crop Protection* 14(5), 379–387.

Oghiakhe, S., Jackai, L.E.N., Makanjuola, W.A. and Hodgson, C.J. (1992) Morphology, distribution, and the role of trichomes in cowpea (*Vigna unguiculata*) resistance to the legume pod borer, *Maruca testulalis* (Lepidoptera: Pyralidae). *Bulletin of Entomological Research* 82(4), 499–505.

Peter, A.J. and Shanower, T.G. (1998) Plant glandular trichomes chemical factories with many potential uses. *Resonance* 3(3), 41–45.

Randhawa, M.A., Sahi, S.T., Ilyas, M.B., Ghazanfar, M.U. and Javed, N. (2009) Comparative assessment of density of glandular hairs, population and size of aperture of stomata in resistant and susceptible cultivars of chickpea to *Ascochyta* blight disease. *Pakistan Journal of Botany* 41(1), 121–129.

Romeis, J., Shanower, T.G. and Zebitz, C.P.W. (1999) Why *Trichogramma* (Hymenoptera: Trichogrammatidae) egg parasitoids of *Helicoverpa armigera* (Lepidoptera: Noctuidae) fail on chickpea. *Bulletin of Entomological Research* 89(1), 89–95.

Sabaghpour, S.H., Mahmodi, A.A., Saeed, A., Kamel, M. and Malhotra. R.S. (2006) Study on chickpea drought tolerance lines under dryland condition of Iran. *Indian Journal of Crop Science* 1(1–2), 70–73.

Sen, D., Weaver, R.W. and Bal, A.K. (1986) Structure and organization of effective peanut and cowpea root nodules induced by rhizobial strain 32H1. *Journal of Experimental Botany* 37(3), 356–363.

Singh, B.B., Mai-Kodomi, Y. and Terao, T. (1999) Relative drought tolerance of major rainfed crops of the semi-arid tropics. *Indian Journal of Genetics and Plant Breeding* 59(4), 437–444.

Stenglein, S.A., Arambarri, A.M., Vizgarra, O.N. and Balatti, P.A. (2004) Micro-morphological variability of leaf epidermis in Mesoamerican common bean (*Phaseolus vulgaris*, Leguminosae). *Australian Journal of Botany* 52(1), 73–80.

Stockwell, V. and Hanchey, P. (1983) The role of the cuticle in resistance of beans to *Rhizoctonia solani*. *Phytopathology* 73(12), 1640–1642.

Sunitha, V., RangaRao, G.V., Vijaya Lakshmi, K., Saxena, K.B., RameshwarRao, V. and Reddy, Y.V.R. (2008) Morphological and biochemical factors associated with resistance to *Maruca vitrata* (Lepidoptera: Pyralidae) in short-duration pigeonpea. *International Journal of Tropical Insect Science* 28, 45–52.

van der Maesen, L.J.G. (1972) *A monograph of the genus with special reference to the chick pea (*Cicer arietinum*), its ecology and cultivation.* Communication of Agricultural University, Wageningen, Dordrecht, the Netherlands.

Wood, J.A., Knights, E.J. and Choct, M. (2011) Morphology of chickpea seeds (*Cicer arietinum* L.): comparison of desi and kabuli types. *International Journal of Plant Sciences* 172(5), 632–643.

Yin, X.-L., Xu, X.-Y., Long, R, Li, S.-C. and Wang, H. (2004) The preliminary anatomy of the vegetative organ *Cajan muscajan*. *Journal of Hebei Normal University of Science & Technology* 2004(3), 40.

Yoshida, M., Cowgill, S.E. and Wightman, J.A. (1997) Roles of oxalic and malic acids in chickpea trichome exudate in host-plant resistant to *Helicoverpa armigera*. *Journal of Chemical Ecology* 23(4), 1195–1210.

7

Oil Crops

7.1 Sunflower

7.1.1 Introduction

As a result of sunflower (*Helianthus annuus*) being an edible oil of high commercial value, sufficient research activities have been undertaken on various fields of this economically important crop. The sunflower is also valuable as ornamental plant. It forms one of the well-known crops in Russia, Spain, France, Germany, Italy, Egypt, India, Manchuria and Japan. The average acre will produce about 50 bushels of merchantable seeds, and each bushel yields approximately 1 gallon of oil for which there is a whole series of important uses.

The sunflower has been developed into an important agricultural crop. With the decline in world supplies of hydrocarbons, there has been a particular interest in the direct use of sunflower seed oil as biodiesel fuel and in catalytic conversion to other liquid fuels.

Nutrient uptake

In comparison with other crop species, sunflower (*H. annuus* L.) has been found to be very tolerant of high manganese (Mn) concentrations in nutrient solution. The first symptom of excess Mn supply (ca. 30 µM Mn in solution) was the appearance of small, dark-brown to black spots (<0.5 mm in diameter) on lower stems and on petioles and blades of the lower leaves. Electron microprobe techniques demonstrated an accumulation of Mn in and around the trichomes. A compartmentation mechanism is suggested, whereby sunflower is able to tolerate high Mn concentrations in its tissues through localization of Mn in a metabolically inactive form. Dark brown lesions (>2 mm in size) developed on the lower leaves, especially along the veins. A concentration of 2205 µg Mn/g in the tops was associated with a 10% reduction in plant dry matter yield (Blamey *et al.*, 1986).

7.1.2 Origin and domestication

Sunflower is one of the few crop species that originated in North America. It was probably a camp follower of several of the western Native American tribes who domesticated the crop and carried it eastward and southward of North America. The first Europeans observed sunflower cultivated in many places from southern Canada to Mexico.

Sunflower was probably first introduced to Europe through Spain, and spread through Europe as a curiosity until it reached Russia where it was readily adapted. Selection for high oil in Russian began in 1860 and was largely responsible for increasing oil content from 28% to 50%. The high producing lines from Russia were introduced into the USA after the Second World War, which rekindled interest in the crop.

Carbonized seeds of domesticated sunflower (*H. annuus* var. *macrocarpus* Ckll.) recovered from the Hayes site in middle Tennessee yielded an accelerator date of 4265 +/– 60 BP. This is the earliest date for domesticated sunflower, extending the known age of this eastern North American domesticate by 1400 years. The process of domestication has changed plants from being multi-stemmed with a number of flower heads to ones that are tall and unbranched, with a single large head.

7.1.3 Utilization

Every part of the plant may be utilized for some economic purpose: the leaves can be used as cattle feed; the stems contain a fibre, which may be used successfully in paper making; and the flowers contain a yellow dye.

The whole seed of sunflower is sold as a snack and can be processed into a delicious peanut butter alternative, sun butter. It is also sold as food for birds and can be used directly in cooking and salads. The seed is rich in oil, which is said to approach more nearly to olive oil than any other vegetable oil known and is largely used as a substitute. Sunflower oil, extracted from the seeds, is used for cooking, as carrier oil and to produce biodiesel, for which it is less expensive than the olive product.

7.1.4 Morphological description

The sunflower is an annual plant in the family Asteraceae (synonym Compositae), daisy or sunflower family. The sunflower inflorescence or flower head reaches 30 cm in diameter (also, capitulum) and is typical of other Asteraceae family members, and represents a contracted raceme composed of numerous individual sessile florets, all

sharing the same receptacle. As in other Asteraceae, it looks superficially like a single flower. The sunflower's 'seed' is actually an achene: a monocarpellate (formed from one carpel) and indehiscent (they do not open at maturity) fruit containing a single seed that nearly fills the pericarp, but does not adhere to it. *H. annuus* is diploid with $2n = 34$ chromosomes.

Vegetative characters

The plant is a mesophytic herb, with a taproot system with highly branched lateral roots. The stem is aerial, erect, greenish, rough, and hard, pubescent and monopodial in branching and grows up to 3 m. Leaves are simple, exstipulate, alternate, pubescent petiolate, ovate shape, with a serrate margin, acute apex, and rough, reticulate venation (Fig. 7.1).

Floral characters

The inflorescence is a terminal, head inflorescence or capitulum or anthodium, and the

Fig. 7.1. Sunflower plant morphology showing monopodial branching head inflorescence (heterogamous head).

head consists of a highly condensed disc-like peduncle on which numerous small sessile flowers called florets arise. The florets are surrounded and protected by involucres of bracts. The florets are of two types (ray and disc florets), hence it is called a heterogamous head (Fig. 7.2). Ray florets are present at the periphery of the head, and are bracteate, sessile, incomplete, unisexual, zygomorphic and epigynous. The disc florcts are present at the centre of the head, and are bracteate, sessile, complete, bisexual, actinomorphic and epigynous. Sepals are modified into a persistent pappus. There are five gamopetalous petals. In disc florets the corolla (petals) is regular and tubular. In ray florets the corolla is zygomorphic and legulate because the two posterior petals are reduced. The androecium is absent in the ray florets; in disc florets there are four stamens, epipetalous and syngenesious, the filaments are free and anthers connate laterally to form a tube around the style. The anthers are dithecous, basifixed and hooded. The gynoecium is present in both ray and disc florets, and consists of a bicarpellary, syncarpous, inferior, unilocular ovary having only one ovule with basal placentation. The style is terminal, simple and long, and the stigma is forked and re-curved.

Sunflower is generally a cross-pollinated crop. If cross pollination does not occur it can self-pollinate by the use of a safety mechanism. Pollination occurs through entomophily, i.e. the pollen is spread by insects, normally honeybees. The sunflower's 'seed' is actually an achene: a monocarpellate (formed from one carpel) and indehiscent (they do not open at maturity) fruit containing a single seed that nearly fills the pericarp, but does not adhere to it.

7.1.5 Anatomical description

Root anatomy

A transverse section of matured root shows a thin layer of bark at the outer side of the root that is produced due to rupture of the epidermis. Below the epidermis a two- to five-layered cortex is present in compressed form (Fig. 7.3). Xylem is exarch with continuously formed secondary xylem vessels. Pith is absent at the centre. Xylem vessels are endarch with prominent secondary xylem vessels present at the centre of the root.

Significance of variations in root anatomy

Cultivar differences are expressed in a number of root characteristics: (i) the presence or absence of sclerenchymatous exodermis

Fig. 7.2. Inflorescence (heterogamous head) showing petals, disc and ray florets.

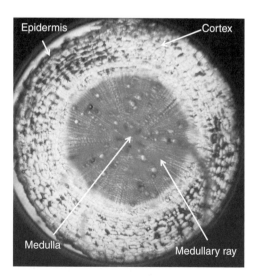

Fig. 7.3. Transverse section of root (100×) showing secondary growth. Primary structure of root is disorganized due to secondary growth and the cortical cells are compressed by enlargement of the stele.

in the cortex and the size of vascular bundle, which may be related to drought resistance; and (ii) the thickness of endodermal cell walls.

Stem anatomy

Stem anatomy of sunflower is typical of that of the dicots. The mature stem shows secondary growth. The primary stem shows three regions, the epidermis, cortex and stele.

The epidermis is the outermost region surrounding the stem, and is uniseriate; having a single layer of compactly arranged barrel-shaped living cells. The outer walls of the epidermal cells are cutinized. The epidermis is also covered by a separate layer of cutin called cuticle on its outer surface, which reduces transpiration. Stomata are present in the epidermis for the exchange of gases and transpiration. Multicellular hairs are present on epidermis.

The cortex is the middle region present between the epidermis and stele, and contains hypodermis, middle cortex and endodermis. The hypodermis is multilayered (with two to six layers) and collenchymatic (Fig. 7.4). Collenchyma may be in the form of complete cylinder or in the form of discrete strands. The hypodermis gives considerable strength, flexibility and elasticity to the young stem. The endodermis is the innermost layer of the cortex. In the young stem the innermost layer of the cortex contains abundant starch, hence this layer is called the starch sheath.

Pericycle is the vascular region of the stele. It is multiseriate and sclerenchymatic. Sclerenchyma fibres may be in the form of bundle caps. Wedge-shaped vascular bundles are arranged in one or two rows. The vascular bundles are described as conjoint, collateral and open type. The xylem consists of many vessels and is endarch with protoxylem present towards

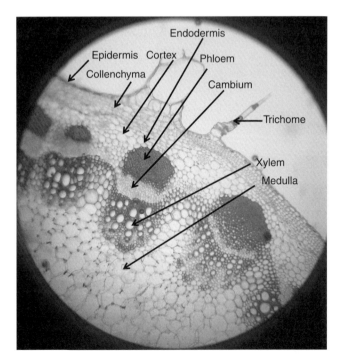

Fig. 7.4. Transverse section of stem (100×) showing epidermis, multilayered collenchyma, cortex, endodermis, xylem, phloem and medulla. Note: collenchyma layers are highly compactly arranged and long trichomes are visible on the epidermis.

the centre. The extensions of parenchymatous pith between vascular bundles are called primary medullary rays, which help in lateral conduction and may give rise to secondary meristem.

Significance of variations in stem anatomy

Epidermal study reveals the following variability between genotypes:

- Density of trichomes may vary.
- Cuticle thickness of the epidermis varies among genotypes.
- Presence or absence of multicellular trichomes; in some genotypes there may be unicellular and multicellular trichomes, which may be related to insect resistance.
- Number of hypodermal layers may differ among genotypes, imparting flexibility and varying the shape of hypodermal cells (collenchyma cells).
- Distribution of collenchyma cells (Fig. 7.5).
- Number of pericycle layers and the shape of sclerenchyma cells differ.
- Sclerenchyma fibres may be in the form of bundle caps in some genotypes, offering strength.

The size of the vascular bundles and the number of vascular bundles reveals the capacity of conduction in sunflower cultivars. It can be observed that the anatomy of young stem shows the presence of a thin cuticle and multicellular trichome; this is subtended by several layers of collenchyma, imparting flexibility of the sunflower stem.

Leaf anatomy

The leaf shows a dorsi-ventral nature, i.e. the leaf blade consists of distinct dorsal and ventral surfaces. The epidermis is single layered, having compactly arranged barrel-shaped (tabular) cells. The cell walls are cutinized and also covered by a thin cuticle. The cutinized outer walls and cuticle reduce the loss of water due to transpiration. Anisocytic type of stomata are present in both the epidermal layers, but the stomata are usually more frequent in the lower epidermis; generally the stomatal index of upper epidermis is 33 and lower epidermis is 39. Stomata facilitate the exchange of gases between the leaf and environment. The epidermis contains two types of trichomes: they are small and unicellular and are either spine or hook shaped. The mesophyll is chlorenchymatic and differentiated into an upper palisade

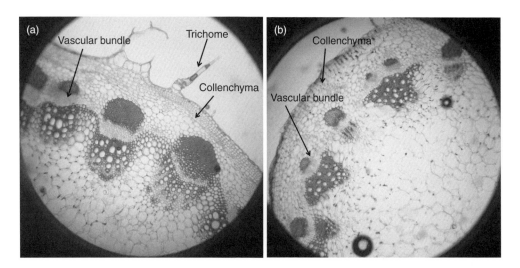

Fig. 7.5. Transverse section of petiole (100×) showing the variation in collenchyma layers and size of the vascular bundles: (a) five to six layers of collenchyma and medium size bundles and (b) two to three layers and large size bundles.

tissue and lower spongy tissue (Fig. 7.6). The palisade tissue is thick and compact in drought-resistant genotypes. Cells of the spongy tissue are thin-walled, irregular in shape, and arranged loosely with large continuous intercellular spaces. Stomata open into large intercellular spaces called substomatal chambers. The cells of spongy parenchyma contain a smaller number of chloroplasts, hence the lower surface of the leaf is light green in colour. Collateral and closed vascular bundles (veins) occur between the palisade and spongy tissue (median in position).

Studies have been undertaken on the development of leaf and cuticle.

Anatomical changes during leaf development

Leaves of *H. annuus* L. were divided into five parts: palisade, spongy, vascular and epidermal tissue, and intercellular space. The percentage of whole-leaf volume that each compartment occupied and the relative surface area of compartment cells in contact with intercellular space and adjacent cells for plants grown under low light were measured. During development all tissue parts

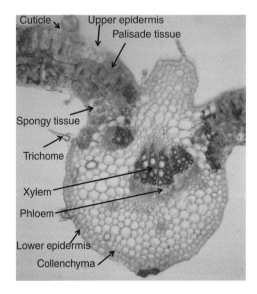

Fig. 7.6. Transverse section of leaf (100×) showing cuticle, upper epidermis, lower epidermis, palisade tissue, spongy tissue, xylem, phloem and collenchyma.

decreased in relative volume, while intercellular space increased. There was a proportional decrease in the amount of contact between mesophyll and vascular tissue, while the surface area of mesophyll cells and vascular tissue in contact with intercellular space increased (Fagerberg and Culpepper, 1984).

Cuticle development

Cuticle development depends on moderate water deficit (MWD). MWD generated from early to late anthesis had an effect on quantitative development of the cuticle, and qualitative and quantitative development of the cuticular waxes (CW) of the pericarp in two hybrids grown under field conditions. At harvest maturity (HM), plants grown under MWD showed higher CW content (31 to 47%) and thicker cuticles (13%) in both experiments compared to controls. Epicuticular wax (ECW) crystals showed a granular morphology. They determined a reduction of CW of 29% (mg CW/g pericarp) during the development of the fruit, from reproductive stage (stage when ray flowers have lost their turgidity) to HM (stage when water content of the fruit was 11%). These results show how internal mechanisms and external variables regulate pericarp wax content, and that fruit dehydration affects both the quantity and quality of wax formation from the time of fertilization to maturity (Franchini *et al.*, 2010).

Significance of variations in leaf anatomy

Arrangement of palisade tissue varies among the genotypes. Genotypes having thick, long and compact palisade tissue can be drought resistant. This is usually observed in thick leaves. Both the type and arrangement of vascular bundle both in the lamina and midrib varies. The number of stomata, trichome density and types of trichomes varies among genotypes (Fig. 7.7). The density of trichomes is related with drought and insect resistance.

Petiole anatomy

The epidermis contains multicellular trichomes (Fig. 7.8). A multilayered (two to

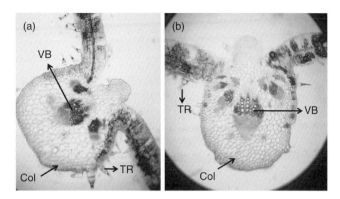

Fig. 7.7. Transverse section of leaf (100×) illustrating variation in trichome (TR) density, distribution of vascular bundles (VB) and collenchyma layers (Col) at the midrib; (a) high density of trichomes, three narrow vascular bundles and more layers of collenchyma; (b) low density of trichomes, smaller amount of collenchyma and five vascular bundles (widely spread).

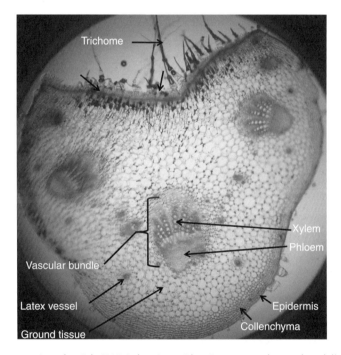

Fig. 7.8. Transverse section of petiole (100×) showing epidermis, cortex, xylem and medulla. Note the high density of long and short trichomes covering the epidermis; latex cavities are distributed in the cortex; and vascular bundles are large with six to eight xylem rows.

six layers) hypodermis of collenchyma cells is found immediately beneath the epidermis and, having chloroplasts, it may carry out photosynthesis. Just beneath the hypodermis ground tissue is found, which consists of thin-walled parenchymatous cells having well-defined intercellular spaces among them. Vascular bundles are arranged in a half ring and scattered in ground tissue. The wedge-shaped vascular bundles are of various sizes in the same petiole. In the petiole, xylem is always found towards the upper side whereas phloem is towards the lower side.

The number of collenchyma layers differs in different genotypes functioning as mechanical support to the petiole and is related to drought resistance. Trichome density varies, which offers insect resistance (Fig. 7.9). The number and size of vascular bundles differs among genotypes, which determines the efficiency of translocation of mineral water (Fig. 7.10). It is postulated that the presence of trichomes and thick cuticle on the leaf surface followed by compactly arranged palisade tissue may confer resistance to drought and insects. In addition, the presence

of thick collenchyma imparts flexibility to the petiole (Maiti *et al.*, 2007).

Insect and drought resistance

It is observed that the presence of trichomes and thick cuticle on the leaf surface followed by compactly arranged palisade tissue may confer resistance to drought and insects. In addition, the presence of thick collenchyma imparts flexibility to the petiole (Maiti *et al.*, 2007).

Trichome – insect resistance

Local species of sunflower are resistant to major insect pests of cultivated sunflower.

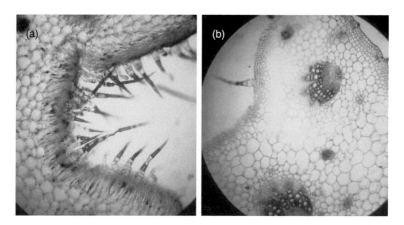

Fig. 7.9. Transverse section of petiole (100×) showing variation in trichome density among the different genotypes: (a) high trichome density; and (b) low trichome density.

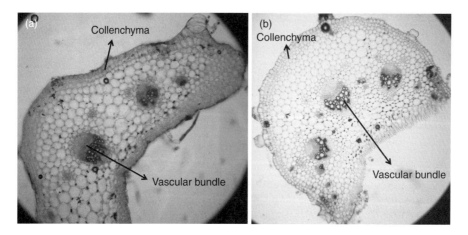

Fig. 7.10. Transverse section of petiole (100×) showing the variation between collenchyma layers and size of the vascular bundles: (a) one to two layers of collenchyma and medium size bundles; and (b) three to four layers and large size bundles.

The density of trichomes has been implicated as a factor in this resistance. The important role of trichomes in the deterrence of feeding by *Cylindrocopturus adspersus* LeConte, a sunflower stem weevil (SSW), was examined. Trichomes were randomly associated with feeding activity. The morphology and feeding behaviour of this insect enables it to avoid sunflower trichome defences. The results indicate that selective sunflower breeding for increased trichome density would have minimal benefit for increased resistance to *C. adspersus* feeding and oviposition (Barker, 1990).

A great variability occurs among sunflower genotypes in trichome density (Maiti, unpublished). This reveals that there is a great necessity to screen and identify sunflower genotypes with greater trichome density and utilize these in breeding for insect and drought resistance.

Drought and salt tolerance

The identification of sunflower genotypes exhibiting tolerance to moisture stress as well as those showing least percentage reduction in their growth and yield attributes under stress is necessary for the improvement of productivity under rainfed environment. Several seedling, physiological, morphological and biochemical traits have been shown to play a significant role in crop adaptation to water stress among which cuticle thickness, compact palisade cells, and a stout petiole with thick collenchyma layers are important (Geetha *et al.*, 2011; Maiti, unpublished).

Application of silica prevents membrane damage (membrane permeability) and increases leaf relative water content. Catalase activity was significantly decreased by drought stress, but supplemental Si increased it. In general, superoxide dismutase and ascorbate peroxidase activities of the cultivars were increased by drought and decreased by application of Si. Under drought stress the non-enzymatic antioxidant activity of the cultivars was significantly increased by Si. Based on this, it can be concluded that applied Si alleviates drought stress in sunflower cultivars by preventing membrane damage (Gunes *et al.*, 2008).

The salt-tolerant line was superior to the salt-sensitive one in growth performance under drought conditions. Stomatal conductance, transpiration and gas exchange in the salt-tolerant line were not affected by water deficit, whereas they were affected in the salt sensitive line. The salt tolerant line had significantly greater values for all these variables than the salt sensitive line. The lines did not differ in water-use efficiency. From this study it is clear that salt tolerance and drought tolerance of the two lines examined share osmotic effect (Ashraf and O'Leary, 1996).

Seed coat anatomy

Integuments of fertilized ovule become the seed coat. The seed coat covers the embryo and provides protection. The seed coat consists of two layers, the testa and the tegman.

The testa is the outermost layer of the seed coat and is formed from the outer integument of the ovule. Generally it is hard and impermeable to water and sometimes shows ornamentations. The testa contains macro- and microsclereids. Macrosclereids are arranged very compactly and are dumb-bell/boat/bone-shaped (osteosclereids). Microsclereids are small in size with a square shape and dentate edges. The hardness of the seed coat depends on the thickness of the testa and the arrangement of sclereids. The rate of water absorption (imbibition) also depends on the thickness of testa. Very hard and thick testa causes seed dormancy in some plants.

The tegman is the inner layer of the seed coat and is formed from the outer integument. It is thin and soft and covers the embryo.

The macro- and microsclereids are not very compactly arranged; therefore water enters easily into the seed and on imbibition it ruptures easily. As a result of this the time of imbibition is less, therefore germination time is about 36 h at 25–30°C temperature in laboratory conditions.

Sunflower is considered as highly susceptible to drought, but some anatomical traits
such as thick cuticle, high density of trichomes, long compactly arranged palisade
tissue and silica in the leaf, thick cuticle,
dense trichomes and thick collenchyma in
the petiole contribute to drought and insect
resistance in spite of the fact there exists
great variability among sunflower genotypes
for these traits. Therefore there is a great
necessity to screen genotypes for these traits
and select genotypes for a combination of
drought and insect resistance traits and utilize them in breeding for tolerance to insects
and drought.

7.2 Groundnut (Peanut)

7.2.1 Introduction

Grain legumes play an important role as
sources of protein and minerals. Among
them, groundnut is a very important legume
for millions of people in semi-arid and tropical regions of the world. Groundnut is a
small erect or trailing herbaceous legume,
about 15–60 cm high. The fruit is a pod with
one to five seeds that develops underground
within a needle-like structure called a peg.
The seeds are rich in oil (38–50%), protein,
calcium, potassium, phosphorus, magnesium
and vitamins. Groundnuts also have considerable medicinal value. They are reported to
be useful in the treatment of disease such as
haemophilia, stomatitis and diarrhoea. Groundnut is used for different purposes, as food
raw, roasted or boiled, as cooking oil, animal
feed and industrial raw materials. Groundnut
seeds contain 40–50% fat, 20–50% protein
and 10–20% carbohydrate.

7.2.2 Origin and domestication

Groundnut originated in South America in
the area from southern Bolivia to north-
western Argentina. The Portuguese apparently took them from Brazil to West Africa

and then to south-western India in the 16th
century. Africa is now regarded as a secondary centre of diversity. Groundnuts are now
grown in most tropical, subtropical and
temperate countries between 40°N and 40°S
latitude, especially in Africa, Asia and
North and South America.

The cultivated groundnut was probably
first domesticated in the valleys of the
Paraguay and Parana rivers in the Chaco
region of Paraguay and Bolivia. The domesticated groundnut is an amphidiploid or
allotetraploid (i.e. has two sets of chromosomes from two different species). The wild
ancestors of the groundnut were thought to
be *Arachis duranensis* and *A. ipaensis*, a
view recently confirmed by direct comparison of the groundnut's chromosomes with
those of several putative ancestors.

Archaeologists have dated the oldest
specimens found in Peru to about 7600 years.
Cultivation was spread by the Spanish
conquistadors of Mexico as far as Mesoamerica. The plant was later spread worldwide by European traders (Maiti and
Wesche-Ebeling, 2002).

7.2.3 Utilization

Most of the world production of groundnuts
is crushed for oil, which is used mainly for
cooking. The press cake from oil extraction
is a feed rich in protein but is also used to
produce groundnut flour, which is used in
many human foods. The seeds or kernels
are eaten raw, boiled or roasted, made into
confectionery and snack foods, and are used
in soups or made into sauces to use on meat
and rice dishes. The vegetative residues
from the crop are excellent forage.

7.2.4 Morphological characters

Vegetative characters

Groundnut is a mesophytic annual herb and
well adapted for semi-arid tropics. It grows
to a height of up to 40 cm, and requires 75°F
and annual rainfall of 50–125 cm. It has a

taproot system with lateral branches containing root nodules formed by a symbiotic association of *Rhizobium* bacteria, which fixes atmospheric nitrogen by mean of biological nitrogen fixation – nitrogenase enzyme. The stem is weak, aerial, erect, greenish, rounded and pubescent. The leaves are compound (Fig. 7.11), exstipulate and alternate, with two to three pairs of leaflets on a rachis. Each leaflet is ovate shape, with an acute apex and unicostate reticulate venation.

Floral characters

Flowers are born on a solitary, axillary inflorescence, pedicillate, stipulate, complete, bisexual, zygomorphic, pentamerous and perigynous. There are five sepals, green, free and pubescent. The corolla consist of five yellow petals, free, in a papilionaceous arrangement, and the posterior petal is large, called a vexillum. Wing petals are present on both sides of the vexillum; keel petals are present on the anterior side. The androecium consists of ten stamens in a monadelphous condition, dimorphic, dithecous, introse, and out of the ten stamens two are staminode. The gynoecium consists of a unicarpellary, unilocular ovary with one to three ovules present on marginal placentation. The stigma is simple with hairy. The leguminous fruit usually contain one to

Fig. 7.11. Groundnut plant morphology showing compound alternate leaves and solitary, axillary inflorescence.

three seeds per fruit. The seeds are rounded and usually pink in colour.

7.2.5 Anatomical description

Root anatomy

The epidermis forms the outermost region of the root and is uniseriate; having a single layer of compactly arranged thin-walled living cells. Epidermal cells produce unicellular root hairs, which are useful in increasing the surface area for absorption of water and helps in obtaining water from soil. The root epidermis is also called the rhizodermis, epiblema and siliferous layer

The cortex is multiseriate and contains relatively homogeneous parenchymatous tissue between epidermis and stele. The symbiotic association with *Rhizobium* bacteroids occurs in the cortex region. The endodermis forms the innermost layer of the cortex and has uniseriate barrel-shaped cells.

The pericycle is the outermost region of the stele, and is uniseriate and parenchymatous. Cells of the pericycle retain meristematic activity. Lateral roots arise endogenously from the pericycle. The lateral roots help in absorption of water.

Xylem and phloem are arranged in the form of separate strands on different radii. They alternate with each other.

Bundle cap length and number of xylem vessels per vascular bundle showed a significant positive correlation with peg strength. Analysis revealed that bundle cap length had a direct positive effect on peg strength whereas number of xylem vessels per vascular bundle played only a minor role. Thus bundle cap length emerged as a reliable anatomical character, which can serve as a selection criterion in breeding cultivars with strong pegs to reduce harvesting losses in groundnut (Tiwari *et al.*, 1988).

Due to the proliferating cell divisions derived from the pericycle (incomplete second state of endodermal development), distinct types of lateral roots can be recognized in groundnut: long, first-order lateral roots, forming the skeleton of the root system, and thin and short second- and higher-order lateral roots, 'feeder roots'. The formation of root nodules at the base of the lateral roots was the result of proliferating cell divisions derived originally from the pericycle (Tajima *et al.*, 2008).

Stem anatomy

The outline of the stem in transverse section shows ridges and grooves. A transverse section of primary stem shows three regions, the epidermis, the cortex and the stele (Fig. 7.12).

The epidermis is the outermost region surrounding the stem. It is uniseriate, having a single layer of compactly arranged barrel-shaped living cells. The outer walls of the epidermal cells are cutinized, and the epidermis is covered by cuticle on its outer surface.

The cortex is smaller than the stele. The hypodermis is multilayered (two to three) and collenchymatic and gives considerable strength, flexibility and elasticity to young stems. The endodermis is innermost layer of the cortex and consists of uniseriate compactly arranged barrel-shaped cells.

There are from ten to 15 wedge-shaped vascular bundles arranged in one ring called a eustele. Each vascular bundle consists of xylem towards the centre of the axis and phloem towards the periphery. In the vascular bundle xylem and phloem are present together (conjoint) on the same radius (collateral), with a strip of cambium (fascicular cambium) between the xylem and phloem (open), hence the vascular bundle is described as conjoint, collateral and open type. The xylem consists of many vessels and is endarch with protoxylem present towards the centre.

The presence of the thick cuticle and collenchyma imparts strength and flexibility to the stem.

Leaf anatomy

The leaf shows a dorsi-ventral nature, i.e. the leaf blade consists of distinct dorsal and ventral surfaces. The epidermis is mainly meant for protection and is present on both surfaces of the leaf blade. The epidermis is single layered with compactly arranged barrel-shaped (tabular) cells. A thin cuticle layer covers the epidermis and it reduces the loss of water due to transpiration. Anisocytic stomata are present in both epidermal layers. The stomatal index of the upper epidermis is 26 and of the lower epidermis is 32 under 40× magnification.

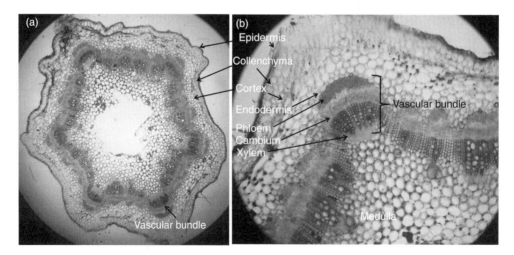

Fig. 7.12. (a) Transverse section of stem (40×) showing ridges and grooves in the outline (hexagonal), vascular bundles arranged in the form of a ring (eustele) and the central hollow region. (b) Sector enlarged illustrating organization of various parts. The vascular bundles are wedge shaped and the cambial ring is formed by the union of inter- and intra-fascicular cambium (secondary growth has just initiated).

The ground tissue of the leaf, present between the upper and the lower epidermal layers, is called mesophyll. Long palisade tissue is present below the upper epidermis, and is compactly arranged (Fig. 7.13). Palisade tissue is the most highly specialized type of photosynthetic tissue. The upper surface of the leaf is dark green in colour due to palisade tissue. The lower part of the mesophyll present towards the lower epidermis is the spongy tissue. Cells of the spongy tissue are thin walled, irregular in shape, and arranged loosely with large continuous intercellular spaces. Vascular bundles (veins) occur between the palisade and spongy tissue (median in position), and are collateral and closed. In the vascular bundle, xylem is present towards the upper epidermis and phloem towards the lower epidermis.

The presence of the thick cuticle and compactly arranged palisade tissue reduce transpiration loss thereby imparting drought resistance, but there exists variability among genotypes.

In an anatomical study, two cultivars showed different characteristics: stomatal density, leaf thickness, relative volume of water storage cells, cell sizes, and number of mesophyll cells per unit leaf surface varied significantly between them (Ferreyra *et al.*, 2000).

Few research activities have been undertaken on the contribution of anatomical characters to insect and disease resistance, but literature is rare on drought resistance.

ROLE OF TRICHOME IN INSECT RESISTANCE The role of long trichomes was studied in groundnut. Groundnut (*A. hypogaea* L.) cultivars of different susceptibility to the jassid, *Empoasca kerri* Pruthi, were studied to determine the inheritance of trichomes on the adaxial surface of the leaf, leaf midrib and petiole, and their association with resistance to *E. kerri*. Genotypes or crosses with long trichomes on the leaves and petioles showed a high level of resistance to jassids (leafhoppers) as evidenced by a very low percentage of yellowed foliage (hopper burn). The presence of long trichomes increases resistance to leafhoppers (Dwivedi *et al.*, 1986).

DISEASE RESISTANCE Leaves from resistant and susceptible groundnut genotypes (*Arachis hypogaea* L.) were observed after inoculation

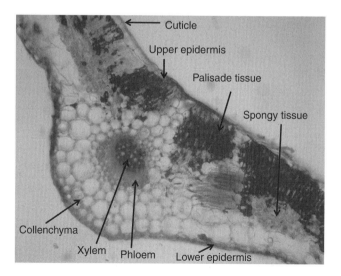

Fig. 7.13. Transverse section of leaf (100×) showing the upper and lower epidermis and vascular bundle at the centre (vein). Note: the thick cuticle on the epidermis and the long and very compactly arranged palisade tissue helps in reducing the transpirational loss of water thereby offering drought resistance.

with *Cercospora arachidicola* Hori and *Cercosporidium personatum* (Berk. and Curt.) Deighton. Both the pathogens induced almost similar anatomical responses in the inoculated leaves. The epidermal and mesophyll cells were shrunken and there was more damage to cell walls. This is due to depletion of polysaccharides, proteins, ascorbic acid and nucleic acids from the diseased host tissue at the site of contact with the pathogen. It was not observed in healthy tissues of these genotypes (Kaur and Dhellon, 1988).

Pod rotting-resistant cultivars (PEANUT) showed the following anatomical characters: the palisade mesophyll cells of 50-day old plants were arranged more compactly in pod rot-resistant than in susceptible genotypes. An index representing total width (μm) of palisade cells/mm leaf blade was more discriminative in distinguishing among genotypes than average of either cell width or cell number alone. The distribution of lignin in groundnut shells was correlated with pod rot resistance. Resistant lines possessed lignified cell walls in the epicarp and sclerenchymatous mesocarp than in the susceptible genotypes. The relationship between anatomical traits of stems, pegs, roots, or juvenile plant leaflets and field pod-rot reaction was not consistent among all genotypes.

Lignin distribution in pods, and an index representing μm palisade cells/mm of leaf blade individually or in combination, might be used effectively to supplement field evaluations in screening breeding lines for pod disease reaction (Godoy *et al.*, 1985).

Petiole anatomy

The petiole is covered by thick cuticle. Two-layered hypodermis of angular collenchyma cells is found immediately beneath the epidermis. The hypodermis gives considerable strength, flexibility and elasticity to young petioles and contains chloroplasts for performing photosynthesis. Just beneath the hypodermis ground tissue is found. It consists of thin-walled parenchymatous cells with well defined intercellular spaces among them. Wedge-shaped vascular bundles are of various sizes in the same petiole. In the petiole, the xylem is always found towards the upper side whereas phloem is found towards the lower side (Fig. 7.14).

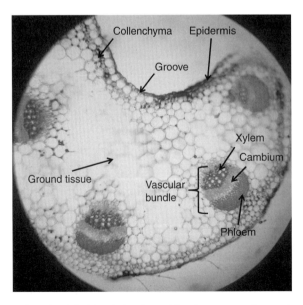

Fig. 7.14. Transverse section of petiole (100×) showing epidermis, cortex, xylem and medulla. The upper side of the petiole has a groove.

Research needs

In the context of the above review on anatomy of various organs of groundnut, it is assessed that groundnut possesses several anatomical traits mentioned that contribute to insect and drought resistance suitable for adaption to semi-arid conditions. There is still a necessity to screen and select cultivars for the combination of resistance traits.

7.3   Mustard

7.3.1   Introduction

Brassica nigra (black mustard) is an annual weedy plant cultivated for its seeds, which are commonly used as a spice. *Brassica juncea*, popularly known as mustard rye, is also called mustard greens, Indian mustard, Chinese mustard and leaf mustard, and is a species of mustard plant.

7.3.2   Utilization

Mustards comprise several plant species in the genera *Brassica* and *Sinapis* whose small mustard seeds are used as a spice and, by grinding and mixing them with water, vinegar or other liquids, are turned into the condiment known as mustard. The seeds are also pressed to make mustard oil, and the edible leaves can be eaten as mustard greens.

7.3.3   Morphological description

Vegetative characters

Mustards are an mesophytic annual herb, growing up to 1 m (Fig. 7.15), with a taproot system. The stem is weak, aerial, erect, greenish, rounded and smooth, with lateral branches emerging from the fourth or fifth leaf. The leaves are simple, exstipulate, alternate, lobed, petiolate and ovate shaped, with a lacerate margin, acute apex and unicostate reticulate venation.

Floral characters

The inflorescence is a corymb, with a small yellow flower, pedicillate, stipulate, complete, bisexual, actinomorphic, tetramerous and hypogynous. In the calyx are found four sepals, yellow in colour and free, with the outer sepals larger in size. The corolla contains four yellow petals in a cross arrangement, and a claw structure is present on petals; the anterior petals are larger than the posterior petals. The androecium is tetradynamous,

Fig. 7.15. Mustard plant morphology.

dithecous and introse. The gynoecium is bicarpellary, syncarpous, with many ovules present on parietal placentation due to a false septum ovary having two chambers. The fruit is a siliqua, 1.25–6.25 cm in length. Seeds are numerous, small, rounded, brown in colour and arranged in rows. The crop is of economic importance for the mustard oil and rapeseed oil that can be extracted from the seeds of brassicas. Mustard seeds are used as a condiment.

7.3.4 Anatomical description

Root anatomy

A transverse section of the root shows three main parts, the epidermis, the cortex and the stele (Fig. 7.16).

The epidermis consists of compactly arranged thin-walled living cells covering the root. Epidermal cells produce unicellular root hairs, which are used for absorption of water. The cortex is multiseriate and relatively homogeneous and consists of loosely arranged thin-walled parenchyma cells with prominent intercellular spaces. The cells

are usually round or oval, colourless and store starch. The epidermis is single layered with compactly arranged barrel-shaped cells. Endodermal cells are characterized by the presence of Casparian strips. Pericycle forms the outermost region of the stele and is uniseriate and parenchymatous. Cells of the pericycle retain meristematic activity. The vascular strands are radial; the xylem is exarch, of the closed type.

Stem anatomy

Epidermis is the uniseriate and compactly arranged outermost region surrounding the stem. Hairs are absent on the epidermis. Below the epidermis three to four layers of collenchyma are present. The inner layer of the cortex consists of endodermis (Fig. 7.17). Xylem and phloem form the vascular tissues, organized into vascular bundles. Each vascular bundle is wedge-shaped, conjoint,

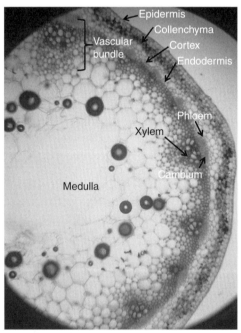

Fig. 7.17. Transverse section of stem (100×) showing epidermis, three to four layers of collenchyma and a smaller amount of cortex. Vascular bundles are arranged in the form of a ring. A large volume of medulla occupies the central portion of the stem.

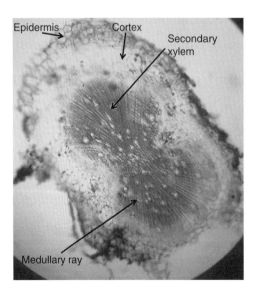

Fig. 7.16. Transverse section of root (100×) illustrating various features. The epidermis and endodermis have been ruptured due to secondary growth in the stele. The cortex is also compressed.

collateral, open and endarch; 15–20 vascular bundles are arranged in one ring called eustele. The central part of the stem is occupied by parenchymatous pith or medulla. The cells are round or oval with intercellular spaces. Medulla stores food materials and aids in lateral conduction.

The presence of the medium-thick collenchyma layer contributes to the flexibility to the stem, offering resistance to breakage.

Leaf anatomy

The epidermis is single layered with compactly arranged barrel-shaped (tabular) cells on both sides of the leaf. The outer walls of the epidermal cells are cutinized. Anisocytic stomata are present in both of the epidermal layers. The stomatal complex consists of three subsidiary cells covering two kidney-shaped guard cells (Fig. 7.18). the stomatal index of upper epidermis is 28 and lower epidermis is 31. The presence of epicuticular wax on both sides of the epidermis is an important character. The mesophyll is chlorenchymatic and concerned mainly with photosynthesis. The mesophyll is differentiated into palisade tissue and spongy tissue

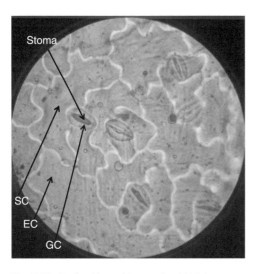

Fig. 7.18. Leaf epidermal impression (400×) showing stomatal complex. Two kidney-shaped guard cells (GC) are covering stoma (pore), and three subsidiary cells (SC) are adjacent to the guard cells – anisocytic stomata.

(Fig. 7.19). Palisade cells are compactly arranged and occupy more area when compared to spongy tissue, which is absent below the veins. Vascular bundles occur between the palisade and spongy tissue (median in position), and are collateral and closed. In the vascular bundle, xylem is present towards the upper epidermis and phloem towards the lower epidermis.

The presence of a thin cuticle and loose palisade associated with large stomata on the leaf surface imparts susceptibility to drought.

Petiole anatomy

Epidermis covers the petiole and is the outer layer. It is uniseriate with a single layer of compactly arranged barrel-shaped living cells. Two layers of collenchyma are present below the epidermis. Collenchyma is living mechanical tissue consisting of chloroplasts (chlorenchyma) below the corners of the collenchyma cells. Just beneath the hypodermis ground tissue is found (Fig. 7.20), consisting of thin-walled parenchymatous cells with well-defined intercellular spaces among them. There are three to four wedge-shaped vascular bundles arranged in a 'C' shape in the ground tissue, and these bundles can be various sizes in the same petiole.

The presence of the thin cuticle and medium-thin collenchyma contributes to the fragile nature of the petiole.

Conclusion

The anatomical characters of mustard reveal that the plant is susceptible to drought; thereby there is a necessity to select genotypes with desirable drought-resistance traits.

7.4 Sesame

7.4.1 Introduction

Sesame (*Sesamum indicum*) is a flowering plant in the genus *Sesamum*. Numerous wild relatives occur in Africa and a smaller

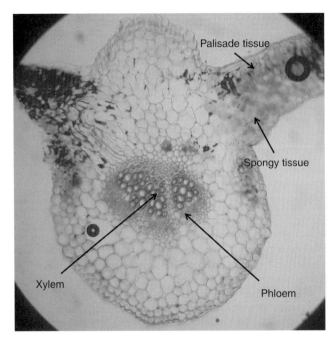

Fig. 7.19. Transverse section of leaf (100×) showing palisade and spongy tissue, and a vascular bundle at the centre of the leaf.

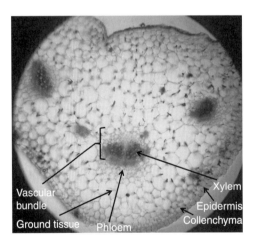

Fig. 7.20. Transverse section of petiole (100×) showing epidermis, cortex, xylem and ground tissue.

number in India. It is widely naturalized in tropical regions around the world and is cultivated for its edible seeds, which grow in pods. The flowers of the sesame seed plant are yellow, though they can vary in colour with some being blue or purple.

Resistance was found to be associated with tropical provenance, high oil content of the seed and high root:top ratio; it was negatively correlated with resistance to the toxic action of $KClO_3$ solution, germinability and rate of water absorption.

7.4.2 Origin

Despite the fact that the majority of the wild species of the genus *Sesamum* are native to sub-Saharan Africa, sesame was first domesticated in India. This is revealed by morphological and cytogenetic affinities between domesticated sesame and the south Indian native *S. mulayanum* Nair., as well as archaeological evidence that it was cultivated at Harappa in the Indus Valley between 2250 and 1750 BC, and a more recent find of charred sesame seeds in Miri Qalat and Shahi Tump in the Makran region of Pakistan.

7.4.3   Utilization

Sesame seeds are used for oil extraction and also used in dishes.

7.4.4   Morphological description

Vegetative characters

Sesame is a mesophytic herb, growing up to 60–150 cm and has a taproot system. The stem is aerial, erect, quadrangular, greenish, smooth, and monopodial in branching. Leaves are simple, stipulate, alternate, glabrous, long petiolate and ovate shaped, with a dentate margin, acute apex and reticulate venation (Fig. 7.21).

Floral characters

Sesame has a racemose bracteates inflorescence, with complete, bisexual, zygomorphic, epigynous flowers. There are five sepals and petals, which are gamopetalous and bilipped. There are four stamens, which are

Fig. 7.21. Sesame plant morphology.

didynamous and epipetalous, the anthers are dithecous, base fixed and introse. The gynoecium is bicarpellary and bilocular, with a syncarpous ovary, gynobasic style and bifurcated stigma. The fruit is a carcerulus.

7.4.5   Anatomical description

Root anatomy

The epidermis forms the outermost region of the root, and is uniseriate with a single layer of compactly arranged thin-walled living cells. The epidermis ruptures after secondary growth. The cortex is the middle region, lying between the epidermis and the stele. It is multiseriate and relatively homogeneous and consists of loosely arranged thin-walled parenchyma cells with prominent intercellular spaces. The cortex either ruptures or is compressed due to secondary growth (Fig. 7.22). Endodermis is very distinct in roots and is single layered with compactly arranged barrel-shaped cells. Endodermal cells are characterized by the presence of Casparian strips or Casparian bands. The endodermis is also called the starch sheet layer, as starch grains are present in the endodermal cells. The pericycle is uniseriate and parenchymatous, and cells of the pericycle retain meristematic activity. Lateral roots arises endogenously from the pericycle, and these help in the absorption of water.

Xylem and phloem are the vascular tissues. Xylem is used for the conduction of water and salts and phloem for food materials. Xylem and phloem are arranged in the form of separate strands on different radii and alternate with each other, hence the vascular strands in root are described as separate and radial. Xylem is exarch, having protoxylem towards the pericycle and metaxylem towards the centre and uniformly distributed secondary xylem vessels. Cambium is absent and hence vascular bundles are described as closed type.

Stem anatomy

The outline of the stem in transverse section is round in shape, and shows epidermis,

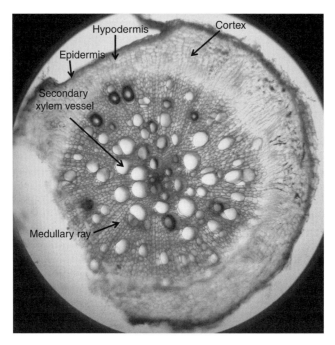

Fig. 7.22. Transverse section of root (100×) showing secondary growth. Secondary xylem vessels are larger in diameter and medullary rays are noticeable. The cortex is compressed due to secondary growth.

cortex and vascular bundles. Epidermis is the outermost region surrounding the stem (Fig. 7.23) and is uniseriate, having a single layer of compactly arranged barrel-shaped living cells. The epidermis is covered by thick cuticle, which reduces the transpiration. Multilayered (three to four layers) collenchymatic hypodermis is present below the epidermis. The presence of two layers of chlorenchyma indicates the photosynthetic activity of the stem. The endodermis is uniseriate and this compactly arranged layer covers the stele. Xylem and phloem form the vascular tissues, which consist of xylem towards the centre of the axis and phloem towards the periphery. The xylem consists of many vessels arranged in rows and is endarch with protoxylem present towards the centre. Medulla and medullary rays are distinct, present at the centre of the stem and usually parenchymatous. The cells are round or oval with intercellular spaces. The medulla stores food materials.

Leaf anatomy

The leaf shows a dorsi-ventral nature, i.e. the leaf blade consists of distinct dorsal and ventral surfaces. In transverse section the leaf shows three distinct regions, the epidermis, mesophyll and vascular system.

The epidermis is present on both surfaces of the leaf blade, with that covering the upper surface of the leaf termed upper epidermis or ventral epidermis or adaxial epidermis. The epidermis present towards the lower side of the leaf is called lower epidermis or dorsal epidermis or abaxial epidermis. Lower epidermis contains a greater number of unicellular trichomes. The epidermis is single layered, with compactly arranged barrel-shaped (tabular) cells. The outer walls of the epidermal cells are cutinized, and the epidermis on its outer surface is covered by a continuous layer of cutin, termed cuticle. The cutinized outer walls and cuticle reduce the loss of water due to transpiration. Anisocytic stomata facilitate

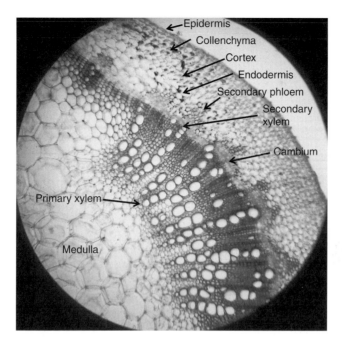

Fig. 7.23. Transverse section of stem (100×) illustrating various parts. Secondary growth has started, with secondary xylem larger in size and diameter than primary xylem.

the exchange of gases between the leaf and environment (Fig. 7.24) and are present in both epidermal layers. The stomatal index of the upper epidermis is 27 and lower epidermis is 31. All the epidermal cells except guard cells are colourless. Palisade tissue is present below the upper epidermis (Fig. 7.25). The palisade cells are thin walled, cylindrical and contain numerous chloroplasts. The cells are compactly arranged in one layer and perpendicular to the upper epidermis. Narrow intercellular spaces are present in the palisade tissue. Palisade tissue is the most highly specialized type of photosynthetic tissue and the upper surface of the leaf is dark green in colour due to palisade tissue. Cells of the spongy tissue are thin walled, irregular in shape, and arranged loosely with large continuous intercellular spaces. Stomata open into large intercellular spaces termed sub-stomatal chambers. The cells of spongy parenchyma contain a lower number of chloroplasts, hence the lower surface of the leaf is light green in colour. The vascular tissue occurs in the form of discrete bundles called veins, which are

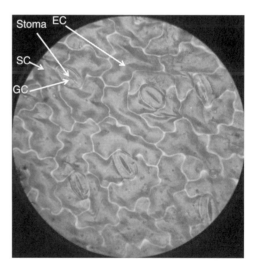

Fig. 7.24. Leaf epidermal impression (1400×) showing epidermal cell (EC), subsidiary cell (SC), guard cell (GC) and stoma.

interconnected to form reticulate venation. Vascular bundles (veins) occur between the palisade and spongy tissue (median in position) and are collateral and closed. In the

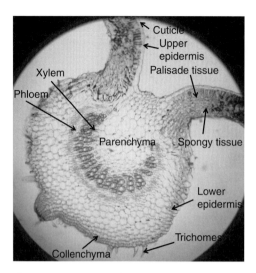

Fig. 7.25. Transverse section of leaf (100×) showing arrangement of tissues. Note: lower epidermis is covered by trichomes, the vascular bundle is in a half-moon shape, and three to four collenchyma layers are present above the lower epidermis.

vascular bundle, xylem is present towards the upper epidermis and phloem towards the lower epidermis. The vascular bundle is surrounded by a parenchymatous bundle sheath. The cells of the bundle sheath are called border parenchyma.

The presence of thin, loose palisade cells and large stomata imparts susceptibility to drought in the mustard crop.

Drought stress – cuticle

A study was conducted at post-flowering stage by withholding irrigation 15 days compared to a control. The objective of this study was to evaluate leaf cuticular wax constituents across a diverse selection of sesame cultivars, and the responses of these waxes to drought-induced wilting. Compared to well-irrigated plants, drought treatment caused an increase in wax amount on most cultivars. Drought treatments caused a large decrease in seed yield per plant, but did not affect the mean weight of individual seeds, showing that sesame responds to post-flowering drought by reducing seed numbers, but not seed size. Seed yield was inversely correlated with the total wax amount ($r=-0.466^*$)

significant at 0.05 level, indicating that drought induction of leaf wax deposition does not contribute directly to seed set (Kim *et al.*, 2007).

Petiole anatomy

Epidermis is uniseriate with a single layer of compactly arranged barrel-shaped living cells. The outer walls of the epidermal cells are cutinized and cuticle further reduces the transpiration. Unicellular hairs are present on the epidermis. The hypodermis is formed of multilayered (Fig. 7.26) angular collenchyma cells and is found immediately beneath the epidermis. Collenchyma is living mechanical tissue. The hypodermis gives considerable strength, flexibility and elasticity to young petioles and, having chloroplasts, it may carry out photosynthesis. Due to the presence of the thick layer of collenchyma the petiole is strong, stiff and given strength. The vascular bundles are of various sizes in the same petiole; the xylem is always found towards the upper side whereas phloem is towards the lower side.

7.5   Castor

7.5.1   Introduction

The castor oil plant, *Ricinus communis*, is a species of flowering plant in the spurge family, Euphorbiaceae. It belongs to a monotypic genus, *Ricinus*, and subtribe, Ricininae. The evolution of castor and its relation to other species is currently being studied.

Its seed is the castor bean, which, despite its name, is not a true bean. Castor is indigenous to the south-eastern Mediterranean basin, eastern Africa and India, but is widespread throughout tropical regions (and widely grown elsewhere as an ornamental plant).

Castor seed is the source of castor oil, which has a wide variety of uses. The seeds contain between 40% and 60% oil, which is rich in triglycerides, mainly ricinolein. The seed contains ricin, a toxin, which is also present in lower concentrations throughout the plant.

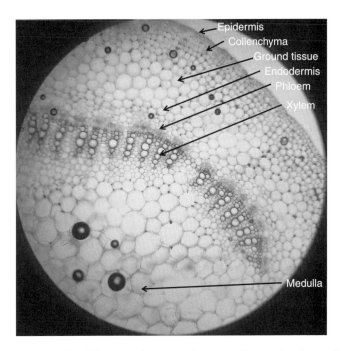

Epidermis
Collenchyma
Ground tissue
Endodermis
Phloem
Xylem
Medulla

Fig. 7.26. Transverse section of petiole (100×) showing various parts. Three to four layers of collenchyma and multilayered ground tissue can be observed.

Global castor seed production is around 1 Mt/year. Leading producing areas are India (with over 60% of the global yield), China and Brazil, and it is widely grown as a crop in Ethiopia.

7.5.2 Utilization

In Brazil, castor oil is now being used to produce biodiesel. In rural areas, the abundant seeds are used by children for slingshot balls, as they have the right weight, size and hardness. The use of castor bean oil in India has been documented since 2000 BC in lamps and in local medicine as a laxative, purgative and cathartic in Unani, Ayurvedic and other ethnomedical systems. Traditional Ayurvedic medicine considers castor oil the king of medicinals for curing arthritic diseases.

The attractive castor seeds are used in jewellery, mainly necklaces and bracelets. This practice is highly dangerous, due to the extreme toxicity of the seeds, especially if the seed coat is drilled for stringing. Castor oil in a processed form, called polyglycerol polyricinoleate or PGPR, is currently being used in chocolate bar manufacture as a less expensive substitute for cocoa butter.

7.5.3 Morphological description

Vegetative characters

The castor oil plant can vary greatly in its growth habit and appearance. The variability has been increased by breeders who have selected a range of cultivars for leaf and flower colours, and for oil production. It is a fast-growing, suckering perennial shrub, which can reach the size of a small tree, but it is not cold hardy.

The glossy leaves are 15–45 cm, long, long-stalked, alternate and palmate with 5–12 deep lobes with coarsely toothed segments (Fig. 7.27). In some varieties they start off dark reddish purple or bronze when young, gradually changing to a dark green, sometimes with a reddish tinge, as they mature (Fig. 7.27). The leaves of some other varieties are green practically from the start, whereas in yet others a pigment masks the green colour of all the chlorophyll-bearing parts, leaves, stems and young fruit, so that they remain a dramatic purple to reddish-brown throughout the life of the plant. Plants with the dark leaves can be found growing next to those with green leaves, so there probably is only a single gene controlling the production of the pigment in some varieties at least. The stems (and the spherical, spiny seed capsules) also vary in pigmentation. The fruit capsules of some varieties are more attractive than the flowers.

Floral character

The flowers are borne in terminal panicle-like inflorescences of green or, in some varieties, shades of red monoecious flowers without petals. The male flowers are yellowish green with prominent creamy stamens and are carried in ovoid spikes up to 15 cm long; the female flowers, borne at the tips of the spikes, have prominent red stigmas (Fig. 7.28).

The fruit is a spiny, greenish (to reddish purple) capsule containing large, oval, shiny, bean-like, highly poisonous seeds with variable brownish mottling. Castor seeds have a warty appendage called the caruncle, which is a type of elaiosome. The caruncle promotes the dispersal of the seed by ants (myrmecochory).

7.5.4   Anatomical description

Root anatomy

Epidermis forms the outermost region of the root. It is a uniseriate, having a single layer of compactly arranged thin-walled living cells. The epidermis ruptures after secondary growth (Fig. 7.29). Cortex forms the middle region, lying between the epidermis and the stele. It is multiseriate and relatively homogeneous and consists of loosely arranged thin-walled parenchyma cells with prominent intercellular spaces. The cortex ruptures or is compressed due to secondary growth. The endodermis is very distinctive in roots and is single layered with compactly arranged barrel-shaped cells. Endodermal cells are characterized by the presences of Casparian strips or Casparian bands. Endodermis is also called a starch sheet layer, as starch grains are present in the endodermal cells. The pericycle is uniseriate and parenchymatous. Xylem and phloem are

Fig. 7.27. Castor plant morphology showing glossy leaves with long-stalked, alternate and palmate leaves with 5–12 deep lobes with coarsely toothed segments and panicle-like inflorescence.

arranged in the form of separate strands on different radii and alternate with each other, hence the vascular strands in the root are described as separate and radial. The xylem is exarch, having protoxylem towards the

Fig. 7.28. Castor inflorescence, with female flowers at the top and male flowers at the bottom.

pericycle and metaxylem towards the centre and uniformly distributed secondary xylem vessels. Cambium is absent and hence vascular bundles are described as closed type. The central part of the root is occupied by a small amount of parenchymatous pith.

Significance of variations in root anatomy

Cultivar differences are expressed in a number of root characteristics: (i) presence or absence of sclerenchymatous exodermis in the cortex; (ii) thickness of endodermal cell walls; and (iii) variability in size of the xylem vessels (Fig. 7.30).

Anatomy of stem

The presence of thick cuticle, hypodermal collenchyma layers and weak sclerenchyma in the wood imparts flexibility to the stem. The outline of the stem in transverse section shows round in shape. The epidermis, cortex and vascular bundle are visible (Fig. 7.31). Epidermis forms the outermost region surrounding the stem. It is uniseriate, having a single layer of compactly arranged barrel-shaped living cells. The epidermis is covered by a thick cuticle and reduces transpiration. Multilayered (from four to five layers) collenchymatic hypodermis is present below

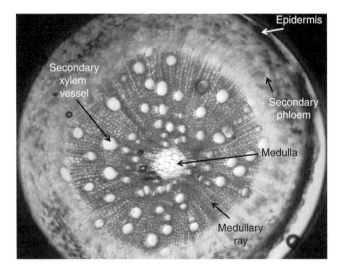

Fig. 7.29. Transverse section of mature root (100×) illustrating secondary growth. The epidermis, secondary phloem, secondary xylem, medullary rays and medulla are clearly visible.

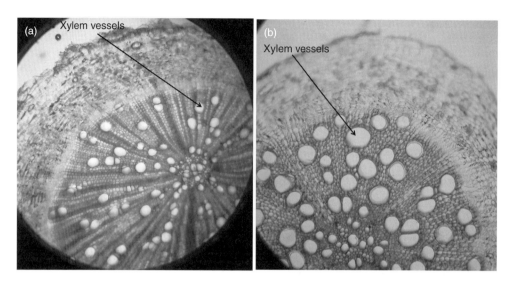

Fig. 7.30. Transverse section of mature root (100×) showing variability in size of the secondary xylem vessels: (a) secondary xylem is smaller in size; (b) secondary xylem is larger in size.

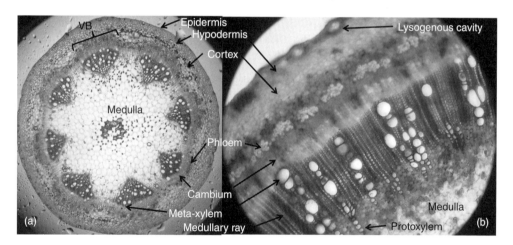

Fig. 7.31. (a) Transverse section of stem showing different parts (100×). (b) Sector enlarged 400×: resin-filled lysogenous cavities, and metaxylem showing larger vessels.

the epidermis. The presence of two layers of chlorenchyma indicates the photosynthetic activity of the stem. The endodermis is uniseriate, and the compactly arranged layer covers the stele. Xylem and phloem form the vascular tissues and are organized into vascular bundles with eight to ten wedge-shaped vascular bundles arranged in one ring. Each bundle consists of xylem towards the centre of the axis and phloem towards the periphery. The xylem consists of many vessels arranged in rows, and is endarch with protoxylem present towards the centre. Medulla and medullary rays are distinctive, present at the centre of the stem and usually parenchymatous. The cells are round or oval with intercellular spaces. The medulla stores food materials.

Significance of variations in stem anatomy

Epidermal study reveals variability between genotypes in the following:

- Cuticle thickness of the epidermis shows variability.
- Presence or absence of multicellular trichomes; in some genotypes there may be unicellular and multicellular trichomes.
- Number of hypodermis layers may be differ between genotypes.
- Distribution of collenchyma cells in hypodermis (compact or loosely arranged).
- Number of pericycle layers; shape of sclerenchyma cells differ in different genotypes.
- Size of the vascular bundles (large, medium and small), size of the xylem vessels ranging from small to large and number of vascular bundles also shows variability (Fig. 7.32).
- The secondary xylem contains xylem vessels mostly isolated or in pairs interspersed with medullary rays and scanty sclerenchyma thereby imparting fragility.

Leaf anatomy

The leaf shows a dorsi-ventral nature, i.e. the leaf blade consists of distinct dorsal and ventral surfaces. In transverse section the leaf shows three distinct regions, the epidermis, mesophyll and vascular system (Fig. 7.33).

The epidermis is mainly for protection and is present on both surfaces of the leaf blade. The epidermis covering the upper surface of the leaf is termed the upper epidermis or ventral epidermis or adaxial epidermis; epidermis present towards the lower side of the leaf is called the lower epidermis or dorsal epidermis or abaxial epidermis. The epidermis is single layered, having compactly arranged barrel-shaped (tabular) cells. The outer walls of the epidermal cells are cutinized and on its outer surface is covered by a continuous layer of cutin called cuticle. The cutinized outer walls and cuticle reduce the loss of water due to transpiration. Anisocytic stomata are present in both epidermal layers. The stomatal index of the upper epidermis ranges from 29 to 31 and of the lower epidermis is 36–39.

The mesophyll is differentiated into palisade and spongy tissues. The palisade

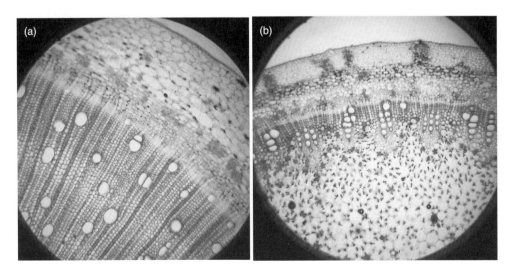

Fig. 7.32. Transverse section of stem showing variability in different parts: (a) large xylem vessels and collenchyma layers; (b) small xylem vessels.

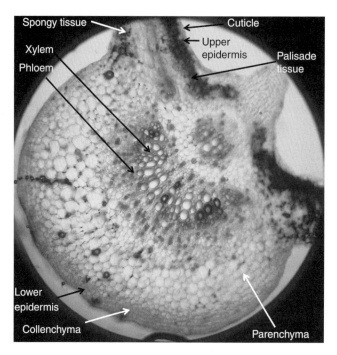

Fig. 7.33. Transverse section of leaf (100×) showing arrangement of tissues.

tissue is present below the upper epidermis. Palisade cells are thin walled, cylindrical and contain numerous chloroplasts. The cells are compactly arranged in one layer and perpendicular to the upper epidermis. Narrow intercellular spaces are present in the palisade tissue. Palisade tissue is the most highly specialized type of photosynthetic tissue. The upper surface of the leaf is dark green in colour due to palisade tissue. Cells of the spongy tissue are thin walled, irregular in shape and arranged loosely with large continuous intercellular spaces. Stomata open into large intercellular spaces called sub-stomatal chambers. The cells of spongy parenchyma contain a smaller number of chloroplasts, hence the lower surface of the leaf is light green in colour.

Vascular bundles (veins) occur between the palisade and spongy tissue (median in position). They are collateral and closed. In the vascular bundle, xylem is present towards the upper epidermis and phloem towards the lower epidermis. Vascular bundles are surrounded by parenchymatous

bundle sheath, the cells of which are termed border parenchyma.

Significance of variations
in leaf anatomy

The arrangement of palisade tissue is different in different genotypes (Fig. 7.34) as is the number of stomata, trichome density and types of trichomes. The type and arrangement of vascular bundles both in the lamina and midrib (Fig. 7.35) contains variability. The characteristic size and shape of bundle sheath chloroplasts also varies among genotypes.

Drought-resistant genotypes possess thick, compactly arranged long palisade cells and a thick cuticle, which reduces transpiration loss compared to the drought susceptible varieties.

Anatomy of petiole

A transverse section of petiole shows four regions, the epidermis, hypodermis, ground tissue and vascular bundles (Fig. 7.36).

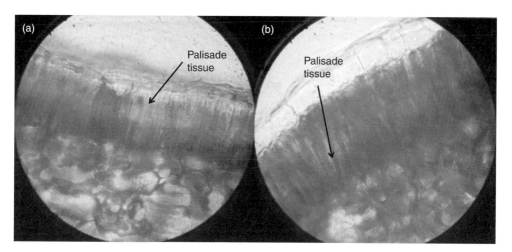

Fig. 7.34. Transverse section of leaf (400×) showing variability in palisade tissue: (a) short and loosely arranged palisade cells; (b) long and compactly arranged palisade cells.

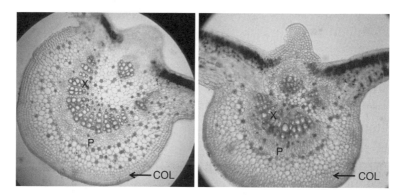

Fig. 7.35. Transverse section of leaf (40×) showing variability in collenchyma layers (COL) and the distribution of xylem (X) and phloem (P) tissue.

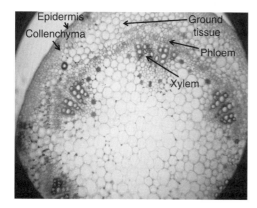

Fig. 7.36. Transverse section of petiole (40×) showing organization of various tissues.

The epidermis is uniseriate, having a single layer of compactly arranged barrel-shaped living cells. The outer walls of the epidermal cells are cutinized. Cuticle reduces the transpiration. Unicellular hairs are present on the epidermis.

Multilayered (two to six layers) hypodermis of collenchyma cells is found immediately beneath the epidermis. Collenchyma is living mechanical tissue. The hypodermis gives considerable strength, flexibility and elasticity to young petioles. Due to the presence of the thick layer of angular collenchyma, the petiole is strong and stiff. Several layers of collenchyma below the epidermis impart flexibility to the petiole, which is the characteristic of drought resistance.

The wedge-shaped vascular bundles are of various sizes in the same petiole. In the petiole, the xylem is always found towards the upper side with phloem towards the lower side.

Significance of variations in petiole anatomy

Number of collenchyma layers differs in different genotypes (Fig. 7.37) as does the number and size of vascular bundles. The presence of a thick cuticle and thick layers of collenchyma in the petiole imparts flexibility, avoiding breakage.

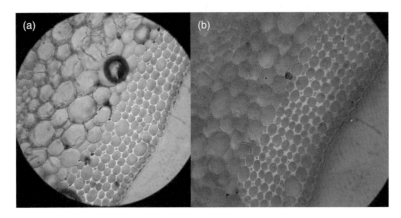

Fig. 7.37. Transverse section of petiole showing variation in collenchyma layers: (a) five layers; (b) seven layers.

References

Ashraf, M. and O'Leary, J.W. (1996) Effect of drought stress on growth, water relations, and gas exchange of two lines of sunflower differing in degree of salt tolerance. *International Journal of Plant Sciences* 157(6), 729–732.

Barker, J.F. (1990) Sunflower trichome defenses avoided by a sunflower stem weevil, *Cylindrocopturus adspersus* LeConte (Coleoptera: Curculionidae). *Journal of the Kansas Entomological Society* 63(4), 638–641.

Blamey, F.P.C., Joyce, D.C., Edwards, D.G. and Asher, C.J. (1986) Role of trichomes in sunflower tolerance to manganese toxicity. *Plant and Soil* 91(2), 171–180.

Dwivedi, S.L., Amin, P.W., Rasheedunisa, Nigam, S.N., Nagabhushanam, G.V.S., Rao, V.R. and Gibbon, R.W. (1986) Genetic analysis of trichome characters associated with resistance to jassid (*Empoasca kerri pruthi*) in peanut. *Peanut Science* 13(1), 15–18.

Fagerberg, W.R. and Culpepper, G. (1984) A morphometric study of anatomical changes during sunflower leaf development under low light. *Botanical Gazette* 145(3), 346–350.

Ferreyra, R.A., Pachepsky, L.B., Collino, D. and Acock, B. (2000) Modeling peanut leaf gas exchange for the calibration of crop models for different cultivars. *Ecological Modelling* 131(2–3), 285–298.

Franchini, M.C., Hernández, L.F. and Lindstrom, L.I. (2010) Cuticle and cuticular wax development in the sunflower (*Helianthus annuus* L.) pericarp grown at the field under a moderate water deficit. *International Journal of Experimental Botany* 79, 153–161.

Geetha, A., Saidaiah, P., Suresh, J. and Siva Sankar, A. (2011) Morpho-physiological and biochemical basis of drought tolerance in sunflower – a review. *Agricultural Reviews*. College of Agriculture, Acharya N.G. Ranga Agricultural University, Rajendranagar, Hyderabad, India.

Godoy, R., Smith, O.D., Taber, R.A. and Pettit, R.E. (1985) Anatomical traits associated with pod rot resistance in peanut. *Peanut Science* 12(2), 77–82.

Gunes, A., Pilbeam, D.J., Inala, A. and Cobana, S. (2008) Influence of silicon on sunflower cultivars under drought stress, i: growth, antioxidant mechanisms, and lipid peroxidation. *Communications in Soil Science and Plant Analysis* 39, 13–14.

Kaur, J. and Dhellon, M. (1988) Histological and histochemical changes in groundnut leaves inoculated with *Cercospora arachidicola* and *Cercosporidium personatum. Phytoparasitica* 16(4), 327–335.

Kim, K.S., Park, S.H. and Jenks, M.A. (2007) Changes in leaf cuticular waxes of sesame (*Sesamum indicum* L.) plants exposed to water deficit. *Journal of Plant Physiology* 164(9), 1134–1143.

Maiti, R.K. and Wesche-Ebeling, P. (2002) *Advances in Peanut Science.* Science Publishers, USA, 376 pp.

Maiti, R.K., Singh, V.P., Purohit, S.S. and Vidyasagar, P. (2007) *Research Advances in Sunflower* (Helianthus annus *L.).* Agrobios (International), ISBN 81-904309-2-0, 512 pp.

Tajima, R., Abe, J., New Lee, O., Morita, S. and Lux, A. (2008) Developmental changes in peanut root structure during root growth and root-structure modification by nodulation. *Annals of Botany* 101(4), 491–499.

Tiwari, S.P., Murthy, T.G.K., Johnson, G.K. and Varmoda, D.L. (1988) Path-coefficient analysis of anatomical characters affecting peg strength in groundnut. *Euphytica* 39(2).

8

Fibre Crops

_____

8.1   Cotton

8.1.1   Introduction

Cotton (*Gossypium* spp.) is the world's most industrialized crop. The term cotton is used both for the plant and the fibres obtained from the lint of the seed, which is derived from the Arabic word 'Qutn'. Cotton has been in use for the last 5000 years and is indigenous to Southeast Asia and the Americas. The domestication of the cotton plant for commercial cultivation for clothing and for other forms of human utilization is considered to have begun from the Harappa civilization in the Indian subcontinent, using diploid or Asiatic cottons (*Gossypium herbaceum* L. and *Gossypium arboreum* L.). Cotton, also known as 'white gold', is grown in more than 80 countries all over the world. China, India and the USA are the leading cotton growers, accounting for 60% of the world cotton production. Other major cotton growing countries are Pakistan, Brazil, Uzbekistan and Turkey.

8.1.2   Origin, evolution and domestication

There are four cultivated species of cotton. Of these, *G. arboreum* (Asiatic/Indian cotton) and *G. herbaceum* (African cotton) are diploids (2*n*=2*x*=26) and *Gossypium hirsutum* (Mexican cotton) and *Gossypium barbadense* (Egyptian/sea island/South American cotton) are tetraploids (2*n*=4*x*=52). The diploid species of cotton are considered to have been domesticated in the Old World. The Indus valley civilization of India and Pakistan cultivated *G. arboreum*, from which it spread to the Mediterranean region and Africa. Cotton seeds and fibres dating back to 6000 BC have been identified in Mehrgarh, an initial establishment of Indus valley civilization. A few centuries later, *G. herbaceum* was domesticated in northern Africa and the Near East. The New World cottons, *G. hirsutum* and *G. barbadense*, are considered to have been domesticated in Mexico and Peru in 3500 BC and 3480 BC, respectively.

8.1.3   Genetic resource

There are about 45 diploid and five tetraploid species in the genus *Gossypium*, which belong to eight genomic groups. The cultivated diploid species *G. arboreum* (A$_1$) and *G. herbaceum* (A$_2$) are closely related. *G. hirsutum* (AD$_1$) is the most widely commercially cultivated cotton species. *G. barbadense* (AD$_2$) is preferred for long staple, high quality cotton fibre. As in other crops, little diversity is present among cultivated genotypes within each species, because farmers tend to grow a few improved varieties with higher yield. However, considerable variability is present in the genetic resource collections of different countries, including China, the USA and India.

8.1.4 Utilization

Cotton fibres are the raw materials for the textile industry. Cloth and yarn are manufactured from cotton. Gun cotton (nitrocellulose) is obtained by mixing cotton with concentrated nitric acid. Cotton seed is used as stock feed, because it contains fats, proteins and vitamins. Oil is extracted from seed, hydrogenated and used as vanaspathi (cooking oil) and also in soap making. Seed cake is also used as cattle feed. Bark is used for the manufacture of low grade paper. Cotton gin by-products are utilized for the manufacturing of fuel pellets.

8.1.5 Morphological description

Vegetative characters

The cotton plant is a mesophytic herb with a taproot system with inclined deep lateral roots. The stem is aerial, erect and branched, with stellate trichomes. The leaves are simple, heart shaped (Fig. 8.1), with three to five lobes, an entire margin and acute apex, stipulate, petiolate, palmately reticulate venation, alternate and pubescent, with stellate trichomes, extra floral nectarines, and exhibit monopodial and sympodial branching.

Floral characters

Cotton plants have a continuous flowering habit with a raceme, axillary and solitary inflorescence. The flower is bracteate, pedicellate, presence of epicalyx, complete, bisexual, pentamerous, hypogynous and actinomorphic. The flowers are normally referred to as 'squares'. The calyx consists of five gamosepalous (united) bowl-shaped valvate sepals, and trichiferous nectar is present at the base of calyx. The corolla consists of five petals, which are polypetalous, large, attractive and twisted. Stamens are numerous,

Fig. 8.1. Cotton plant morphology: simple, heart shaped, three- to five-lobed, alternate and pubescent leaves; flowers born on axillary and solitary inflorescence.

monadelphous and show protandry. Anthers are monothecous, kidney shaped and extrose; pollen grains are spherical in shape. The gynoecium is multicarpellary, syncarpus, and has a superior ovary with axile placentation. The stigma is lobed, with the terminal style and number of stigma equal to the number of locules. Botanically, the fruit is called a loculicidal capsule (pericarp breaks vertically along the middle of each locule) and is rounded or oblong in shape. The fruit is usually known as a 'boll'. The seed is round in shape, grey or ash colour with a rough surface, 2.5–3 mm in diameter and consists of outgrowths of cellulosic fibres (lint).

8.1.6 Anatomical description

Root anatomy

Transverse section of the young root shows the tetrach condition of xylem (Fig. 8.2). Numerous root hairs arise from the epidermis, which helps in absorbing water and minerals. Transverse section of a matured root shows a thin layer of bark at the outer side of the root, which is produced by rupture of the epidermis. Below this, the two- to five-layered cortex (Fig. 8.2) band is compressed. Xylem is exarch with continuously formed secondary xylem vessels in mature root and pith is absent at the centre. Endarch xylem vessels with prominent secondary xylem vessels are present at the centre of the root.

Significance of variations in root anatomy

Cultivar differences are expressed in a number of root characteristics: (i) presence or absence of sclerenchymatous exodermis in the cortex; and (ii) thickness of endodermal cell walls.

A high density of vascular tissue (xylem) is essential under drought condition for efficient translocation of water to the plant parts. Large xylem vessels help in efficient conduction of water and minerals to the shoot system (Fig. 8.3).

Root anatomy in applied field (biotic and abiotic stress)

PATHOGEN BARRIER   Catechin (3,5,7,3',4'-flavanpentol) and gallocatechin (3,5,7,3',4', 5'-flavanhexanol) were isolated from methanol extracts of roots of 1-week-old 'Acala 4-42' cotton seedlings and identified by infrared spectrophotometry. Root chemical composition has a direct impact on pathogen penetration. The localization of the catechins is secreted by the

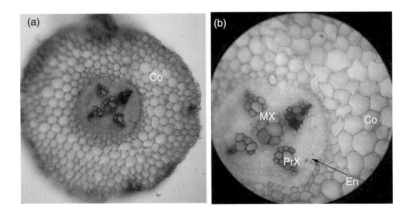

Fig. 8.2. Transverse section of young root. (a) Ground plan (40×) showing exarch, tetrarch, closed radial vascular bundles. (b) Sector enlarged (100×) showing endodermis (En), cortex (Co), protoxylem (PrX), and metaxylem (MX).

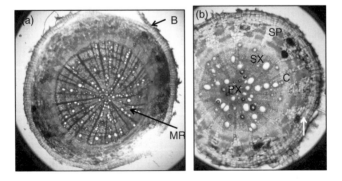

Fig. 8.3. Transverse section of mature root showing secondary growth: (a) medullary rays (MR) and bark (B) are visible; (b) secondary xylem vessels (SX) are clearly visible in the picture, along with bark (B), secondary phloem (SP), cambium (C) and secondary xylem (SX).

pathogen in the hypodermis, endodermis, scattered cells of the xylem parenchyma, and the proximal portion of the root cap (Mace and Howell, 1974).

WATER FLOW   The intensity of root anatomical components influences water flow within the root system. In a study, the comparison of water flow and xylem anatomy between roots with tetrarch and pentarch vascular bundle arrangements showed no significant differences in the measured values of water flux for the primary root. Water flux, estimated using Poiseuille's equation and measured xylem dimensions, was greater for the tetrarch roots, primarily because of the larger diameter of individual vessel elements. The increased number of vessel elements in the pentarch primary root did not result in any apparent decrease in axial resistance to water flow (Oosterhuis and Wullschleger, 1987).

DROUGHT STRESS   Drought affects root anatomy in arid zones. The original phloem may be killed and become functionless. New phloem is produced, which may contain the same elements although in varying proportions. The proliferated cells are generally thin-walled parenchyma cells varying somewhat in size and form. Tyloses were present in both cases, but do not seem to be related to acid injury (Gore and Taubenhaus, 1931).

An increased number of vascular elements under stress of a drought resistant cotton cultivar determines its resistance to drought. In a study, the vascular 'bundle' arrangement in all the commercial entries and in the T169 strain were tetrarch (four distinct bundles), while the vascular bundle arrangement of strain T25 was pentarch (five distinct bundles). A stereological analysis of the root vascular systems showed that the total cross-sectional vessel area per root was 45% greater in T25 than in T169. These differences were primarily because of an increased number of vessel elements as a result of an additional vascular bundle. There was an increase in lateral root development (increased number and length) in T25. The increased number of vessel elements in T25 suggested a decrease in root axial resistance to water flow (McMichael *et al.*, 1985).

Stem anatomy

The stem anatomy of cotton is typical of that of the dicots. The mature stem shows secondary growth. The primary stem shows three regions, the epidermis, cortex and stele (Fig. 8.4).

The epidermis is the outermost region surrounding the stem. It is uniseriate, having a single layer of compactly arranged barrel-shaped living cells. The outer walls of the epidermal cells are cutinized. The epidermis is covered by a separate layer of cutin called cuticle on its outer surface, which reduces transpiration. Stomata are present in the epidermis for exchange of gases and transpiration. Unicellular hairs are present on epidermis.

The cortex is the middle region present between the epidermis and stele. It comprises three sections, the hypodermis, middle cortex and endodermis. Hypodermis is the outermost region of the cortex and is multilayered (with from two to six layers) and collenchymatic. Collenchyma may be in the form of a complete cylinder or in the form of discrete strands. Hypodermis gives considerable strength, flexibility and elasticity to the young stem.

The middle cortex is multilayered and parenchymatic and consists of mucilaginous cavities. The cells may be round or oval with prominent intercellular spaces. This region also stores food materials temporarily. Endodermis

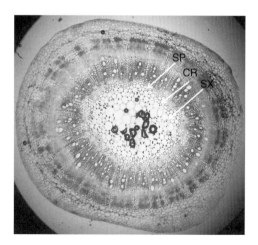

Fig. 8.4. Transverse section of stem showing initiation of secondary growth in vascular tissue, cambium ring (CR) forming outer secondary phloem (SP) and inner secondary xylem (SX).

is the innermost layer of the cortex. In young stems the innermost layer of the cortex contains abundant starch.

The stele is the central region of the stem and consists of pericycle, vascular bundles and medullary rays. Pericycle is the non-vascular region of the stele, and is multiseriate and sclerenchymatic. Wedge-shaped vascular bundles are arranged in one or two rows. The vascular bundles are described as conjoint, collateral and open type. The xylem consists of many vessels and is endarch with protoxylem present towards the centre. Medullary rays are the extensions of parenchymatous pith between vascular bundles and are called primary medullary rays, which help in lateral conduction and may give rise to secondary meristem.

The presence of several layers of collenchyma and strong secondary xylem contributes mechanical strength to the root tissue.

SECONDARY GROWTH   After the secondary growth the primary structure undergoes disturbance. Increasing secondary tissues produce pressure on the primary tissues causing compression. Secondary growth initially starts in vascular system. During the secondary growth cambial strips are formed from medullary rays. These are called inter-fascicular cambium. Fascicular cambium (cambium of vascular bundles) and inter-fascicular cambium combine and form a cambial ring (Fig. 8.5). The meristematic cells of the cambial ring form secondary xylem towards the inner side and a smaller amount of secondary phloem towards the outside. Phellogen or cork cambium is formed from the cells of the hypodermis or middle cortex. Cork cambium forms secondary cortex towards the inner side and cork tissue towards the outer side (Fig. 8.6). A larger amount of cork tissue is produced than secondary cortex. Secondary phloem and secondary xylem are larger in size (Fig. 8.7). Cotton genotypes show much variation in secondary growth and products of secondary growth, including the size of the distribution and size of the xylem vessels and thickness of bark (Fig. 8.8).

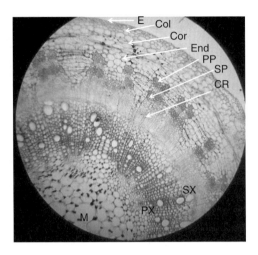

Fig. 8.5. Transverse section of stem (sector enlarged) showing secondary growth in vascular tissue: the primary structures are also clearly distinguishable; epidermis (E), collenchyma (Col), cortex (Cor), endodermis (End), primary phloem (PP), secondary phloem (SP), cambial ring (CR), secondary xylem (SX), primary xylem (PX) and medulla (M).

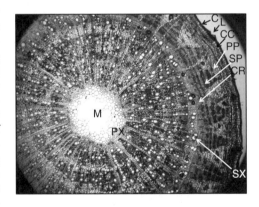

Fig. 8.6. Transverse section of stem showing complete secondary growth: cork tissue (C), lenticel (L), cork cambium (CC), primary phloem (PP), secondary phloem (SP), cambium ring (CR), secondary xylem (SX), primary xylem (PX) and medulla (M).

Significance of variations in stem anatomy

Cotton genotypes exhibit a good deal of variation in cuticle thickness, number and types of trichomes, shape and size of hypodermis layer, distribution of collenchyma cells and lignification, numbers of pericycle layers,

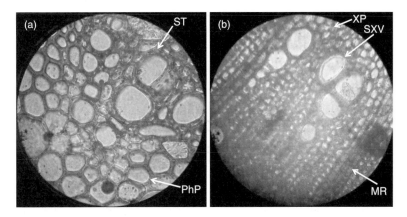

Fig. 8.7. Transverse section of stem showing secondary phloem; (a) phloem parenchyma (PhP) and sieve tubes (ST) can be observed in the picture; (b) xylem parenchyma (XP), secondary xylem vessels (SXV) and medullary rays (MR).

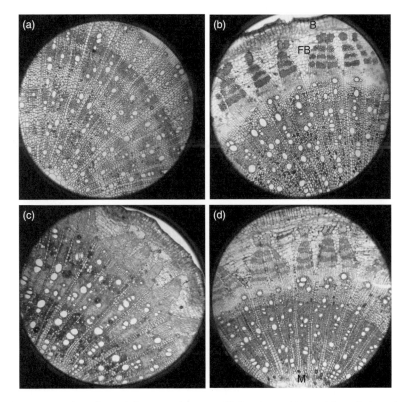

Fig. 8.8. Transverse section of stem (after secondary growth) showing variations: (a) small size xylem vessels with uniform distribution; (b) thick bark (B), compressed primary phloem, uniformly spread small to medium size secondary xylem vessels and fibre bundles (FB, phloem fibres); (c) medium thick bark and medium size secondary xylem vessels; (d) large size secondary xylem vessels and compressed primary xylem vessels at the centre. Medulla is prominent at the centre of the stem.

shape of sclerenchyma cells, number of secondary xylem vessels and number and size of vascular bundles. Variability in the density of mucilaginous canals in the cortex region is also noticed (Fig. 8.9). A large volume of sclerenchyma gives mechanical support and avoids the loss of water from internal parenchyma tissue by evapo-transpiration. It is expected that medium to large size and a high number of xylem vessels are required for efficient translocation of water under drought conditions (Fig. 8.8c). The presence of large vessels associated with dense lignified sclerenchyma offers strength to the stem, as well as drought resistance facilitating efficient translocation of water and photosynthates through phloem to the growing bolls.

Leaf anatomy

The leaf shows a dorsi-ventral nature, i.e. the leaf blade consists of distinct dorsal and ventral surfaces. The epidermis is single layered, having compactly arranged barrel-shaped (tabular) cells. The cell walls are cutinized and also covered by a thin cuticle, which reduce the loss of water due to transpiration. Anisocytic stomata are present on both the epidermal layers (Fig. 8.10). The stomata are usually more frequent in the lower epidermis (Fig. 8.11). The stomatal density varies among cotton cultivars, but on average the stomatal index of the upper epidermis is 25 and the

lower epidermis is 33. Stomata facilitate the exchange of gases between the leaf and environment. The epidermis contains two types of trichomes: small unicellular spine-like and hook-like trichomes.

The mesophyll is chlorenchymatic and differentiated into an upper palisade tissue and lower spongy tissue. The palisade tissue is thick and compact in drought-resistant genotypes, while in susceptible ones these are loose leading to greater loss of water by transpiration. Cells of the spongy tissue are thin-walled, irregular in shape and arranged loosely with large continuous intercellular spaces. Stomata open into large intercellular spaces called sub-stomatal chambers. The cells of spongy parenchyma contain lower numbers of chloroplasts, hence the lower surface of the leaf is light green in colour.

Collateral and closed vascular bundles (veins) occur between the palisade and spongy tissue (median in position). The vascular bundle is surrounded by parenchymatous bundle sheath.

Significance of variations in leaf anatomy

Ample genotypic variability is observed for the leaf anatomical structures under different adaptive environments. Variations in the arrangement of palisade tissues (Figs 8.12, 8.13), number of stomata, leaf trichome density (Fig. 8.14), density of gossypol glands

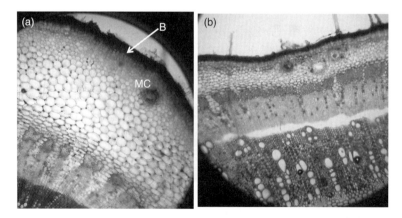

Fig. 8.9. (a) Transverse section of stem showing very thick bark, and a smaller number of mucilage cavities (MC) in the cortex. (b) Transverse section of stem showing medium thick bark, and a greater number of mucilage cavities in the cortex.

(Fig. 8.15), type and arrangement of vascular bundles both in the lamina and midrib and size and shape of bundle sheath chloroplasts are found in cotton germplasm. A number of mechanisms providing resistance to dehydration are observed in genotypes adapted to water-limited environments. Sclerenchyma gives mechanical support and avoids the loss of water from internal parenchyma tissue by evapo-transpiration.

The presence of a thick cuticle on the epidermis of the leaf reduces transpirational loss of water. Thick, compactly arranged palisade tissue also helps reduced stomatal density to minimize water loss. Some additional mechanisms such as high trichome density reduce the direct sunlight effect and minimize leaf temperature, thus finally reducing transpirational loss. The high density of trichomes also offers resistance to sucking insect pests. Insect pest resistance is also enhanced by the presence of a higher number of glands producing gossypol.

There exists a large variation in the compactness of palisade cells among cotton hybrids. Compactly arranged palisade cells in the leaf contribute to reducing transpiration loss and in drought resistance; on the other hand, loosely arranged palisade cells leads to a higher rate of transpiration.

Cotton hybrids show a large variation in the types and intensity of trichomes, which are related to insect pest tolerance. Three types of trichome are present on cotton leaf surface: (i) stellate; (ii) unicellular simple; and (iii) bifurcate types varying among genotypes (Fig. 8.16).

The leaf surface of cotton shows the presence and intensity of gossypol glands, which are related to insect tolerance. There exists a large variation in dermal morphology among cotton hybrids.

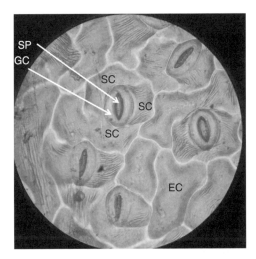

Fig. 8.10. Stomatal complex of the leaf epidermis (400×) showing two kidney-shaped guard cells (GC) covering the stomatal pore (SP), with three unequal subsidiary cells (SC) surrounding the guard cells. This type is called an anisocytic stomata. Epidermal cells (EC) are irregular in shape and unequal in size.

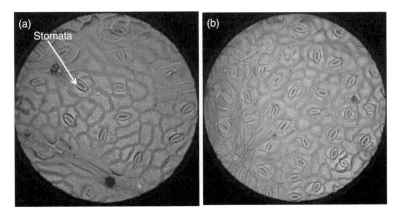

Fig. 8.11. Upper epidermis (a) of the leaf (400×) showing a lower frequency of stomata compared to lower epidermis (b).

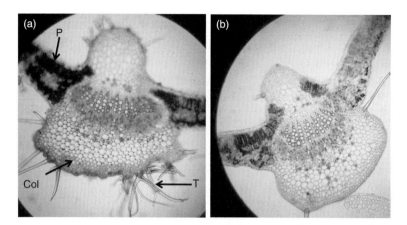

Fig. 8.12. Transverse section of leaf showing: (a) high density of trichomes (T) on both the epidermis, thick compactly arranged palisade (P) tissue, thick cuticle, and three to four layers of collenchyma (Col) tissue above the lower epidermis; and (b) very low density of trichomes on both the epidermis, thin, loosely arranged palisade tissue, thin cuticle, and two to three layers of collenchyma tissue above the lower epidermis.

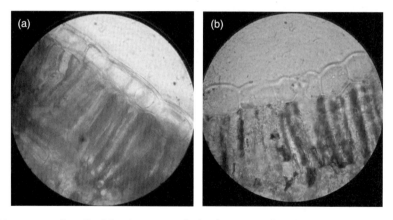

Fig. 8.13 Transverse section of leaf showing portion of palisade tissue: (a) thick, long compactly arranged and (b) thin, loosely arranged palisade cells.

Leaf anatomy in applied field (biotic and abiotic stress)

PALISADE AND STOMATA   The mature leaves of *G. barbadense* are larger and thinner compared to those of *G. hirsutum*, with a thinner palisade layer. *G. barbadense* leaves also show significant cupping or curling, which allows for a more even absorption of insolation over the course of the day and much more light penetration into the canopy. Although *G. barbadense* leaves have a 70–78% higher stomatal density on both the abaxial and the adaxial surfaces, its stomata are only one-third the size of those of *G. hirsutum*. This results in

G. barbadense having only about 60% of the stomatal surface area per leaf surface area compared to *G. hirsutum*. These results are indicative of the anatomical and physiological differences that may limit the yield potential of *G. barbadense* in certain growing environments (Wise *et al.*, 2000).

TRICHOME INTENSITY   In a study, differences in density of trichomes between leaf pubescence ratings were found. Pubescence ratings and trichome counts were highest for upper canopy leaves. Trichomes became less dense as leaves enlarged, then tended to abscise as leaves aged, therefore

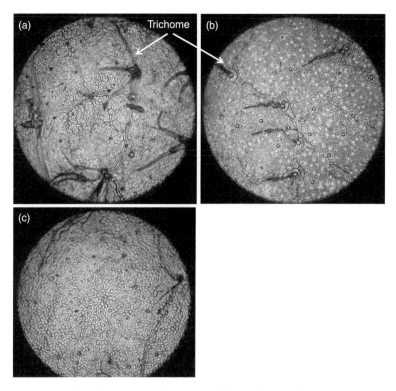

Fig. 8.14. Leaf upper epidermal impression showing variability in density of trichomes: (a) high, (b) medium and (c) low density.

leaf pubescence should be rated using the youngest, fully expanded main-stem leaves. Within each field trial, leaf pubescence ratings varied significantly among cultivars and were consistent for cultivars at different locations and across years. These data indicate that the rating system has a morphological basis and can be used to characterize the leaf pubescence of cotton cultivars (Bourland *et al.*, 2003).

CUTICLE ABRASION IN RELATION TO WATER POTENTIAL MEASUREMENTS   Transverse sections viewed with a transmission electron microscope indicated substantial direct damage to the cuticle with large sections of cell wall devoid of a cuticular layer. Although the exposed cell walls were intact, the lateral cell walls were physically compressed and distorted during abrasion. In addition, the cytoplasmic and vacuolar membranes of the epidermal cells were also frequently ruptured. Evaluation of the damage following abrasion indicated that the release of turgor

by the affected cells may contribute to increased sample variability and possibly to errors in osmotic potential (ψ) measurements (Wullschleger and Oosterhuis, 1987).

ANATOMICAL CHANGES UNDER INCREASED ULTRAVIOLET RADIATION   Enhanced UV-B radiation increased epicuticular wax content on adaxial leaf surfaces, and stomatal index on both adaxial and abaxial leaf surfaces. Leaf thickness was reduced following exposure to UV-B owing to a decrease in thickness of both the palisade and mesophyll tissue, while the epidermal thickness remained unchanged. The reproductive parameters were reduced at both ambient and high UV-B levels. The study shows that cotton plants are sensitive to UV-B at both the whole plant and anatomical level (Kakani *et al.*, 2003).

GOSSYPOL GLANDS   The intensity and the volume of the gossypol amount is an important part of defence against plant insects and diseases. The number of gossypol glands on the

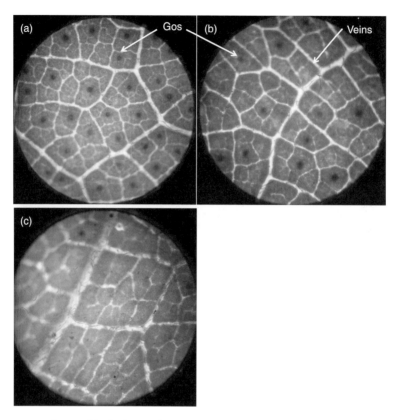

Fig. 8.15. Lower epidermis of the leaf (10×) showing variation in density of gossypol glands (Gos): (a) high, (b) medium and (c) low density.

stem, boll walls (carpels), leaves and seeds were visually counted and ranged from 0 to 142/cm² for stem, 0 to 135/cm² for leaf, 0 to 85/cm² for boll wall and 0 to 16/mm² for embryo (kernel). The number of gossypol glands varied for the genotypes and years but this variance was too low in some genotypes. Measured parameters were positively and significantly correlated to each other. Correlations for number of gossypol glands between stem and leaf ($r = 0.77**$), boll wall ($r = 0.65**$) and seed ($r = 0.32**$) were significant (Bolek *et al.*, 2010).

THE ROLE OF TRICHOME INTENSITY ON INSECT RESISTANCE   This study revealed that pubescence provides a mechanism of resistance to movement of newly hatched larvae. The study observed movements of newly hatched tobacco budworm larvae, *Heliothis virescens* (F.), on upper and lower leaf surfaces and petioles of four cotton (*G. hirsutum* L.) strains (Ramalho *et al.*, 1984).

Trichome density on leaf surface to jassid tolerance was observed. The existence of a positive relationship between trichome density and jassid tolerance was also confirmed through artificial screening (Kannan *et al.*, 2006).

Trichome density on lower leaf surface is found to be related to sucking pest resistance (Vibha Seeds).

The intensity of gossypol glands influences insect and disease resistance.

Petiole anatomy

There is a close similarity between petiole and stem with regard to the structure of epidermis. The ground parenchyma of petiole is similar to that of the stem cortex in the arrangement of cells and in the number of chloroplasts. The supporting tissue is collenchyma. In relation to the arrangement of vascular tissues in the stem, the vascular

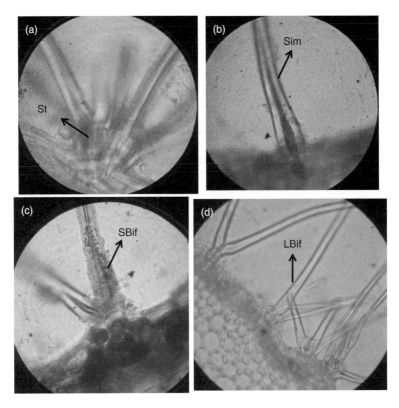

Fig. 8.16. Transverse section of petiole (portion of epidermis) showing different types of trichomes: (a) stellate (St), (b) simple unicellular (Sim), (c) short bifurcated (SBif) and (d) long bifurcated type (LBif).

bundles of the petiole are collateral. The epidermis contains unicellular trichomes. Multilayered (two to six layers) hypodermis of collenchyma cells is found immediately beneath the epidermis and, having chloroplasts, it may carry out photosynthesis. Just beneath the hypodermis ground tissue is found. This consists of thin-walled parenchymatous cells having well defined intercellular spaces among them. Vascular bundles are arranged in a half ring and scattered in the ground tissue. The wedge-shaped vascular bundles are of various sizes in the same petiole. In the petiole, xylem is always found towards the upper side whereas phloem is found towards the lower side.

Significant variation on petiole anatomy

Plant water deficits induce both leaf and boll abscission from cotton under field conditions. Petiole, being near the abscission zone, plays

an important role in hormonal regulation of leaf abscission. Cotton petioles with a higher number of vascular bundles will thus help increasing the longevity of leaves, thus increasing photosynthesis and biomass. Drought-resistant cotton cultivars possess a thick cuticle (Fig. 8.17), high density of trichomes (Fig. 8.18), high number of collenchyma layers (Fig. 8.19) and larger sized vascular bundles, thereby giving flexibility and strength to the petiole. Drought-resistant cultivars also have a long and stout petiole. The petiole is also commonly used for nitrogen status analysis in cotton, as the nitrate taken up by root moves to the leaves through the petiole. Hence, the higher number of vascular bundles and presence of thick collenchyma for stiffness and flexibility of the petiole will be helpful for efficient translocation of photosynthates, particularly under water stress conditions. The presence of cuticle thickness subtended by several layers of collenchyma contributes to

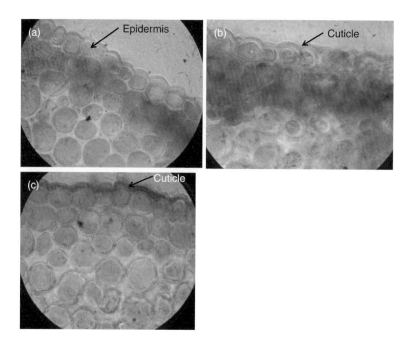

Fig. 8.17. Transverse section of petiole showing variation in cuticle thickness: (a) thin layer of cuticle (Col) on the epidermis; (b) moderately thick; and (c) very thick layer of cuticle.

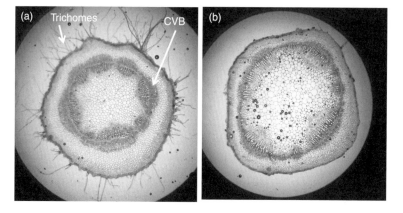

Fig. 8.18. (a) Transverse section of petiole showing high density of trichome and medium size vascular bundles (CVB) and small amount of centre ground tissue. (b) Low density of trichomes and high amount of centre ground tissue.

drought resistance. It has been observed that the variation in leaf epidermal and leaf anatomical traits may be used in specific characteristics of a cotton cultivar.

Research trends

In the context of the above mentioned literature it can be assessed that leaf surface anatomy with special reference to the intensity of trichomes and gossypol glands contributes greatly to the tolerance or susceptibility of cotton cultivars to some insect resistance. There exists a large variability in the intensity of the leaf anatomical components, thereby giving enormous opportunity in the selection of genotypes for tolerance to jassids and sucking pests. On the other hand, the cuticular

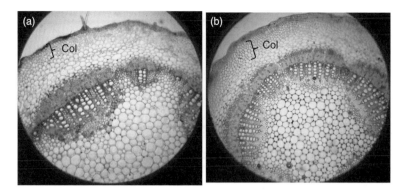

Fig. 8.19. Transverse section of petiole showing variability in collenchyma and distribution of vascular bundles: (a) four to five layers of collenchyma tissue (Col) and irregularly distributed xylem vessels in the vascular bundle; (b) six to seven layers of collenchyma tissue, uniformly distributed xylem vessels in vascular bundle.

thickness and the compact long palisade tissue associated with thick collenchyma layers contribute to drought resistance. All these traits may be effectively utilized for tolerance to biotic and drought stress.

Pedicel anatomy

The pedicel and the receptacle have a typical structure, with a normal vascular cylinder (Fig. 8.20); the cylinder may be unbroken or it may contain a ring of vascular bundles. In the region where floral organs are borne, the pedicel expands into the receptacle. The vascular cylinder also expands and the vascular bundles increase somewhat in number, and finally traces begin to diverge. The appendage traces are derived from the receptacular stele.

Parts of the flower and their arrangement

Longitudinal sections of the flower at different stages of development give a clear idea of the arrangement of floral parts and its development (Fig. 8.21). The flower consists of an axis, also known as receptacle, and lateral appendages. The appendages are known as floral parts or floral organs. The sepals and petals that constitute the calyx and corolla respectively are the sterile parts. The stamens and the carpels are the reproductive parts. The stamens consist of androecium and united carpels compose the gynoecium.

The flower shows limited growth. In flower the apical meristem ceases to be active after the formation of floral parts.

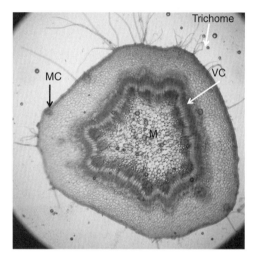

Fig. 8.20. Transverse section of pedicle illustrating central medulla surrounded by vascular cylinder (VC). The epidermis has long trichomes and mucilaginous canals (MC) are distributed in ground tissue (parenchyma).

A whorled arrangement of flower parts is observed in cotton.

Calyx

The sepals resemble leaves in their anatomy. Each sepal consists of ground parenchyma, a branched vascular system and an epidermis. The chloroplasts are found in the green sepals but there is usually no differentiation into palisade and spongy parenchyma. They may contain cells, laticifers, tannin cells and other

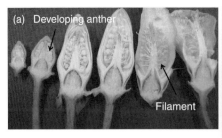
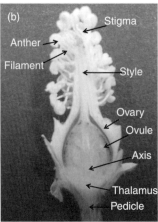

Fig. 8.21. (a) Longitudinal sections of flowers at different stages of development. (b) Flower showing arrangement of floral parts on thalamus. Floral parts (sepals, petals and stamens) develop below the level of the gynoecium (hypogynous flower).

idioblasts. The epidermis of sepals may possess stomata and trichomes.

Corolla

Petals contain ground parenchyma, more or less branched vascular system and epidermis. They may also have crystal-containing cells, tannin cells, laticifers and certain other idioblasts. They contain pigments containing chromoplasts. The epidermis may also contain stomata and trichomes. Some genotypes contain volatile oils. The presence of various types of trichomes may also be observed.

Androecium

The anthers contain two lobes (Fig. 8.22), and are found to be situated on a slender filament that bears a single vascular bundle. The structure of the filament is quite simple. The vascular bundle is amphicribral and remains surrounded by parenchyma. The epidermis is cutinized and bears trichomes. Stomata may also be present on the epidermis of both anther and filament. The vascular bundle is found throughout the filament and culminates blindly in the connective tissue situated in between the two anther-lobes.

The outermost wall layer of the anther is the epidermis. Just beneath the epidermis

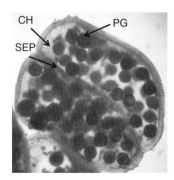

Fig. 8.22. Longitudinal section of anther showing two chambers (CH) (monothecous), septum (SEP) and matured pollen grains (PG).

there is endothecium, which usually possesses strips or ridges of secondary wall material mainly on those walls that do not remain in contact with the epidermis. The innermost layer is composed of multinucleate cells; this is nutritive in function and known as tapetum. The wall layers that are located between the endothecium and tapetum are often destroyed during the development of pollen sacs. On the maturation of the pollen the tapetum disintegrates and the outer wall of the pollen sac now consists of only the epidermis and endothecium. At the time of dehiscence of the anthers the pollen is released out through the stomium.

Gynoecium

The ovary is multicarpellary, syncarpus, with superior axile placentation (Fig. 8.23). The stigma is lobed, with a terminal style consisting of numerous hairs (Fig. 8.24). The ovary contains two to six carpels with five locules (seed chambers) on axile placentation corresponding to the number of carpels. Depending upon the species of cotton, each locule can have 8–12 ovules, five to nine of which usually mature.

A thermo-tolerant cultivar was characterized by having higher fertilization efficiencies under heat stress, higher pre-stress levels of antioxidant enzyme activity (glutathione reductase and superoxide dismutase) and higher levels of ATP, water-soluble calcium and total calcium in the pistil. Reproductive thermo-tolerance was also associated with higher photosynthetic thermostability due to higher pre-stress antioxidant enzyme activity in the subtending leaf (Snider, 2010).

Boll

After pollination occurs the cotton boll begins to develop (Fig. 8.25a). Under opti-

mum conditions it requires approximately 50 days for a boll to 'open' after pollination occurs. Boll development can be characterized by three phases: enlargement, filling and maturation.

The enlargement phase of boll development lasts approximately 3 weeks. During this time the fibres produced on the seed are elongating and the maximum volume of the boll and seed is attained. During this period, the fibre is basically a thin-walled tubular structure, similar to a straw. The boll consists of the axis at the centre and seeds covered by pericarp (Fig. 8.25b).

The filling phase of boll development begins during the fourth week after flowering. At this time, fibre elongation ceases and secondary wall formation of the fibre begins. This process is also known as fibre filling, or deposition. Cellulose is deposited inside the elongated fibre every 24 h, filling the void space of the elongated fibre.

The boll maturation phase begins as the boll reaches its full size and maximum weight. During this phase, fibre and seed maturation take place and boll dehiscence occurs. The capsule walls of the boll dry, causing the cells adjacent to the dorsal suture to shrink unevenly. This shrinking causes the suture between the carpel walls to split, and the boll opens.

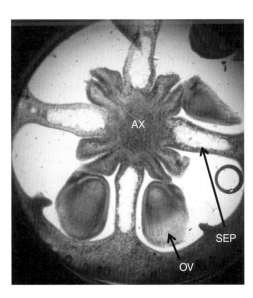

Fig. 8.23. Transverse section of ovary consisting of centre axis (AX), septum (SEP) and ovules (OV). The ovules are attached to the central axis (axial placentation).

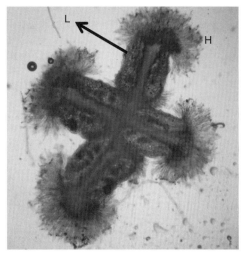

Fig. 8.24. Transverse section of style showing four lobes (L). Each lobe has numerous minute hairs (H).

Fig. 8.25. (a) Longitudinal sections of boll (fruit) at different stages of development, showing axis (AX) and developing seeds. (b) Fully developed boll showing pericarp (P), axis (AX), seeds (S) and fibres (F).

Fibre development

Cotton fibre development is divided into four distinct phases: initiation, elongation, secondary thickening and maturation (Jasdanwala *et al.*, 1977). The rate of fibre elongation and rate of water content show close parallels. Maturation occurs after boll opening and describes the drying of the mature, metabolically inactive fibre; the first three stages occur while the fibre is alive and actively growing (Fig. 8.26).

INITIATION   Fibre initiation involves the initial isodiametric expansion of the epidermal cell above the surface of the ovule (Fig. 8.26a). This stage may last only a day or so for each fibre. Because there are several waves of fibre initiation across the surface of the ovule (Stewart, 1975), one may find fibre initials at any time during the first 5 or 6 days post anthesis. Cytoskeletal elements may play separate roles in fibre morphology and initiation mechanisms. During the early development of fibre cells, a micronucleolus in the nucleus of fibres is produced. This developmental marker appears at or a little before 4 days after anthesis (DAA) in about 10% of the fibres and increases thereafter to nearly 80% provided the fibres are growing on fertilized ovules. Micronucleoli are neither seen in nuclei of fibres at 0–2 dpa nor in nuclei of non-fibre cells. Consequently, it is postulated that they are the product of specific developmental genes associated with fibre growth. Plasmolysis studies showed that the fibre initials and adjacent non-initiating ovule epidermal cells have similar osmotic potential. Initiation and rapid elongation of these fibres requires the expression of sucrose synthase (Sus) and, potentially, a transient closure of plasmodesmata.

ELONGATION   Each cotton fibre is a single cell that elongates to 2.5–3.0 cm from the seed coat epidermis within about 16 days after anthesis (DAA). Fibre elongation (Fig. 8.26b) is initially achieved largely by cell wall loosening and finally terminated by increased wall rigidity and loss of higher turgor. Depending on genotype, this stage may last for several weeks post-anthesis. During this stage of development the fibre deposits a thin, expandable primary cell wall composed of a variety of carbohydrate polymers. As the fibre approaches the end of elongation, the major phase of secondary wall synthesis starts. In cotton fibre, the secondary cell wall is composed almost exclusively of cellulose. During this stage, which lasts until the boll opens (50 to 60 DAA), the cell wall becomes progressively thicker and the living protoplast decreases in volume. There is a significant overlap in the timing of the elongation and secondary wall synthesis stages. Thus, fibres are simultaneously elongating and depositing secondary cell wall throughout the initiation and early elongation phases of development, and cotton fibre expands primarily via diffuse growth (Seagull, 1995; Tiwari and Wilkins, 1995). Later in fibre development, late in cell elongation, and well into secondary cell wall synthesis (35 DAA), the organization of cellular organelles is consistent with continued diffuse growth (Seagull *et al.*, 1998). Many cells that expand via diffuse growth exhibit increases in both cell length and diameter; but cells that

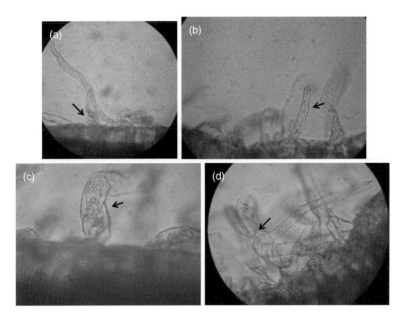

Fig. 8.26. Developmental stages of fibre cell (1000×): (a) initiation, (b) elongation, (c) secondary thickening and (d) maturation. The arrow points to the specific fibre cell being described.

exhibit tip synthesis do not exhibit increases in cell diameter (Steer and Steer, 1989).

EXPANSION   Cell expansion is also regulated by the extensibility of the cell wall. Regardless of whether cell expansion occurs via tip synthesis or diffuse growth, the wall in the region of expansion must yield to turgor pressure if the cell is to increase in size (Fig. 8.26c). For this reason, cell expansion most commonly occurs in cells that have only a primary cell wall (Cosgrove, 1997). Primary cell walls contain low levels of cellulose. Production of the more rigid secondary cell wall usually signals the cessation of cell expansion. Secondary cell wall formation is often indicated by the development of wall birefringence. Fibre diameter significantly increases as fibres grow and develop secondary cell walls. Fibre cells show increases in diameter; however, the specific rates of change differed; fibres continue to increase in diameter during the secondary wall synthesis stage of development, indicating that the synthesis of secondary cell wall does not coincide with the cessation of cell expansion.

MATURATION   During the maturation stage the fibres are thickened, i.e. diameter is increased

and the water content decreases drastically (Fig. 8.26d). Increases in fibre dry weight are the result of increases in the synthesis of cellulose (Meinert and Delmer, 1977). The cell wall that develops during this phase has been termed secondary cell wall and is composed almost exclusively of high DP (degree of polymerization) cellulose (Marx-Figini, 1982). Unlike secondary cell walls of other plant cells, the cotton fibre secondary cell wall contains little non-cellulosic component and no lignin. The secondary cell wall must be plastic enough to allow for the observed increases in fibre diameter. As lignin is one of the wall components thought to strengthen and make the wall rigid, the lack of lignin is consistent with the observed increase in fibre length and diameter (Fig. 8.26).

There exists a large variability in morphology of fibre cells, in length, breadth and wall thickness, which may be related to fibre quality.

Significance of variations in fibre anatomy

Cotton cultivars shows variation in fibre characters, which includes fibre length (staple length), diameter, wall thickness (Fig. 8.27),

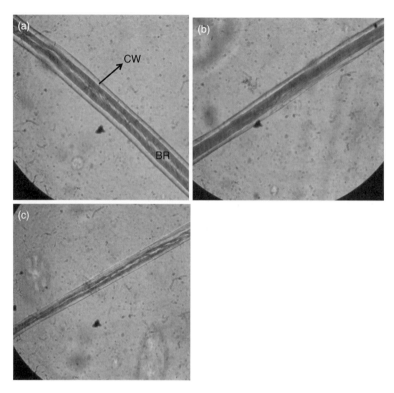

Fig. 8.27. Variability in fibre cell morphology in relation to wall thickness and breadth. (a) Coarse fibre showing high cell wall (CW) thickness and high breadth (BR). (b) Medium-fine fibre showing medium cell wall thickness and breadth. (c) Fine fibre showing less cell wall thickness and less breadth.

length/breadth (L/B) ratio and percentage of uniformity. These characters determine the quality of the fibre.

It is generally observed that genotypes having highest fibre length, less fibre cell diameter, less wall thickness and high L/B ratio give finer and stronger fibre. The rate of fibre cell elongation is an important parameter in determining the longest fibre, with the higher rate of fibre elongation giving fine quality fibre. Fine and long fibres have the higher rate of cell elongation per day. Fibre with the highest uniformity gives good spinning quality. Fibres having higher fibre cell diameter and high wall thickness produce coarse fibres.

In the above parameters, the rate of fibre cell elongation is very important for obtaining long staple fibres, which have commercial demand.

Differentiation of lint and fuzz is believed to be a continuous process, beginning just

before the flower opens, and if fertilization occurs, continuing up to the 28th day after flowering. The fibres at the base of the seed often develop first, and are longer than those that arise at the tip of the seed. The elongation of the fibre occurs during 25–30 days after flowering. Towards the end of this period there is a deposition of the secondary wall, which may continue up to the 78th day after flowering. The growth in length and the deposition of the secondary wall depend on the variety of the cotton plant and the environmental conditions (Flint, 1950).

Seed anatomy

The cotton seed is composed of embryo (Fig. 8.28), endosperm, perisperm, inner pigment layer, palisade (Malpighian) layer, colourless layer, outer pigment layer and epidermis, including lint hairs; traces of starch, in addition to oil and protein, occasionally occur in the cells of both young and mature embryos. The palisade layer is a part of the inner rather than of the outer integument. The ovules are campylotropous. Endosperm is present and usually is fairly abundant. The embryo and endosperm contain an abundance of oil and protein, but very little starch at maturity. In developing ovules, however, starch is commonly found in the integuments, nucellus,

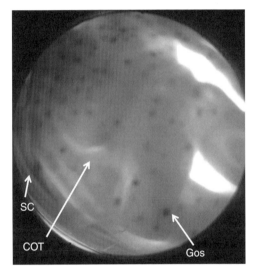

Fig. 8.28. Transverse section of cotton seed showing seed coat (SC), cotyledons (COT) and gossypol glands (Gos).

endosperm and embryo. Epidermal outgrowths are in the form of cellulosic fibres, called lint (Reeves and Valle, 1932).

Integuments of fertilized ovule become the seed coat, which covers the embryo and gives it protection. The seed coat consists of two layers, the testa and tegman. Testa is the outermost layer of the seed coat and is formed from the outer integument of the ovule. Generally it is hard and impermeable to water and sometimes contains ornamentations. The testa contains macro- and microsclereids (Fig. 8.29). Macrosclereids are arranged very compactly. Macrosclereids are dumb-bell/boat/bone shaped (osteosclereids). Microsclereids are small in size with a square shape and dentate edges (Fig. 8.29). The hardness of the seed coat depends on the thickness of testa and the arrangement of sclereids. The rate of absorption of water (imbibition) depends on the thickness of testa. Two types of fibres are produced from epidermal layers: short fibres called 'fuzz' and longer fibres known as 'lint'.

The tegman is the inner layer of the seed coat and is formed from the outer integument. It is thin and soft and covers the embryo. This thin seed coat and the fact that the macro- and microsclereids are not very compactly arranged, means that water enters easily into the seed and on imbibition it ruptures easily. The time of imbibition is less so germination time is about 36–40 h at 25–30°C under laboratory conditions. The outer integument is

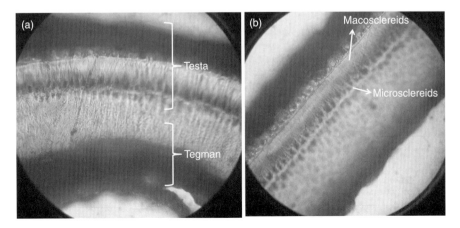

Fig. 8.29. (a) Transverse section of seed coat illustrating testa and tegman (100×); (b) the testa shows macro- and microsclereids (400×).

thin, usually being only two or three cells in thickness. The inner integument is much thicker, but is partially reabsorbed during development of the embryo.

In seed of tetraploid cotton (*G. hirsutum* and *G. barbadense*), approximately 30% of the seed coat epidermal cells develop into cellulose-enriched fibres, while the embryos synthesize oils and proteins. Hence, both the maternal and filial tissues of the cotton seed are of significant economic value. After initiation from the ovule epidermis at or just before anthesis, the single-celled fibres elongate to 2.5–6.0 cm long in the tetraploid species before they switch to intensive secondary cell wall cellulose synthesis. A number of candidate genes and cellular processes that potentially regulate various aspects of fibre development have been identified (Ruan, 2005).

The presence of a thick cuticle and thick macroscleOreids delays the time of initiation of germination in cotton seeds.

8.2 Bast Fibre Crops

8.2.1   Introduction

Next to food-producing plants, fibre plants have played a significant role in early and modern civilization. Primitive people learned to use plant fibre for clothing and other domestic use. In spite of the advent of modern synthetic fibres, vegetable fibres have great demand and compete with wool, silk and synthetics for quality, resistance, durability, colour and lustre. A great number of plant fibres are used in the fabrication of clothing, rope and paper. Vegetable fibre is an anatomical structure obtained from stems, leaves, roots, fruits and seeds. Among the bast fibres, jute (*Corchorus olitorius, Corchorus capsularis*), kenaf (*Hibiscus cannabinus, Hibiscus sabdariffa*) and other species are mostly of secondary origin (cambium), while ramie (*Boehmeria nivea*) and flax (*Linum usitatissimum*) are mostly of primary origin (pro-cambium). Fibres are also obtained from different plant parts such as cotton from seeds, sisal (*Agave*) from leaves, coir from fruits, *Muhlenbergia* from roots and so on.

Fibre crops have played a very important role in the world since primitive times to meet daily needs, and in modern civilization to meet industrial needs. The studies on fibre crops deal with distribution, origin, botany, ecological conditions, agronomy, harvesting method, quality and processing of different species of fibres. In addition, the different aspects of fibre plants grown in wild conditions, the methods of extraction, processing and uses of different species vary widely. This book provides a critical analysis of prospects and problems of fibre crops. It also discusses the technological aspects and market potentials of several fibres.

Although fibre science does not attract the special interest of many scientists, it deserves special attention for the mysteries of its origin, its structure, its productivity and its great economic importance in modern civilization.

8.2.2   The development of fibre filament and its anatomical structure

In plants, development of fibre is a determined and unidirectional biological process. A stem fibre is an anatomical structure present in the stem and derived mainly from the cambial activity where fibre filaments grow protophloeic in origin. The sieve tubes of stem and companion cells are obliterated by the expansion and elongation of the surrounding cells. The fusiform initials of the cambium are elongated cells that run lengthwise along the axis of the stem. The ray initials produce ray parenchyma in between the tiers of fibre bundles as observed in a transverse section of the stem. Fibre cells, after full development, become cemented at the ends by cementing materials such as hemicellulose and pectin to form the fibre strands. The fibre strands make their way through the stem in a zigzag pattern and become cemented here and there, thereby forming the meshy structure of jute and other bast fibres except in ramie, flax and sunnhemp (*Crotalaria juncea*).

In the case of ramie, fibre filament is composed of one fibre cell starting from the base of the plant up to the top of the plant and it is the strongest fibre of vegetable origin.

The formation of a meshy network is the main characteristics of these bast fibres. The ultimate fibre cells derived from fusiform initials, which form the building skeleton of the fibre filament, are pointed at the apex and have a lumen, the morphology of which varies among fibre crop species. The general characteristics of fibre bundle structure and ultimate fibre cells are depicted in the following figures. A large variation is observed in fibre bundle structure among crop species and among varieties of the same species with respect to shape of fibre wedge, number of layers, number of fibre bundles and their structures.

The fibre wedge is in general tapering to blunt with gradual reduction of fibre bundles towards the tip. The interconnection of fibre bundles forms the meshy structure. For example, the structure of fibre bundles and their cross-sectional area varies among the varieties of *capsularis* jute.

A breeder has to select a plant on the basis of its morphological structure, as fibre retting will destroy the crop without providing seed for progeny population. Moreover, during the retting process microbial organisms degenerate the parenchymatous tissues without affecting ligno-cellulosic fibre filaments, thereby liberating fibre strands after completing retting process and washing with water. Over-retting or under-retting reduces fibre quality, therefore it is difficult to assess the fibre quality if the fibre is not retted properly. In this respect anatomical traits can give correct prediction of the potential quality of the genotype before retting, which may be used in the breeding process. Considering these limitations, anatomy of the fibre filament depicts clearly the genetic potential of the fibre quality of crop plants as discussed herein.

Large variability occurs among bast fibre crop species in the anatomical structure of fibre bundles and their orientation and intensity in the fibre wedge. In the following section, the variability of fibre bundle anatomy of a few fibre crop species and their varieties is depicted.

The irregular contour of fibre bundles produces fibre filaments with irregular surface of poor fibre quality. The morphology of ultimate fibre cells, and tip and size vary widely among crop species. It has been reported that longer fibre cells contribute to greater strength. In this respect *C. olitorius* (tossa jute) produces good quality fibre for a desirable fibre bundle structure compared to *C. capsularis* (white jute), which has an undesirable fibre bundle structure.

In *C. olitorius* there exists large variations in fibre bundle intensity and fibre bundle structure, thereby offering scope for selection of varieties with high yield potential and high fibre quality. There is enough scope for selection of genotype for better qualities, such as fineness and bundle with uniform surface, and use them for genetic improvement of quality. The same is true in the case of *C. capsularis*. The main defects of this jute are the irregular fibre bundle and more meshy structure compared to that of *C. olitorius*. Therefore, emphasis should be given to improve fibre quality using anatomical traits. A greater number and more layers of fibre bundles per fibre wedge may also be considered for yield improvement (Maiti, 1973).

The fibre filaments with highly meshy structure offer obstacles during the carding process, while fibre filaments with minimal meshy structure are easier to handle during the carding process. A good example of better fibre type is *Malachra capitata*, which produces ideal fibre bundles for the spinning process. The fibre bundles of this plant are more or less rectangular with uniform surface structure and are scarcely meshy, revealing good fibre quality. The fibre filament tenacity is also high. *Hibiscus cannabinus* produce better quality fibre compared to that of *H. sabdariffa*. Different species of minor *Hibiscus* spp. produce poor quality fibres with irregular fibre bundles and highly meshy fibre strands such as *H. panduraeformis*. On the other hand, *H. suratensis* produces uniform fibre bundles with less meshy structure. Similarly, *Sida rhombifolia* produce strong and good quality fibre. In a study (Maiti, 1969), it was reported that *Hibiscus vitifolius* possesses uniform rectangular fibre bundles, producing good quality fibre of high fibre strength. The fibre bundles of *Abutilon indicum* are united laterally producing flat ribbon-like fibre filaments after retting. Therefore, the anatomical structures contribute greatly to the quality determination of fibre filaments (Maiti, 1971a).

In the case of some stem fibres, such as ramie, flax and sunnhemp, which are mainly of primary origin derived from procambium, fibre anatomy is quite different compared to those of other bast fibres of secondary origin mentioned above. The fibre cells are oval and grouped in the cortical regions and are cellulosic in nature. The fibres cannot be extracted by biological retting process owing to the fact that the bacteria responsible for retting directly feed on cellulosic fibre, unlike the ligno-cellulosic fibres of other bast fibres such as jute and kenaf. Ramie fibre is extracted by chemical retting methods using specific chemicals.

In the case of ramie (*B. nivea*), which produces the strongest unicellular fibre filaments, the oval or egg-shaped fibre cells are derived from procambial activity and are arranged in groups in the cortical regions. The intensity and cross-sectional area of fibre cells varies greatly among varieties. A single fibre cell produces the fibre filament, unlike other bast fibres where the fibre filament is composed of many ultimate fibre cells cemented at the tips. Therefore there is great scope in the selection of varieties with higher yield potential and better fibre quality.

Flax (*L. usitatissimum*) fibres are similar to ramie fibre cells, being ovoidal, and are present in patches in the cortical regions. Sunnhemp (*Cr. juncea*) also produces fibre cells in the cortex in a similar manner. There are variations among varieties in the size and abundance of fibre cells, thereby giving scope for selection of varieties for high yield and fibre quality.

Yield potentials

A greater number of fibre bundles in fibre wedges per unit area and fibre bundle area is associated with greater plant height and basal diameter for long fibres of secondary origin (jute, kenaf, other *Hibiscus* spp.).

In the fibres of primary origin viz. ramie, flax and sunnhemp, a greater number of fibre cells in transverse section of stem, and a greater number of fibre cell layers in fibre patches will predict higher fibre yield of the particular fibre crop before retting.

8.2.3   Quality parameters

Bast fibres of secondary origin

CROSS-SECTIONAL AREA   Fibre bundles with greater cross-sectional area produce coarse and strong fibres suitable for gunny bags, whereas fibres with low cross-sectional area produce fine fibres used in fabrics, and finer textiles.

SURFACE STRUCTURE   Fibre bundles with irregular surfaces are not desirable and produce bad quality fabric, on the contrary fibre bundles with uniform surface produce uniform filament for the fabrication of good quality fabric.

ULTIMATE FIBRE CELLS   Long ultimate fibre cells with uniformity in surface and fibre cell tips confer higher fibre strength/tenacity.

Fibres of primary origin

Low cross-sectional area of fibre cells and uniformity in cell surface produce finer textiles.

It is suggested that there is a great necessity for the evaluation and selection of genotypes with desirable yield potentials and quality parameters of the main fibre crop species such as jute (*C. olitorius, C. capsularis*), kenaf (*H. cannabinus, H. sabdariffa*), ramie (*B. nivea*) and flax (*L. usitattisimum*), both for yield potential and quality parameters. There is a great amount of germplasm available, which should be assessed for the potential fibre characteristics mentioned above, which can be stored in a germplasm bank for identification of the species and their possible utilization.

Once selected, the desirable genotypes could be included in crossing programmes for genetic improvement for yield and quality parameters along with plant height and basal diameter.

8.2.4   Fibre bundle anatomy determines the yield potentials and fibre quality of bast fibre (long fibre)

Vegetable fibres consist of cellulose, lignocelluloses, pectin and hemicelluloses. In the case of bast stem fibre, fibre is derived from meristematic tissue of primary or secondary origin depending on the species. Fibres of ramie and flax are mostly cellulose.

Apart from seed fibres (cotton), yield and quality determinations of vegetable fibres are difficult to assess unless the fibre is extracted by the retting process. This involves a series of biological and physical changes in fibre structure induced by microorganisms. Once the fibre retting is complete, its quality is determined by several parameters including fibre tenacity, fineness, surface contour, rigidity etc., assessed by various testing machines. During the retting process the fibre quality is affected severely by poor- or over-retting. Plant breeders use indirect methods to estimate potential fibre yields. In general in the case of stem fibres, selection criteria for genetic improvement of fibre yield are plant height and basal diameter. It is expected that the greater plant height and basal diameter contribute to higher fibre yield, but a greater basal diameter combined with a lower number of fibre bundles usually produces poor yields. On the contrary, a greater basal diameter with a greater number of fibre bundles increases fibre yield. Therefore, the selection by greater basal diameter may be misleading in few cases. For example, for soft stem mutants, the stem diameter is large, but with scanty fibre bundles. There are large variations among crop species and within the same species in the intensity of fibre bundles and their structure. It is evident that the anatomy of bast fibre plays a great role in determining fibre yield (Maiti, 1971a).

It may be stated that until now there have been no direct selection criteria for fibre yield and quality, which is only possible after extraction of fibre by the retting process. The yield and fibre quality are highly affected by retting methods and condition. At present, plant height and basal diameter of stem are indirect selection criteria, but these selection criteria are influenced by the amount of fibre bundles or cells in the stem bark and the quality is affected highly by poor retting conditions. Specific examples may be given in case of 'Soft Stem Mutant' and 'Halomehara', which possess scant amounts of fibre bundles.

Therefore, it may be hypothesized that anatomical traits of fibre bundles may be reliable selection criteria for genetic improvement of both fibre yield and quality.

In the context of the above discussions it is observed that great variability was observed in anatomical traits of fibre bundles of different long fibres such as jute, kenaf and other bast fibres among species and among varieties of the same species. It is suggested that the following parameters may be considered as selection criteria for yield and quality improvement.

8.2.5 Simple technique for screening varieties of main bast fibre crops for yield and quality potentials

This technique is suitable for main bast fibre crops such as jute, kenaf, ramie and flax. A small quadrant (piece) of bark is cut at the base of the stem of the respective species at flowering stage. Then transverse section of the bark is cut with a sharp razor blade and viewed under a microscope by adding a drop of safranin in the case of jute and kenaf for distinguishing lignocellulosic fibre strands and by adding bismark brown or light green in the case of ramie or flax for observing cellulose fibre cells in transverse section. By using this method, a large number of plants can be screened and selected within a short time for desirable yield-contributing and quality traits. This technique gives an opportunity to utilize the selected plants in the crossing programme for possible genetic improvement of yield and quality. The genetics of these traits need to be investigated for understanding the inheritance of these traits.

Taking a small portion of bark at the base of stem will not affect the plant growth but give an opportunity to the breeder to utilize a selected plant in crossing with another desirable selection. The selection procedure will continue in F_1, F_2 to F_7 to obtain homozygous plants for the combination of desirable yield and quality parameters. For example, there is a necessity to improve the quality of white jute, *C. capsularis*, owing to the irregular fibre bundle and highly meshy structure. As mentioned earlier there exists large variability in fibre bundle structure among *capsularis* varieties. The jute variety 'Sudan' is finer in bundle, which could be utilized in improving fineness of the fibre filament in *capsularis*. The same technique needs to be adopted in the case of *H. sabdariffa* for improving fibre quality. The plant breeders may use this simple technique in the genetic improvement of yield and quality of bast fibre crops following the schematic programme given in Fig. 8.30.

8.2.6 Research needs

Some investigational needs should be considered here, albeit only in general as the particular research needs relevant to reach the specific crop will be considered under the discussion of that crop. For seed-produced fibres such as cotton, the principal focus will be the establishment of new criteria for the improvement of resistance and length of the fibre, and more clearly associating these fibres with the quality not the yield.

In Mexico and Latin America, the principal leaf-fibre sources in the arid zones are from henequen, lechuguilla and palm, but there is insufficient research base concerning improvements in quality, resistance, durability, the number of fibres per leaf, percentage of fibres per leaf, product improvement, secondary product or basic anatomical studies.

For stem-origin fibre crops, investigation parameters should consider thickness and characteristic of bark, mainly the number, length, and distribution of cells and fibre bundle structure; as well as the extraction methodology, fibre percentage and fibre-to-wood ratio of the product. Lastly, the fibre quality, extraction times and methods of improvement of the extraction process should be considered.

For vegetable fibres obtained from root sources, it may be necessary to explore new agricultural machinery for extraction of roots from the soil. Root-source fibres represent a minor crop, the exploitation of which is limited in most advanced countries, but the root-fibre crop essentially remains a hand-labour operation in the developing countries.

Few attempts have been made to identify selection criteria for genetic improvement of fibre quality. The author strongly feels that some of the anatomical characters mentioned above could be used in the crop improvement programme of different fibre crops. The study of the fibre structure is important in determining or predicting fibre quality. In this sense, the principal objective of the breeders of vegetable fibres should

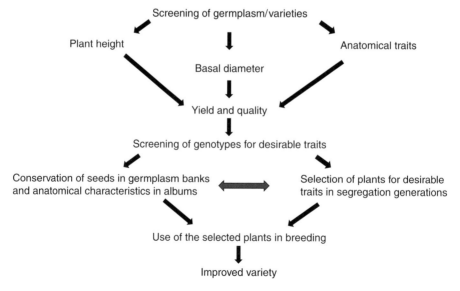

Fig. 8.30. Schematic hypothetical diagram for possible genetic improvement of yield and quality.

be to incorporate the component of yields and quality in order to elevate them in a new genotype or hybrid.

A concerned research effort should be made to promote the diversification and new uses of different vegetable fibres and expand the national and international market possibilities.

8.3   Jute

8.3.1   Introduction

Jute fibre, which is obtained from stem, is of great commercial importance in the world in the manufacture of bags, carpets, curtains, fabrics, and in the manufacture of paper. The fibre is obtained mainly from the species *Corchorus capsularis* and *C. olitorius*. These two species are commercially called white and tossa jute, respectively. It is cheaper than synthetic fibre, and contributes largely to the income of countries where it is cultivated. Jute is cultivated in developing countries of central and South-east Asia, including India, Bangladesh, China, Thailand, Myanmar, Vietnam, Uzbekistan and Nepal. It is also cultivated to a lesser extent in African and South American countries such as Sudan, Egypt, Zimbabwe and Brazil. Jute belongs to order Malvales and family Teliaceae. The number of chromosomes in these two species are $2n = 4$.

8.3.2   Origin and distribution

In the early 1950s, Kundu (1956) studied the distribution pattern of *Corchorus* sp. and noted that they are mostly distributed throughout the tropical regions. He proposed Africa as the primary centre of origin, and India and the Indo-Burma region as two secondary centres of origin for *C. olitorius*. Several workers later observed that the members of genus *Corchorus* are distributed in Africa, North and South America, China, Australia, India, Bangladesh, Nepal, Sri Lanka, Taiwan, Myanmar, Indonesia, Thailand, Malaysia, the Philippines, Japan and in some European countries. More precisely, south-east Africa is the centre of origin and differentiation of *Corchorus*; south-east Africa is the centre of origin and the first differentiation centre of wild *C. olitorius*, while the region of India-Burma-China is the second differentiation centre of wild *C. olitorius*, and also the centre of evolution of cultivated *C. olitorius*.

It has recently been proposed that south China, on the other hand, is the centre of origin and differentiation of wild *C. capsularis*, and also the centre of evolution of cultivated *C. capsularis*. It is postulated that the *olitorius* jute has migrated from Africa to India and China in early civilization via the Syrian trade route.

8.3.3   Utilization

Jute fibre is the most important fibre after cotton, and is cheaper. Jute fibre is mainly used in the preparation of gunny bags. Mats, threads, twine, tarpaulin are prepared from jute fibres, as are low-grade blankets, coarse paper and ornamental items. The stems are used as fuel and charcoal is prepared from the stems and used in the preparation of gunpowder.

8.3.4   Morphological characters

Vegetative characters

Jute is a mesophytic annual herb, growing up to 1.5–4.5 m with a taproot system. The base of the stem adheres closely to the bark. The fibre colour is white. The stem is erect, pubescent, generally unbranched and conical from the base, varying in plant height among the varieties with a range between 4 and 6 m, the diameter ranging from 1 to 2 cm. The stems at their apical part could produce ramification. Depending on the variety the stem colour could be green or dark red. The leaves are 10–16 cm in length and 3–6 cm in width, alternate, clear green, oblong, narrow and thick, dorsi-ventral, unicostate with reticulate venation and present small dentate margins. The petiole is green and pigmented with the length varying from 3 to 7 cm. The stipules are of two types, conical and crossed. The angle of leaves is highly variable, from sharp to open.

Floral characters

Flowers are present in the axils in solitary form or in groups of three to four (panicle). Flowers are pedicillate, with a small pedicle, bisexual, actinomorphic, pentamerous and hypogynous. The calyx consists of five sepals, united, green in colour, and with volvate aestivation. The corolla has five yellow petals, free, 4.5 mm in length, with imbricate aestivation.

The androecium consists of numerous stamens, free, short filaments, dithecous, and introrse gynoecium with five carpels, syncarpus, five locules and a superior ovary. Numerous ovules are present in each locule, with axile placentation, a short style and the stigma is flat.

The fruit is a capsule (dehiscent capsule) measuring 5–6 cm in length, 2 cm in diameter with a corrugated appearance. At maturity the capsule opens in five parts without showing partitions between the seeds. The seeds are numerous, small, 2–3 mm, of dark colour and pyramidal form.

8.3.5   Variations between *C. capsularis* and *C. olitorius*

The two cultivated species of jute are distinct in their morphology, but similar in their general characteristics, with the exception of the leaves and the fruit. They differ in leaf shape and size, pod shape, growth habit and fibre quality. Variability in morphological and yield-contributing traits is of great importance in the genetic improvement of a crop. In this respect, varieties and germplasm of *C. capsularis* vary widely in pigmentation and branching nature of the stem, the shape and size of leaf lamina and petiole.

In *C. olitorius*, leaves possess a sweet flavour, the stem is cylindrical and smooth in surface; the capsules are long, slender and cylindrical with a ribbed surface; the seeds are smooth, the fibre has a less meshy structure, and is very little adhered to the bark. The fibres are golden brown in colour.

In *C. capsularis*, leaves have a bitter flavour and vary in morphology among genotypes; the pod is small, globular, rough and corrugated, and seeds are pyramidal; the fibre is highly reticulated.

It is reported that the varieties of *C. capsularis* show wide variations in morphological characters, varying in the nature of pigmentation and branching habit of stem, pigmentation, shape and size of leaf lamina and petiole. This indicates that there is a wide range of diversity in morphological characters, thereby offering great scope to the breeders for the selection of desirable varieties for a certain growth parameter. For example, higher leaf area, leaf angle (45°C) and longer petiole length may lead to greater photosynthetic capacity.

Tossa jute (*C. olitorius*) yields more fibre per unit area compared to white jute (*C. capsularis*). The fibre is finer, softer, more lustrous and less rooty than *C. capsularis*. The leaves are sweet whereas *capsularis* leaves are bitter in taste. The capsules in *C. olitorius* are cylindrical, but are globose in *C. capsularis*.

8.3.6   Anatomy

Meristem

The bud initials are laid in the usual method in the primordial meristem in the case of branching plants of *C. capsularis*. The non-branching character of some varieties is due to the different structural organization of the shoot apex. In the non-branching plants the absence of bud initials is associated with early vacuolation of the meristematic cells. Cases occur of sporadic development of a few axillary and extra-axillary buds (Kundu and Rao, 1954).

Nodal anatomy

The cotyledonary node is unilacunar single trace or two traces, and the first leaf node is unilacunar single trace or trilacunar depending upon the species. The approximation of two traces from the cotyledonary node and three traces from the first leaf node forms a single trace in the lamina region (Thanki *et al.*, 2000).

Fibre anatomy and development of fibre

The fibre is an anatomical structure present in the stem and derived mainly from the cambial activity and the growth in length by the apical meristem. An assessment of fibre is only possible after the extraction of the fibres by the retting process. Therefore, the knowledge of morphogenesis of the fibres could enlighten us on the fibre-producing capacity of jute and other bast fibres.

In jute, 90% of the total fibre of the plant is developed in the secondary phloem by the activity of the cambium, and the remaining 10% is formed in the protophloem region. The vegetative phase represents the period of greatest cambial activity. Cambium maintains a definite ratio in its division with approximately two xylem initials produced for each phloem initial, due to which the secondary wood developed by the cambium forms a much thicker zone and occupies the greater part of the radial growth. The fibre cells developed from cambium undergo rapid change in length and breadth. The secondary wall of the young fibre cells becomes thicker by the gradual deposit of secondary layers of cellulose materials.

It is reported that the outermost layers of the fibre bundles in the fibre wedge are protophloeic in origin. The sieve tubes and companion cells are obliterated by the expansion and elongation of the surrounding cells. The procambium in the protophloeic region undergoes cell division and modification leading to the protophloeic fibres. This process continues towards the inner side, thereby crushing and obliterating all the sieve tubes and the companion cells, and the entire protophloem is covered by the solid fibre patches. The remaining bulk of the fibres are derived from meristematic activity of cambium, which contains two types of fibre initials, the fusiform initials and ray initials. Fusiform initials are elongated cells that turn lengthwise along the axis of the stem and give rise to new fibres by repeated longitudinal divisions. Ray initials are short cells in rows that divide to give rise to ray cells. The fibre initials after being cut off from fusiform initials make their way longitudinally through the tissue by an apical intrusive and symplastic growth. After full development of the fibre cells they become cemented at the ends by cementing materials of hemicelluloses and pectins to form the fibre strands. The fibre strands make their way through the stem in a zigzag pattern and become connected here and there, forming a meshy structure of jute and other bast fibres. The fusiform initials not only cut off fibre initials, but also at times undergo transverse divisions leading to the formation of soft parenchyma lying intermingled between the fibre strands, thus making possible their separation and isolation. This process leads to the formation of networks in bast fibres.

The ray initials in turn cut off derivatives to the formation of the ray cells, which make their way at right angles to the main axis of the stem. Thus, the derivatives of cambial initials have a bidirectional momentum of force, one (fibre initial) along the longitudinal direction and another (ray initial) along the transverse direction. This bidirectional force is associated with a vertical twisting growth by apical meristem leading to the peculiar mode of arrangement in the stem. The formation of a meshy network (Fig. 8.31) is a main characteristic of jute and other bast fibre, depending on the activity of cambium. The degree of meshiness is associated with tangential expansion and radial growth of the stem. The meshes are loose towards the top and compact towards the basal region. At the basal region, the meshes are loose towards the periphery and more compact towards the cambium. The meshiness is a drawback because it hinders the carding process. Coarse meshiness reduces the quality of fibres and weakens the fibre strands during splitting and combing operations in the spinning process.

Fibre filaments as they appear in commerce are developed and oriented in a given manner peculiar to each variety of species. Fibre cells derived from the meristematic activity of cambium are assembled together in a definite pattern. The differential cambial activity that governs the yield and quality of fibres is characteristic of the genotype of the different species or varieties. However, the mode of arrangement of the fibre bundles in

Fig. 8.31. Pattern of anatomical structure of fibre bundles in jute stem: (a) transverse section; (b) single fibre wedge; (c) view of meshy structure of fibre filaments in longitudinal section; and (d) meshy structure (Maiti, 1997).

the secondary phloem fibre differs from one species to another. Generally it is a ribbed meshy structure broad at the base and gradually tapering upwards into free filiform strands, mostly composed of primary cells. In a transverse section these fibre strands appear as an extra-cylinder ring of pyramidal wedges (see Fig. 8.31), having a centripetal broad base and a narrow conical and radially directed

towards the periphery with alternate patches of fibre bundles and thin-walled tissue in between. Fibre bundles are arranged radially in superimposed arcs along radii, tangential in the axis of the stem, and the bundles are traversed by phloem rays and show great variability in the fibre bundle structure among *C. olitorius* varieties. The ultimate fibre cells, which form the building skeleton of the fibre filament, are pointed at the apex and have a lumen at the centre, although they vary between two species. The ideal criterion for the selection of jute for high yield is a combination of three factors: plant height, basal diameter and fibre content (number of fibre bundles), which is influenced by the cambial activity during the vegetative phase.

The secondary cambium forms the phloeic fibre in all peripheral directions, and wood in a centripetal direction towards the stem pith. The contribution of apical meristem leads to the elongation of stem, and its direct contribution to quantity of fibre is very low since the greater quantity of fibre corresponds to the secondary activity of the cambium. Diverse anatomical studies show that the activity and function of the apical meristem finishes when the flowering begins, but the activity of the secondary cambium could continue even after flowering. This study in the quantitative assessment of the cambial activity in relation to the production of fibre bundles in some cultivars of *C. capsularis* indicated that the varieties differed widely in the optimum growth responses. In another study, the cambial activity with respect to the formation of fibre and wood in four varieties

of *C. olitorius* indicated that growth rate of the meristem (intercalary and apical) for the formation of fibre and plant height showed alternate peaks, but the period of activity for both components was different in other varieties (Fig. 8.32). The period of optimum cambial activity among the varieties was similar at 40 days. However, the period of optimum activity of the apical meristem (42 days) was very different from the activity of cambium (Maiti *et al.*, 1974).

Shaikh *et al.* (1980) applied several anatomical methods in order to select mutants of *C. capsularis* with good fibre yield. In this study, two morphological characteristics, plant height and basal diameter, were considered and anatomical components were studied in transverse sections. The coefficients of environmental variability and genetics and their correlations were computed in order to confirm the effects of other agronomic and anatomical characteristics on fibre yield. Some genotypes with good yield have very thick bark and greater area of fibre bundles in each section. The number of cells per fibre bundle in each strip showed a negative influence on the yield of fibre. The same fibre cell area has high variability in the genetic coefficient.

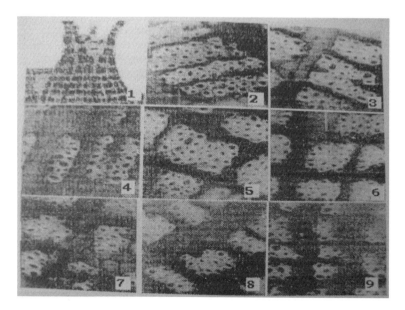

Fig. 8.32. Variability in fibre bundle structure among *C. olitorius* genotypes: 1, fibre wedge; 2–9, fibre bundle morphology (Maiti, 1997).

This indicates an opportunity for selecting these characters. Shaikh *et al.* consider these characteristics the most effective for the determination of a good yield potential, better than the agronomic parameters of plant height and basal diameter. These observations support the findings of Maiti (1980), who emphasizes the importance of anatomical characters in determining yield as well as quality.

The ultimate fibre cell (Fig. 8.33) is wide in the centre and gradually tapering toward the ends; the cell presents lumen that on occasion gets obliterated. In *C. olitorius*, frequently the lumen is wider and uniform in thickness, and sometimes has a constriction on the cell wall in *C. capsularis*. The lumen is irregular with frequent constrictions. The maceration of fibre filaments to individual fibre cells could be carried out in a mixture of 10% chromic acid and 10% nitric acid in a proportion of 1:1, in an incubator at 60°C approximately for 1 h.

In a study by Hao *et al.* (1993) in Fujian, China, correlation coefficients among fibre counts, fibre strength and single fibre cells were determined in 16 accessions of *C. capsularis*. It is suggested that advanced generation for selection of single fibre cell length, small cell cavity and thin cell wall may be effective in improving the quality of the fibre. This has also been suggested by Maiti (1980).

Ali (1989) advocated the use of ultimate fibre cells for selection in *C. olitorius*. He determined the length, thickness, and length to thickness ratio of the ultimate fibre cells from the top, middle and base of the stems of

11 indigenous genotypes. Highest mean fibre cell length and thinnest fibre cell occurred in the upper part of the stem.

The genetic relationship between anatomical characters and fibre yield and quality in varieties of *C. capsularis* was reported by Chen *et al.* (1990) and by Chen (1991). The number of fibre bundles and layers was highly correlated with plant height and stalk diameter. Secondary fibre cells began differentiation 21 days after emergence, but the duration of development varied significantly among the varieties, especially in the number of fibre cells and fibre bundles, mainly at the middle and late growth stages. In mature fibre cells, the cell cross-sectional diameter showed significant positive correlation with cell wall thickness. Fibre cells took 30 days from differentiation to ripening, the growth and ripening rate differing with variety, plant growth stage and temperature.

The number of fibre bundles and number of cells in the bundle are indicators of fibre yield as well as the irregularity and regularity of the bundle surface, the length of the ultimate fibre, the length/breadth ratio and the area of section of bundle. The improvement of fibre quality is imperative where there is competition with synthetic fibres. Several studies have suggested that anatomical traits associated with morphological characters could be considered as selection criteria for the genetic improvement of jute for yield and quality.

Some structural parameters of fibre bundles, as revealed by the microscopic study of the transverse section of jute plant stems, have been found to have high positive correlations with filament fineness of extracted fibres. Equations have been derived by linear regression analysis to predict the fineness of extracted fibres. The study would not only be very helpful in selecting the fine fibre-producing plants in breeding programmes, but will also offer some information on the structure–quality relationship of jute fibre, essential for genetic upgrading of fibre quality (Majumdar, 2002a).

Fibre anatomy in mutants

The anatomical characteristics were studied of one X-ray induced macromutant, undulating

Fig. 8.33. Ultimate fibre cell (Maiti, 1997).

stem. This true-breeding mutant produces only primary phloeic fibres. The secondary phloeic fibres, which are the derivatives of vascular cambium, are found lacking, though there is formation of secondary phloem pyramids from the cambial tissue. Inheritance of fibre (bast) in jute (*C. capsularis*) is monogenic and recessive (Mitra, 1984).

In a study the mutant grew more slowly, had shorter internodes and yielded much less fibre after retting. The fibre of the mutant contained 50% less lignin but comparatively more cellulose than that of the normal type. The lower and earliest developed part of the mutant stem had no lignified fibre cells. This developmental deficiency in lignification of fibre cells was correlated to a similar deficiency in phenylalanine ammonia lyase activity, but not peroxidase activity, in the bark tissue along the stem axis. This mutant may be utilized to engineer low-lignin jute fibre strains (Sengupta and Palit, 2004).

Yield on the basis of anatomy

Correlation studies showed that bark thickness, number of phloem pyramids, number of fibre bundles in phloem pyramids, number of fibre bundle arcs in phloem pyramids and area of phloem pyramids were positively and significantly correlated with fibre yield in the genotypes. Genotypes showed strong association of the component characters with fibre yield except number of fibre bundle arcs in phloem pyramids. Bark thickness showed highly significant correlation with phloem pyramid areas in almost all the genotypes (Ali, 1993).

Fibre quality on the basis of anatomy

Plant height and basal diameter are used as the conventional method of selection for improvement of yield of jute and allied fibres, but to the breeders there are no authentic breeding criteria for quality improvement. Breeders require some suitable and rapid techniques for assessing quality on a single-plant basis. They cannot depend on physical methods for estimation of quality. Anatomists have attempted to correlate some of the anatomical components with quality characters of fibres. Different

varieties of *capsularis* jute indicate that these show much structural variabilities, and it is predicted that some of the anatomical characters are helpful in the selection of a variety in quality breeding. As fibres are in the stem, the fibre yield may be reflected by the attainment of plant height and basal diameter, but the yield may be less if the fibre content is less.

Fibre cells are broad and narrow-lumened type; tips are straight and pointed, notches are present in denticulate manner in some, the lumen is broader than wall thickness, constricted and blocked; and the cell surface is generally regular.

The tenacity values of bundles of jute fibres belonging to different varieties have been predicted from the structure of their fibre bundles as seen in the transverse section of jute stems and the length of the constituent ultimate cells by multivariate regression analysis. Jute plant stems having fibre bundles comprising a greater number of compactly arranged long and fine ultimate cells give fibres of higher strength when retted under standard conditions (Majumdar, 2002b).

There is a simple procedure for identifying the textile bast fibres from flax, nettle/ramie, hemp and jute. The procedure is based on measuring the fibrillar orientation with polarised light microscopy and detecting the presence of calcium oxalate crystals (CaC_2O_4) in association with the fibres (Bergfjord and Holst, 2010).

Fibre quality parameters

The quality of the jute fibre is considered in accordance with five major criteria: force, fineness, lustre, colour and length. The ideal fibres for spinning have high lustre, good colour and good length. For commercial purposes, the most important quality parameters are fibre strength, fibre fineness, nature and extent of defects, root content of the fibre, colour of the fibre and fibre density. Besides the length of the fibre cell, the diameter and distribution of fibre cells are also important quality parameters of jute fibre.

FIBRE STRENGTH   Fibre strength is defined as the amount of force required to break a fibre of unit length and weight. The unit length is generally fixed at 5 cm or 10 cm and a mechanical jerk is given to break the fibre. The fibre strength is expressed as g/tex. The fibres have high initial modulus and low elasticity; therefore, they have small extension capability before breaking.

FIBRE FINENESS   Fineness is described as the width of fibre filaments when fibre filaments are separated. It is also expressed as the width of fibre strands when separated mechanically or manually. The fibre fineness ultimately depends on the fibre cell diameter. Jute fibre cells are not circular, rather elliptical shaped with many irregularities. This makes direct estimation of fineness difficult from fibre cell width.

ROOT CONTENT   The root content in the fibre describes the hard bark regions at the lower end of stem. As the region is hard, retting is not often complete in this region. It is expressed as the percentage of rooty fibre weight to total fibre weight. When fibre is exported, generally the rooty ends are chopped out and sold separately, and are called 'cuttings'.

FIBRE COLOUR AND LUSTRE   Although fibre colour does not affect the end use of jute fibre as hessian, sacking or clothing, it is a primary determinant of jute fibre price in the market, especially for tossa jute. The *capsularis* jute is dull white in colour, while *olitorius* jute turns golden yellow after retting. The colour is also indicative of extent of retting of fibre, thereby serving as a selection criterion for the buyer. The lustre of the fibre also indicates the extent of retting as over-retted fibres will lose lustre and will become dull.

FIBRE CELL LENGTH   The individual length of jute fibre ranges from less than 1 mm to greater than 5 mm, with an average around 2.8 mm. Longer fibre cells indicate better quality fibre and are more suitable for blending in clothes and paper pulp industries.

Yarn quality

The unidirectional thermoplastic commingled yarn composites have been fabricated from three different varieties of yarns in a hot press and then studied for the bending properties on a DCS-500 tester using a three-point loading system. It was observed that the bending properties of composites, especially the transverse bending properties, are affected by the mixing degree of reinforcing fibre/matrix fibres and bulkiness of reinforcing fibres bundle (Li *et al.*, 2002).

Pulp quality

Jute fibre pulp showed better papermaking properties than jute cuttings and caddis. The tear index of these raw materials was similar to softwood. The bleachability of jute fibre pulp was also better than that of cuttings and caddis. Pulp yield and bleachability was higher and kappa number lower for jute fibre than jute cuttings and caddis, but α-cellulose, S10 and S18 values and viscosity were almost similar in these three raw materials (Jahan *et al.*, 2007).

8.4   Kenaf

8.4.1   Introduction

Kenaf (*Hibiscus* spp.) is a bast fibre and is next to the jute fibre in importance and a valuable substitute of jute for sack and rag production. The main kenaf fibre-producing species of economic importance are *H. cannabinus* and *H. sabdariffa* (*H. sabdariffa* var. *altissima*) and often the fibres from these two species are difficult to distinguish in the market. They belong to the family Malvaceae. The major producers of kenaf fibre are India, China, Thailand, Vietnam, Cambodia, Indonesia, Brazil and Cuba. China was the largest producer of kenaf in the first decade of the 21st century.

In India, these two species are also known as 'roselle' (*H. sabdariffa)*, mesta pat or mesta (*H. cannabinus* and *H. sabdariffa*), Deccan jute and ambari. *H. cannabinus* is also known as Java jute and siami jute in South-east Asia, stokroos in South Africa and Guinea hemp in West Africa. In northern Africa, particularly in Egypt, kenaf is known as teal or teel. In the European countries, kenaf was introduced later and is generally named after the country from which it was introduced, such as 'cáñamo de la India' in Spain.

8.4.2 Origin and distribution

H. cannabinus originates from Africa, where it is found in its wild form, and also in the subtropical regions of Asia. There are some controversies with regard to the origin of *H. sabdariffa* var. *altissima*: some authors indicate that it originated in Africa and India; however, it has been reported that *H. sabdariffa* var. *altissima* has been introduced in India through single seed mixture with *Calapogonium muconoides*.

H. cannabinus and *H. sabdariffa* are widely distributed in the USA, Latin American countries, Russia, Sudan, Egypt, India, Australia, the Philippines, Java, Iran, Nigeria, Senegal, Thailand, China and Brazil. It has been introduced into Cuba, Guatemala, El Salvador, Colombia, Mexico, Costa Rica and Haiti as a source of fibre and into Tanzania, Kenya and Australia for the production of paper. It was introduced into Europe early in 20th century.

The crop was introduced into China from India through Taiwan and from Russia in the early 20th century. A cultivar called 'Madras Red' was introduced in Zhejiang province, China. In Russia, kenaf was cultivated on a large scale in the 1930s.

8.4.3 Adaptation

Kenaf is more adaptive than jute under diverse conditions of climate and soil, and is very resistant to drought. Kenaf is a traditional fibre crop with adaptability to both tropical and temperate environments, although a tropical environment is more favourable. It is considered as a fibre crop of marginal farming communities under drought-prone rainfed conditions.

8.4.4 Utilization

Kenaf fibres are used mainly in the fabrication of paper, cordage, fabric, yarn and decorative objects, and also in combination with other synthetic fibres or vegetable fibres such as jute. In some countries the fruit is eaten. The quality of the kenaf fibre is good except for its fineness and semi-meshy structure. Fibre strands are more irregular than jute.

Kenaf seeds contain 16–22% oil and 32% protein, ideally suiting them for use in cooking; they also contain unsaturated fatty acid in a low proportion, which is used in the elaboration of margarine. After extraction of oil, the residue contains approximately 35% protein and is used as livestock feed. The seedlings, leaves and fruits are used to make sauces, jellies and wines.

8.4.5 Morphological description

Vegetative characters

Both species are herbaceous and present two sexes in the flowers present in the leaf axils on the stem. They are annual crops. The taproot is pivotal, and secondary and tertiary roots are derived to give support to the crop and absorb nutrients. The stem could be flat and thorny and reach a height of 2–4 m depending on the environmental conditions and varieties. Roselle (*H. sabdariffa*) is an annual plant. The stem shows several growth patterns and is generally purple. The leaves are polymorphic, palmate, extremely lobulate with three or five lobes, and alternate on the stem with very deep venation (Fig. 8.34).

Floral characters (H. sabdariffa)

The flowers of *H. sabdariffa* are smaller than in *H. cannabinus*, and solitary in the axiles of the upper leaves on a short peduncle. The peduncles are shorter than the petioles; the bracts and the calyx show upward growth, the sepals are triangular in form and acuminate, with more than half an intense purple colour. The flowers are generally cream or yellow, with scarlet to purple in the interior part of the neck. The capsule (fruit) is ovoidal, pointed, hairy, and shorter than the axis; the seeds are reniform, coffee-coloured and smaller than those of *H. cannabinus*.

Fig. 8.34. Kenaf plant morphology: leaves are polymorphic, palmate, extremely lobulate with three or five lobes; the flowers are cream or yellow in colour.

8.4.6 Comparative morphology of *H. cannabinus* and *H. sabdariffa*

The stem of *H. cannabinus* has a higher basal diameter and its wood is soft, while *H. sabdariffa* has a harder wood. The basal leaves in the varieties of *H. sabdariffa* lack bifurcations and are not lobed. The petiole has projections on its surface. The flowers are big and cream coloured, with purple at the centre, reddish and yellow at the mouth. The peduncle is axillar and very short, the sepals are lanceolate and wide in the middle; the flowers are located in the leaf axils and are found generally in the apical part. The corolla is big, wide open and yellow, the capsules are five-locular with four to five seeds in locules. The calyx is pubescent and curved.

8.4.7 Fibre anatomy

The quality of kenaf fibre is good except for its fineness and semi-meshy structure. Fibre strands are more irregular than jute. The fibre is coarser than jute, hence it cannot be used for finer clothing purposes.

The pattern of orientation of the fibre bundles is similar to that of jute, but kenaf varies in the structure and form of the fibre bundle surface (Fig. 8.35). Furthermore, the intensity of reticulation determines the quality of the fibre. In kenaf, the structure of the bundle surface is more irregular and has more intense reticulation than jute. Within kenaf, in *H. sabdariffa* the anatomical structure of the fibre cells has many interconnections among fibre bundles, resulting in a more abundant reticulation; the periderm is thick and the surface of the bundle is irregular; therefore, it has a poorer fibre quality than *H. cannabinus*. A comparative study (Maiti, 1971b) on the anatomy and contents of fibres among ten varieties of *H. cannabinus* in India showed that the number of fibre cells per bundle and the number of fibre bundles per fibre wedge was related to the yield of fibre under similar environmental conditions.

The morphology of ultimate fibre cells varies in several genotypes. In general, it is regular; the lumen varies from wide to narrow in accordance with the genotypes. Also, the tips of the ultimate fibre could be pointed or round, and the cell wall varies in thickness and undulations (Fig. 8.36) The fibre bundle tenacity (force) increases with an increase in the length to breadth ratio (L/B) of the ultimate fibre. A high L/B ratio of fibre cells offers good fibre quality both for fineness and spinning quality (Maiti, 1980).

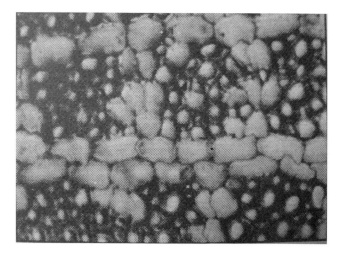

Fig. 8.35. Fibre bundle structure in *Hibiscus sabdariffa* (Maiti, 1997).

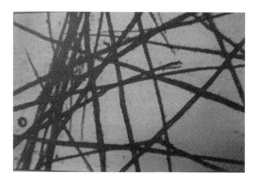

Fig. 8.36. Ultimate fibre cell in *Hibiscus cannabinus* (Maiti, 1997).

Although the quality of *H. cannabinus* and *H. sabdariffa* is inferior to that of jute, varieties exist whose quality is superior on the basis of their anatomical characteristics, e.g. cross-sectional area of the fibre bundle and L/B of the ultimate fibre. These anatomical traits may be used in the selection and genetic improvement of kenaf for both yield and quality.

8.5 Flax

8.5.1 Introduction

Flax, *L. usitatissimum*, is one of the oldest crops cultivated by human beings and its products are greatly useful. Flax belongs to the family Linaceae in subfamily Linoideae. The crop is also known as linseed in English, lin in French, lein, saatlein or flachs in German and lino in Spanish. In India, the common name for *Linum* is tissi, while in China the crop is known as Hu Ma or Ya Ma. Fibre is made into linen, and oil from the seed is made into the drying agent linseed oil. Linen has been made for over 3000 years. It was found in archaeological remains on the river-sides in the Swiss Alps dating to 8000 BC. At the same time, flax was also domesticated in the 'Fertile Crescent' and was one of the earliest crop species to be domesticated. Egyptian civilizations used flax to manufacture cloth, and fed the livestock with flax seeds and seed oil. The chromosome number is $2n = 30$.

8.5.2 Origin and distribution

From ancient periods, flax has been found to be distributed over many continents. The Indian subcontinent is the region where most probably flax originated. Vavilov (1951) suggested flax originated in this region as it represents the greatest diversity of flax. This is also supported by the fact that flax was cultivated in the Indus valley civilization. However, archaeological evidences of seeds of *Linum bienne* have been documented in Syria, Turkey and Iran as old as 8000 BC. It was also known in ancient China, dating back to about 2000–5000 BP. There are three geographical centres of diversity of *Linum*: the Mediterranean area, southern North America and Central America, and South America.

Flax grows in most of the temperate and subtropical countries of the world. Countries that produce fibre and oil are Austria, Germany, Belgium, Finland, Poland, Rumania, Czech Republic, France, Italy, the USA, Argentina, Mexico, Ethiopia, Australia and other countries. As a seed crop, Canada is the world's largest producer of flax, producing about 40% of the world production of 2.7 Mt flax seed, and the USA, China and India produce another 40%. In India, flax is cultivated mainly for seed purposes, and only in a few places on the Himalayan plateau is fibre flax cultivated. However, India is a leading figure in seed flax production, contributing about one-fourth of the total world production of seed flax.

8.5.3 Utilization

Flax is one of the finest quality plant fibres, consumed mainly in the clothing industry. Among the main characteristics of flax fibre the most important is its strength, elasticity, lustre and flexibility; it is used in the fabrication of canvas, cloth, carpets, yarn, paper and insulating materials. Its utility depends in great part on its quality and other factors during the process of extraction.

Fibre is made into linen, and oil from the seed is made into the drying agent linseed oil. Flax is a high quality plant fibre, being primarily used in the clothing industry. Best quality flax fibres are highly valued in the developed countries, such as the USA, Europe and Japan. The linen fabric is primarily used in household cloths, such as bed linen, furnishing fabrics and interior decoration accessories. Shorter fibres are used for tents, towels, sails, canvas, etc. In recent years, there has been an upsurge in the use of flax fibres as composites in the automobile industries and in the production of thermoplastic resins.

8.5.4 Morphological description

Vegetative characters

The crop is annual and herbaceous; the root system consists of a principal taproot, which could reach a depth of 80–130 cm, from which secondary and tertiary roots are derived. The stem is erect, thin and flat, of green grizzly colour, with a height of between 30 and 120 cm; sometimes two or more tillers can originate from the basal part of the plant. The leaves are narrow, lanceolate, alternate and sessile, varying from 2 to 5 cm in length and 5 to 10 mm in width, although varieties with similar leaves do exist; the lamina is smooth and shining and has no pubescence (Fig. 8.37).

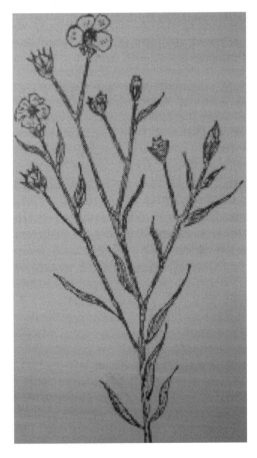

Fig. 8.37. Morphology of flax plant: leaves are narrow, lanceolate, alternate and sessile; the lamina is smooth and shining and has no pubescence; the flowers are assembled in terminal clusters (Maiti, 1997).

Floral characters

Flowers are assembled in terminal clusters. Their five sepals are oval and acuminate and have three principal veins; petals are generally free and persistent. There are five petals of blue, white or rose colour in accordance with the variety and reach senescence before the sepals. They are self-pollinated and present a lower percentage of crossing when the varieties have a long flowering period. Nevertheless, insects and winds influence the amount of crossing. The androecium consists of five stamens, with yellow or rose anthers according to the variety; in some cases, the stamens are present in alternate form, the styles in their terminal part are free, and the stigmas are aligned and nailed. The ovary is five-carpelled with axillar placentation; each carpel presents two divided locules. The fruit is a globular capsule enclosed by a persistent calyx, formed by carpels of ovoidal or spherical form according to the variety. It is of clear coffee colour, dehiscent or indehiscent with ten brilliant yellow seeds with a size of 3.5–5 mm in each capsule; the thousand seed weight ranges from 3.8 to 7 g. Their oil content varies from 30 to 45% according to the variety.

8.5.5 Anatomy

Leaf development

Flax contains a two-layered tunica and a corpus two or three layers in depth. Foliar primordia are initiated in cells of the second layer of tunica located on the flanks of the apical meristem. Apical growth of the primordium occurs by sub-apical initial, constant for only a limited time. Succeeding growth in length of the primordium comes from disperse intercalary divisions. Marginal growth, at the medial level, is established early in the ontogeny of the foliar primordium. Sub-marginal initials by alternating oblique divisions first produce two internal subsurface layers of cells. At varying distances from the leaf margin, the abaxial subsurface layer divides periclinally, thus producing a central layer. In the basal regions of a leaf the middle layer is created by the adaxial subsurface layer. All the vascular elements of small veins are the result of the subdivision of a single cell of the middle layer. Formation of large veins begins from groups of cells. Procambium differentiates in the structure of leaf traces and develops acropetally, being always in continuity with procambium associated with older leaves. Procambium is always present beneath the site of a leaf primordium when it is initiated, possibly even before. Vertical differentiation of procambium appears to become arrested below a primordium until the latter has reached a height of about 50 μm, and then pro-cambialization of the primordium begins. During the development of the primordium, procambium was never noticed any closer than 30–50μm below the subapical initials. Phloem differentiation is acropetal into growing primordia, being always in continuity with older phloem in the axis. The first sieve tube within a leaf primordium appears to be correlated with the cessation of apical growth of the primordium. Xylem is initiated at the base of a leaf, thence undergoes bidirectional differentiation. Xylem differentiation within the leaf is alternating (Girolami, 1954).

Anatomical features of the fibre cells

The flax fibre extends from cotyledonary leaves to the seed capsules at the top. In flax the fibre patches are developed in the region of protophloem of the primary vascular bundle. The fibres are derived from procambium as elements of protophloem. The sieve tubes and companion cells mature first. Later all these are crushed while other elements of the protophloem become wider and more elongated, resulting in the development of fibres. The cells of the fibre are present in the form of fibre patches around the wood, distanced from each other (Fig. 8.38). Although they are few in number; in some varieties the cells of the fibre are present in groups, but during the retting and processing the fibre strips separated in individual form. The fibre cells are of prosenchymatous origin and are presented in small or isolated groups in the pericycle, and the number of cell layers varies from one to four. Anatomical studies carried out by Maiti (1980) in India indicated that there are significant differences for diverse anatomical characteristics in certain varieties of flax, such as the thickness of the fibre patch, the preparation of wood to the region of fibre, the area of cross-section of the fibre cells, the thickness of the fibre cell wall, and the number of lamella in the cell wall.

On the basis of the distribution of the fibre cells, the overall varieties are classified into three groups:

1. Orientation of the fibre into a compact group.
2. Fibre cells in few groups or in isolated form.
3. Fibre cells isolated and distributed.

The fibre cells are cylindrical with a flat surface and pointed termination; they have a very narrow lumen and a very thick cell wall. The lumen is not distinguishable near the tip of the fibre cell. One of the important characteristics of the flax fibre cell is the presence of

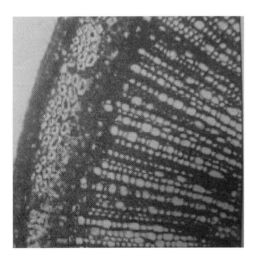

Fig. 8.38. A transverse section of flax stem showing orientation of fibre cells in the cortex (Maiti, 1997).

transverse partition and nodes in an X form (Fig. 8.39). The individual fibre cell in cross-section is polygonal or circular. The fibre cell has a length of 27 mm with a range of 9 to 70 mm and an average diameter of 23 μm.

Cellulose is the main constituent of flax fibre. Cellulose deposition in the fibres of flax hypocotyls were observed by Goubet *et al.* (1993) using $^{14}CO_2$. The glucose was incorporated into storage polysaccharides (probably starch) and used when needed for the secondary deposition of cellulose.

Lignification and other deposition in the fibre cell

The process of lignification in flax fibre has been studied in detail by a number of workers. With the aid of spectroscopic and microscopic observations, a definite relationship of lignin deposition and fibre maturation has been identified. Salnikov *et al.* (1993) studied the ultrastructure of flax development using transmission electron microscopy of ultra-thin section of plant dry material and cytochemical staining of the polysaccharides. Their findings suggest that Golgi apparatus, endoplasmic reticulum, microsatellites and plasmodesmata take an active part in the synthesis of the polysaccharide wall matrix. The orientation of cortical microtubules and microfibrils of secondary cell walls was annular. The two layers of the secondary wall differed in the distribution and content of polysaccharides. It was also observed that the inner electron-dense layer of the fibre cell wall became thinner at fibre maturation.

Enzymes such as peroxidase were actively involved in the process of fibre cell lignifications in flax fibre. Cell wall peroxidase

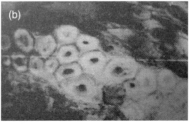

Fig. 8.39. Ultimate fibre cell of flax showing X-shaped cross-section (Maiti, 1997).

reached an optimum level around the time of maximum xylem differentiation. There was an alteration in the type of peroxidase enzymes at the time of active lignification of fibre cells. In a recent study from dissolved fibre bearing tissue from the stem of flax, it was observed that the activities of cell wall peroxidases were correlated with the fibre lignification. The onset of fibre lignification was associated with an increase in the levels of peroxidases both ionic and covalently bound to wall, but the rise was greater in the levels of covalently bound wall peroxidases. From PAGE analysis a number of cationic and anionic enzymes were identified; these enzymes may have specific roles in the lignification of flax fibres (McDougall *et al.*, 1993).

Love *et al.* (1994) suggested that the amounts of phenolic substances reported earlier are overestimated. They estimated deposition of bound phenolic substances in flax fibres by solid state NMR spectroscopy at low field (25 MH$_2$) under conditions to give quantitative responses from the aromatic carbon atoms. This method gave a better estimate for determination of phenolic structures in fibres in comparison with single-pulse excitation used earlier.

Fibre quality

The fibre is soft, lustrous and flexible, and much stronger than cotton or synthetic fibre and wool, but it not as strong as ramie. It is used in spinning factories in the same way as other fibres. It has several uses, alone or in combination with other fibres. Normally, the coarse fibres are used in the manufacture of bags and paper.

QUALITY PARAMETERS   Flax fibre is reputed among natural fibres for its high quality. Two types of fibres are mainly available in the market, the long fibre (also known as line fibre), which is traditionally extracted using dew retting, and the short or tow fibre, which is of relatively poor quality and used for various purposes. The fibre is prepared by mechanical drawing and carding, followed by enzymatic treatments to obtain a better quality.

Chemical composition of flax fibre

Chemical analysis reveals that several monosaccharides can be obtained from flax fibre on decomposition. The fibre carbohydrate contains primarily glucose (72–75%), galactose (3–4%), mannanose (3–4%) and traces of arabinose and xylose. Acid insoluble lignin content varies from 3 to 4%.

8.5.6   Effect of herbicide

In flax, a range of glyphosate treatments was applied at three stages of flowering. During normal maturation, increase in fibre cell wall thickness, lignification of the fibres and differentiation of the secondary xylem continued for 3–5 weeks after the beginning of flowering. This differentiation was halted by the application of glyphosate. In some sections epidermal and cortical cells showed the most damage, this being consistent with herbicide uptake at the stem surface. In other sections phloem and associated parenchyma cells showed the most damage. This resembled the release of fibre bundles that results from conventional post-harvest retting (Fraser *et al.*, 1982).

8.5.7   Biotic stress – anatomical changes

A study was made comparing the effects of two phloem-limited viruses, curly top and aster yellows, using flax, *L. usitatissimum* L., as the host plant.

The first degenerative changes are rapidly followed by hyperplastic divisions, resulting in the production of numerous short, abnormal, sieve elements. The hyperplastic tissue differs strikingly from normal phloem in number, arrangement and size of the sieve elements. Often the centripetal spread of symptoms occurs so rapidly those cells, which in healthy plants would become xylem, differentiate into abnormal phloem elements. Thus there is suppression of xylem differentiation and of cambial activity. Phloem fibres are lacking in the abnormal phloem. Fundamental tissues surrounding the vascular system often show signs of degeneration. The aster-yellows virus induces degenerative changes in the phloem of flax, which are basically similar to those induced by the curly-top virus. In vegetative shoots degenerative changes occur after the differentiation of the first sieve tubes, appear first in cells adjacent to the normally differentiated sieve elements, and consist of hyperplasia and increased chromaticity of cell content. Later, hyperplastic division of cells within the vascular bundles results in the formation of numerous short, peculiarly shaped sieve elements. As in curly-top infections, the vascular tissue located at approximately the level of leaf insertion undergoes the greatest amount of degeneration (Girolami, 1955).

8.5.8 Drought stress on stem anatomy

Under conditions of pronounced stress-induced plant growth retardation, fibre intrusive growth was suppressed relatively softly; their number on the stem transverse sections was reduced only by 16%. However, this determined irreversible diversity in the fibre length in various stem regions. Such insignificant suppression of intrusive growth under osmotic stress indicates the functioning of special mechanisms of its regulation (Chemikosova *et al.*, 2006).

8.6 Ramie

8.6.1 Introduction

Ramie (*B. nivea*) belongs to the family Urticaceae and is known as 'China grass'. It is one of the best vegetable fibres for textile purposes. The crop may be harvested three times per year in China. Ramie was used in the Far East around 3000 BC in the fabrication of cloth and for string in the manufacture of bows. It is also known as nettle fibre in English, Schou, Ch'U Ma or Schou-ma in Chinese, rhea in Assamese, gun in Burmese and kunkura in Bengali. The principal importing countries are Japan, Germany, France and the UK. This fibre was used by the first civilizations of India, central China and Indonesia.

8.6.2 Origin and distribution

Ramie originated from western China and was introduced to many countries of Europe, Africa, the western hemisphere and Asia. When it was taken to Europe in 1733, it was grown only as on ornamental plant in the botanical gardens of Holland and its capacity for fibre production was not known. In China, the species distribution is high, with 32 species and 11 varieties.

It is grown in many countries including China, Brazil, the Philippines, India and South Korea. China is one of the main growers of ramie in the world. One century after the introduction of ramie in Europe, it was introduced into diverse countries of Africa and America as a species of great potential for fibre, mainly in the south-east regions of the USA (Dewey, 1929).

In India, about 19 species of ramie have been reported so far, most of which are distributed in north-eastern India including Meghalaya, Assam, Arunachal Pradesh, Manipur and Sikkim. Some species have also been reported from the Western Ghats, northern parts of West Bengal and Uttaranchal. The principal fibre-producing countries of ramie are China and Japan.

8.6.3 Utilization

Ramie fibre is considered the strongest of the vegetable fibres and is used in the fabrication of cloths, yarns, curtains, tapestry and paper, either alone or in combination with other fibres. Its superior characteristics in respect to fibres of hemp, jute and cotton are greater length, durability, resistance, and resistance to humidity. It also has good lustre and appearance. Ramie is also a good source of protein in the form of flours, which could be used as a part of balanced diet for cattle.

Ramie fibre is highly recommended for use as fish nets or marine ropes. The ramie fibre is used for shoe threads, sewing machine threads, hand sewing threads, mats, and as binding threads for electrical wiring. The most important use of ramie is in clothing and furnishing fabrics, either alone or in combination with other textile fibres. It is used also in making canvas, carpet backing and fibre hoses.

8.6.4 Morphological characters

Vegetative characters

Several varieties and species of ramie grow and develop in diverse tropical and sub-tropical conditions. The plants vary in their morphological characteristics as a result of adaptation. The ramie plant does not produce ramifications; it is tall, herbaceous, and perennial (Fig. 8.40). Its root is profuse and rami-fied; it possesses rhizomes with intermediate colour, scaly leaves of light colour and buds present in the stem that can develop into new plants. The plant has several stems arising from the underground rhizomatous root-stocks. The stems could be thin or thick with a diameter of about 2.5 cm and present nodes. Its height at maturity could be 2 m or more depending on environmental conditions. It is soft, flat without ramifications, and it also presents inconspicuous pubescence. The leaves are green, with a white glistening cover on the abaxial surface due to the presence of profuse pubescences. Leaves are alternate, and very acuminate, with indentations in their margins. They have a long petiole with a variable length of between 5 to 20 cm.

Fig. 8.40. Morphology of ramie plant: the leaves are green, profusely pubescent, alternate, and very acuminate; the flowers are borne on axillary panicles.

Floral characters

The inflorescence is unisexual and mono-ecious in the form of axillary panicles. The panicles appear in groups and are shorter than the petioles. The staminate flowers are present in the lower part of the stem, or well under the pistillate flowers, tubular with one, two or four persistent sepals. The calyx sur-rounds the fruits. There is only one ovary and it possesses one ovule. The fruit is long with thin pubescence. The fruit is an achene, con-stituted by one seed covered with a dry calyx. The seeds are small, coffee coloured and oval.

8.6.5 Fibre anatomy

The fibre cells of ramie are derived from procambium by cell division, followed by pronounced cell elongation in the axis of the plant (Kundu, 1954; Maiti, 1980). In transverse section of the stems, the fibre cells of ramie are arranged in a few layers in the bark of the stem (Fig. 8.41). The indi-vidual fibre cells are elliptical or smooth, oval, isolated or in groups in the cortical region around the stem; the cell possesses a wide lumen and a large cross-sectional area. The anatomical characteristics show varia-tions in different varieties (Maiti and Ghosh, 1974). There exist significant differences in thickness of fibre patches around the wood, the number of cells per unit area, number of layers of fibre cells, diameter of fibre cells, thickness of fibre cell walls, and number of stratifications in fibre cell walls. The number of cells per unit of area probably has good correlation with the fibre yield and could be considered as a criterion for selection in order to increase the fibre yield in ramie (Maiti, 1980).

As for the length of ultimate fibre, ramie possesses the longest filament with respect to other fibres obtained from stems; the length varies from 29 to 280 mm, the width is from 50 to 100 mm, and the proportion of length to width of the fibre cell is 1:3000, approxi-mately (Maiti, 1970, 1980). The fibre cells are cylindrical with rounded or triangular tips; the cell wall is very thick, with longitudinal striations and structures like nodes along their surface (Fig. 8.42). The individual fibre

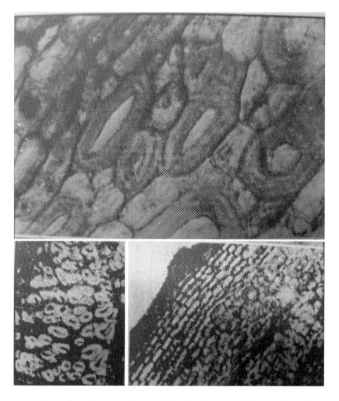

Fig. 8.41. Transverse sections of ramie stem showing the distribution of fibre cells in the cortex (Maiti 1997; World Fibre Crops).

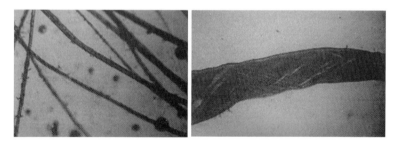

Fig. 8.42. Ultimate fibre cell of ramie (Maiti, 1997).

cells are united at the ends with gums, wax and pectins.

A great deal of information is available on the helical structures, microfibrillar orientation and stratifications in the cell wall of the fibre. For example, there are three stratifications (S_1, S_2, and S_3) forming two helical layers in the cell wall of the fibre, the external layer in the form of the S and the internal one in the form of a Z, with the possibility of the existence of a third cellular covering. Several researchers have studied the structure of the fibre of ramie by means of X-rays and electron micrograph.

Technique for studying secondary wall

A technique was developed by Maiti and Ghosh (1974) and Maiti (1980) in India to study the lamellar orientation of the secondary wall of the ramie fibre. Two or three drops of H_2SO_4 at 70% are added to a transverse section of stem in a Petri dish at 40°C for 5 min. Subsequently, the solution is removed and a few drops of chlorozinc-iodine are added to the sample, and the sample is covered with a cover slip for observation under a microscope. With this treatment, the lamellar structures of ramie fibre cells are visible. On the basis of this study, it was concluded that there are differences among the varieties of ramie with respect to the lamellar orientation in the secondary wall, which is probably related to the strength, rigidity and elasticity. Also, the area of cross-section of the ramie fibre cell indicates the fineness of the fibre in the same study. These authors found that there are large differences in the size of cross-section of fibres among varieties of ramie. The fibre cells of ramie show a spiral structure in the cell wall (Maiti, 1980).

8.6.6 Chemical properties of fibre

Chemically, the fibre consists of cellulose and hemicellulose with very little amount of lignin. Most of the cellulosic material of the fibre is holocellulose. Harvested fibre has a very high gum content, which comprises about 20% of the fibre.

8.6.7 Effect of growth regulators

The effects of two mixtures of four plant growth regulators (choline chloride, gibberellin (GA_3), benzyladenine (6-BA) and $NaHSO_3$) at 20:9:5:800 mg/kg (H1) and 20:42:43:2350 mg/kg (H3) (active ingredients), respectively, were investigated on yield and fibre quality in ramie (*B. nivea* (L.) Gaud.). The mixtures were sprayed over the canopy at two growth stages (10 and 20 days after the previous cut) of field-grown ramie. The treatments increased raw fibre yield by 13–18%, and improved fibre fineness by 57–349 m/g, increased the number of leaves per plant and also improved all yield components. Treatment H1 resulted in a denser distribution, smaller diameters and greater quantity of fibre cells in stem cross-section. Physiological responses included improving leaf water status, increasing net photosynthetic rate and decreasing electrolyte ex-osmosis rate (Liu *et al.*, 2001).

8.6.8 Drought – anatomical changes

Drought-tolerant cultivars of ramie had more fine hairs on abaxial leaf surfaces, produced thicker leaf cuticles, better preserved leaf shape and erectness, and lost fewer leaves. They also generated longer roots with larger root masses and more storage organs, preserved higher root-to-shoot ratios, produced larger diameter stem vessels, and better conserved cell integrity than drought-sensitive cultivars of ramie when plants were grown under drought stress. Fibre yield was better in drought-tolerant cultivars of ramie, because these cultivars had adapted root systems, leaf responses, cellular responses and biochemical activities to allow plants to continue higher levels of photosynthesis and carbon deposition under more stressful environments than the less drought-tolerant cultivars (Liu *et al.*, 2005).

Under drought stress conditions there was a decrease in yield, biomass and agronomic characteristics, i.e. plant height, stem diameter and ribbon thickness. Under osmosis and drought stress conditions, the ramie genotypes resistant to drought had a lower withering rate compared to that of genotypes of lower drought resistance. Under drought stress conditions, drought resistant genotypes had a greater leaf thickness and volume of root system, and a higher dry matter weight of underground parts (Cheng, 2000).

8.7 Hemp

8.7.1 Introduction

Hemp (*Cannabinus sativa*) is used mainly for fibre production from the stem in temperate countries, although it is also used for medicinal purposes. It has been used since the first Chinese civilization. At present, hemp is exploited in temperate climates for fibre and seed oil. As a medicinal plant, it proportionates resin with high stimulating power or drugs known by the name of 'hassis', whose main hallucinogenic narcotic power constituent is 9-tetra-hydro-cannabinol (THC). Generally, the habitat of narcotic species (*C. indica*) is in the tropical regions. In a temperate climate, hemp does not have a narcotic property. In several countries such as India and Mexico, cultivation of the crop is restricted due to its narcotic properties and its profusely branched habit, which makes it unsuitable for fibre production.

8.7.2   Distribution

Hemp is mainly produced in the former USSR, Europe and Chile. It also grows in large areas of the USA as a single crop. About 75% of the worldwide production is located in the former USSR, Italy, Holland, Hungary and Rumania. In the USA, hemp fibre is produced in Kentucky and other central states.

Hemp belongs to the family Urticaceae and is located in the Cannabinaceae tribe, whose principal producer of fibre is *C. sativa* L. The plants used for fibre have long stems and few ramifications, but the plants with medicinal purposes have many ramifications. Several varieties that grow in the warm climates of India and Syria are restricted because of their narcotic properties under semi-arid conditions.

8.7.3   Utilization

Hemp fibres are used in the manufacture of strings, twines, carpets and canvases as well as paper. Oil content from seeds is used in the fabrication of soap and paint. Hemp is suitable for making fine quality paper for books and cigarettes. Hemp produces narcotics in tropical climates, for which reason it is not allowed to be grown in countries with this climate. However, illegal cultivation has been carried out for narcotic purposes in countries such as Mexico and India.

8.7.4   Morphological description

Vegetative characters

Hemp is an erect annual plant that reaches a height of 1.5–4.5 m. Under normal conditions of growth, it presents two forms, flowers and vegetative parts. Under adverse conditions, the plant presents two sexes. The taproot of hemp can reach a great depth under adequate conditions of humidity and soil, but if the prevalent conditions are not ideal, the root system is affected. The stem is thin, erect and hollow, and produces few branches in the apical part of the plant; the diameter varies from 4 to 20 mm in accordance with the climatic conditions and the variety. The leaves are compound and palmate and each leaf has from seven to ten leaflets (Fig. 8.43).

Floral characters

The flowers are dioecious, although they sometimes appear monoecious; the inflorescence occurs first in the stamens, but the pistillate ones remain until the seeds mature. The staminate flowers in the plant appear in a cluster, whereas the pistillate ones are located in the form of spikelets in the spikes in the axils of the leaves. The fruit is classified as an achene; it does not present albumen and has one seed. At maturity, the pericarp breaks and separates from the testa of the seed.

8.7.5   Fibre anatomy

The fibre cells are distributed in the bark of the stem, as in flax and ramie. The fibre cell

Fig. 8.43. Hemp plant morphology: thin, erect and with a hollow stem; leaves palmately compound and seven to ten leaflets; staminate flowers in a cluster and pistillate flowers located in the form of spikelets (Maiti, 1997).

has an average breadth of 22 mm and length of nearly 25 mm. The lumen is broader than in flax and is frequently straight and lineal; at the apex, it is very narrow. The fibre cells are rounded and present projections; the cell wall is very thick. Also, the fibre cell presents some nodulations and sub-terminal ramifications, which do not occur in flax.

In Cairo, comparative morphological and anatomical studies of male and female plants of *C. sativa* were undertaken at different stages in their life cycle. No significant difference was observed in cannabinoid content between sexes; the highest concentrations were found in young stems and leaves, and the concentrations decreased gradually with age.

The dermal sheath of glandular trichomes of *C. sativa* was studied by transmission microscopy. Prior to the secretory activity, cuticle thickened selectively on the outer wall of disc cells of each trichome, whereas thickening was less evident on the dermal cells of the bract. Membraned secretory vesicles that differ in size and appearance were the source of precursor for cuticle synthesis. Vesicle contents were released following the degeneration of the vesicle membrane upon contact with the subcuticular wall and finally contributed to both structural and amorphous phases of cuticle development. The structural phase was evident by deposit and thickening of cuticle at the subcuticular wall–cuticle interface to form a thickened cuticle. In the amorphous phase precursors permeated the cuticle in a liquid phase, as evidenced by the fusion of cuticle and wax layers between contiguous glands.

The development of the secretory activity and the formation of the subcuticular wall of glandular trichomes in *C. sativa* were examined by transmission microscopy. The secretory cavity originated at the wall–cuticle interface in the peripheral wall of the discoid secretory cells. During the pre-secretory phase in the development of the glandular trichome, the peripheral wall of the disc cells became laminated into a dense inner zone adjacent to the plasma membrane and a less dense outer one adjacent to the cuticle. Loosening of wall matrix

in the outer zone initiated a secretory cavity among fibrous wall materials. Hyaline areas accumulated conspicuous electron-dense contents that were released into the secretory cavity, thereby forming rounded secretory vesicles. Fibrous wall material secreted from the surface of the disc cells became distributed throughout the secretory cavities among the numerous secretory vesicles.

8.7.6   Fibre quality

The fibre colour is white or creamy to coffee colour or brown. However, when it is extracted by machinery, it is grey in colour. The fibre length varies from 1.2 to 2.1 m on average and can be used for spinning and fabrics. Fibre quality is measured in terms of force, fineness, uniformity, colour, lustre, length and strength.

The quality of hemp stems as a raw material for paper was assessed at Wageningen, the Netherlands. Bark content decreased during the growing season, and at harvest in September it ranged from 30 to 35% depending on cultivar and plant density. The proportion of secondary bast fibre in the bast fibre fraction increased with stem weight from 10 to 45%. Differences in chemical composition within the sets of bark or core samples were small compared with the difference between bark and core. The bark of a French cultivar contained less cellulose than that of the Hungarian cultivars. Bark quality for paper improved during the growing season, because the cellulose content in the bark increased while the content of lignin and extractives decreased.

8.8   Sunnhemp

8.8.1   Introduction

Sunnhemp (*Crotalaria juncea*) is a fibre crop of Indian and Asiatic origin. It is next to cotton, jute and kenaf in importance. The demand for sunnhemp is increasing in the manufacture of specialized tissue paper and currency notes. The fibre has more tensile strength and durability under exposure to humidity than jute. Sunnhemp is a promising raw material for pulp production in the USA. Little or no nitrogen fertilizer is required to be applied, it is drought resistant and grows on poor soil. The stalks dry out faster than other species after the killing frost. It has a lower yield than kenaf, and the stems are fragile. It is a short-day plant, which restricts the period during which it can be grown. It is cross-pollinated, and seed production is dependent on bees. Sunnhemp belongs to the family Fabaceae or bean or pea family.

8.8.2 Distribution

C. juncea is cultivated in practically all the states of India with the exception of Jammu and Kashmir, Assam and Manipur. The major plantations are in Uttar Pradesh. It is also cultivated in Sri Lanka and southern Asia, mainly in the tropical countries. In India, several varieties exist, and germ plasm is available in the Jute Agricultural Institute, Barrackpore, and the Indian Council of Agricultural Research, New Delhi. The variety 'K-12' includes black and yellow seeds. The plant is grown to a limited extent in other tropical countries.

8.8.3 Utilization

It is used in the making of ropes, strings, twines, high-grade paper, sacks, mattings, tarpaulin, rugs, carpets, insulation, fire-hoses, and soles for shoes, sandals and marine ropes. In the USA it is used in the manufacture of cigarette paper, silk and cordage. The crop can also be used to improve soil in rotation with cereal crops.

8.8.4 Morphological description

It does not show big differences in the morphological characteristics among the several varieties, only in colour and size of seed as well as in the extent of ramification of the stems.

Vegetative characters

The plant is annual, erect, with cylindrical stems and a height of 1.2 to 3 m. The taproot is long and presents great ramification with enough nodules for the fixation of N through bacteria. The stem is extremely green. The leaves are simple, narrow, sessile and lanceolate and covered with tiny pubescences.

Floral characters

The inflorescence is a cluster of yellow flowers; the stamens are united in circles; the anthers are short, versatile, and basifixed. Sunnhemp is a short-day plant, although a day-neutral plant has been reported by Mohan (1974). The plants are cross-pollinated. In India, the flowers open between 2 and 3 p.m. The anthers dehisce between noon and 1 p.m. Sunnhemp is an autogamous crop, as has been verified by covering the crop with bags, through mechanical pollination. The fruit is cylindrical, with 10–16 seeds (Kundu, 1964). The testa of the seeds could be yellow or black.

8.8.5 Fibre anatomy

The bundle of fibre develops in the protophloem of the primary vascular cylinder. The fibres are derived from the procambium as an element of the protophloem. First, the components of the protophloem are lost, while the rest broaden and elongate to give rise to the growth of the fibre. Although most of the fibre is of protophloeic origin, some originates in the vase of the stem due to the secondary activity of the cambium.

In transverse section of a stem, the fibre cells are seen in groups of bundles in the region of the pericycle and they are present in the form of a half moon.

The ultimate fibre cell is long with rounded termination and lumen very enlarged until the terminal part. The fibre cell wall is thick; its surface is irregular because of the presence of transverse and longitudinal depressions. The length of the ultimate fibre varies from 1466 to 9600 µm, the width is 22 to 43 µm, and the relationship of length to breadth of the fibre cell (L/B) is 77, and the width of the lumen is 16.6 µm.

8.8.6 Quality

Sunnhemp fibre is of better quality than jute fibre, since it has good lustre, higher tensile strength and better resistance to environmental exposure. It is coarse, flattened and dark grey in colour. It cannot be woven in the same machinery as jute, because the fibre is too thick (Chaudhuri, 1950; Maiti, 1980).

References

Ali, M.A. (1989) Ultimate fibres in tossa jute (*Chorchorus olitorius* L.) and their utilization in selection. *Proceedings of Indian Academy Of Science, Plant Science* 99(1), 7–41.

Ali, M.A. (1993) Study of some anatomical component characters in white jute (*Corchorus capsularis* L.) and their utilization in selection [in Bangladesh]. *Bangladesh Journal of Jute and Fibre Research* 18(1–2), 13–20.

Bergfjord, C. and Holst, B. (2010) A procedure for identifying textile bast fibres using microscopy: flax, nettle/ramie, hemp and jute. *Ultramicroscopy* 110(9), 1192–1197.

Bolek, Y., Fidan, M.S. and Oglakci, M. (2010) Distribution of gossypol glands on cotton (*Gossypium hirsutum* L.) genotypes. *Notulae Botanicae Horti Agrobotanici Cluj-Napoca* 38(1), 81–87.

Bornman, C.H., Addicott, F.T., Lyon, J.L. and Smith, O.E. (1968) Anatomy of gibberellin-induced stem abscission in cotton. *American Journal of Botany* 55(3), 369–375.

Bourland, F.M., Hornbeck, J.M., McFall, A.B. and Calhoun, S.D. (2003) A rating system for leaf pubescence of cotton. *Journal of Cotton Science* 7, 8–15.

Chemikosova, S.B., Pavlencheva, N.V., Gur'yanov, O.P. and Gorshkova, T.A. (2006) The effect of soil drought on the phloem fibre development in long-fibre flax. *Russian Journal of Plant Physiology* 53(5), 656–662.

Chen, B.Z. (1991) Fibre development of different varieties in *Corchorus capsularis*. *Journal of the Fujian Agricultural College* 20(4), 378–384.

Chen, S.H., Lu, H.R. and Zheng, Y.Y. (1990) The genetic relationship between anatomical characters and fibre yield and quality in jute. *Journal of Fujian Agricultural College* 20(4), 378–384.

Cheng, J.Y. (2000) Morphological, anatomical and agronomical characteristics in different genotypes of ramie (*Boehmeria nivea*) with drought resistance. *China's Fibre Crops* 22(4), 14–22.

Cosgrove, D.J. (1997) Relaxation in a high-stress environment: the molecular basis of extensible cell walls and cell enlargement. *Plant Cell* 9, 1031–1041.

Dewey, L.H. (1929) Ramie a fiber yielding plant. USDA. Miscellaneous Circular, 110.

Flint, E.A. (1950) The structure and development of the cotton fibre. *Biological Reviews* 25, 414–434.

Fraser, T.W., Courtney, A.D. and Harvey, B.M.R. (1982) Pre-harvest retting of flax: a light microscope study of the effects of glyphosate treatment on the maturation of stem tissues. *Annals of Applied Biology* 101, 533–537.

Girolami, G. (1954) Leaf histogenesis in *Linum usitatissimum*. *American Journal of Botany* 41(3), 264–273.

Girolami, G. (1955) Comparative anatomical effects of the curly-top and aster-yellows viruses on the flax plant. *Botanical Gazette* 116(4), 305–322.

Gore, U.R. and Taubenhaus, J.J. (1931) Anatomy of normal and acid-injured cotton roots. *Botanical Gazette* 92(4), 436–441.

Goubet, F., Matini, F., Thoiron, B. and Morvan, C. (1993) Incorporation of D-(U-14C) glucose and 14 Co2 into the cell wall polymers of flax hypocotyls, in the course of fiber differentiation. *Plant and Cell Physiology* 34(6), 841–848.

Hao, J.M., Lu, H.Y., Zheng, Y.Y., Chen, F.Q., Wang, Y.J. and Shen, Y.B. (1993) Study on inter-relationship among three components of fibre cell structure and fiber quality in jute. *China's Fiber Crops* 1, 18–21.

Jahan, M., Al-Maruf, A. and Quaiyyum, M.A. (2007) Comparative studies of pulping of jute fibre, jute cutting and jute caddis. *Bangladesh Journal of Scientific and Industrial Research* 42(4), 425–434.

Jasdanwala, R.T., Singh, Y.D. and Chinoy, J.J. (1977) Auxin metabolism in developing cotton hairs. *Journal of Experimental Botany* 28, 1111–1116.

Kakani, V.G., Reddy, K.R., Zhao, D. and Mohammed, A.R. (2003) Effects of ultraviolet-b radiation on cotton (*Gossypium hirsutum* L.) morphology and anatomy. *Annals of Botany* 91(7), 817–826.

Kannan, S., Ravikesavan, R. and Kumar, M. (2006) Leaf trichome density – an indicator of jassid tolerance in cotton. Conference paper; Plant breeding in post genomics era. *Proceedings of Second National Plant Breeding Congress, Coimbatore, India* Vols 1–3: 137–144.

Kundu, B.C. (1954) Origin, development and structure of important vegetable fibres. Presidential address, Section of Botany. *Proceedings of the 41st Indian Science Congress, Part II.*

Kundu, B.C. (1956) Jute – World's foremost bast fibre. Botany, agronomy, disease and pests. *Economic Botany* 10, 103–133.

Kundu, B.C. (1964) Sunnhemp in India. *Proceedings of Soil and Crop Science Society of Florida* 24, 396–403.

Kundu, B.C. and Rao, N.S. (1954) Origin and development of axillary buds in jute (*Corchorus capsularis*). *Annals of Botany* 18(3), 368–375.

Li, L., Wang, S.Y. and Yu, J.Y. (2002) Mechanical properties of commingled yarn composites. *Indian Journal of Fibre and Textile Research* 27, 287–289.

Liu, F.H., Liang, X.N., Zhang, N.G., Huang, Y.S. and Zhang, S.W. (2001) Effect of growth regulators on yield and fibre quality in ramie (*Boehmeria nivea* (L.) Gaud.), China grass. *Field Crops Research* 69(1), 41–46.

Liu, F.H., Liu, Q.Y., Liang, X.N., Huang, H.Q. and Zhang, S.W. (2005) Morphological, anatomical, and physiological assessment of ramie [*Boehmeria Nivea* (L.) Gaud.] tolerance to soil drought. *Genetic Resources and Crop Evolution* 52(5), 497–506.

Love, G.D., Snape, C.E., Jarvis, M.C. and Morrison, I.M. (1994) Determination of phenolic structures in flax fibre by solid-state 13 C NMR. *Phytochemistry* 35(2), 489–491.

Mace, M.E. and Howell, C.R. (1974) Histochemistry and identification of condensed tannin precursors in roots of cotton seedlings. *Canadian Journal of Botany* 52(11), 2423–2426.

Maiti, R.K. (1970) Fibre microscopy for the study of performance of fibre crops in different fields of research. *Bulletin of Botanical Society of Bengal* 24(104), 37–44.

Maiti, R.K. (1971a) Histo-morphological studies of some long fibre crops in relation to yield and quality. DSc thesis, Calcutta University, West Bengal, India.

Maiti, R.K. (1971b) Study on growth pattern in mesta (*Hibiscus cannabinus* L.) in relation to environmental conditions. DSc thesis, Calcutta University, India.

Maiti, R.K. (1973) Relationship between cross-sectional area and perimeter with the number of cells in the fibre bundles of jute and mesta. *Journal of the Textile Association* 34(4), 202–203.

Maiti, R.K. (1980) *Plant Fibres*. Bishen Singh Mahendra Pal Sing, Dehra Dun, India, 418 pp.

Maiti, R.K. (1997) *World Fibre Crops*. Science Publishers, Lebanon, USA and Oxford & IBH Co Pvt. Ltd, New Delhi, India, 352 pp.

Maiti, R.K. and Asima, L.M. (1974) Quantitative study of cambial function in jute. *Jute Bulletin* 37(7/8) Oct–Nov.

Maiti, R.K. and Ghosh, K.L. (1974) Comparative microscopy of fibre strands of some ramie varieties (*Boehmeria nivea* Gaud.) with special reference to its relation to yield and quality. *Jute Bulletin* 37(5–6).

Majumdar, S. (2002a) Prediction of fibre quality from anatomical studies of jute stem: Part I – Prediction of fineness. *Indian Journal of Fibre and Textile Research* 27, 248–253.

Majumdar, S. (2002b) Prediction of fibre quality from anatomical studies of jute stem: Part II – Prediction of strength. *Indian Journal of Fibre and Textile Research* 27, 254–258.

Marx-Figini, M. (1982) The control of molecular weight and molecular weight distribution in the biogenesis of cellulose. In: Brown, R.M. Jr (ed.) *Cellulose and Other Natural Polymers*. Plenum, New York, pp. 243–271.

McDougall, G., Milium, S. and Davidson, D. (1993) Alterations in surface associated peroxidases in vitro root development of explants of *Linum usitatissimum*. *Plant Cell, Tissue and Organ Culture* 32(1), 101–107.

McMichael, B.L., Burke, J.J., Berlin, J.D., Hatfield, J.L. and Quisenberry, J.E. (1985) Root vascular bundle arrangements among cotton strains and cultivars. *Environmental and Experimental Botany* 25(1), 23–30.

Meinert, M.C. and Delmer, D.P. (1977) Changes in biochemical composition of the cell wall of the cotton fibre during development. *Plant Physiology* 59, 1088–1097.

Mitra, G.C. (1984) Genetic control of secondary phloic fibres in jute (*Corchorus capsularis* L.). *Genetica* 63(1), 9–11.

Mohan, K.V.J. (1974) A day neutral variety in sunnhemp. *Jute Bulletin* 36, 11–12.

Oosterhuis, D.M. and Wullschleger, S.D. (1987) Water flow through cotton roots in relation to xylem anatomy. *Journal of Experimental Botany* 38(11), 1866–1874.

Ramalho, F.S., Parrott, W.L., Jenkins, J.N. and McCarty, J.C. (1984) Effects of cotton leaf trichomes on the mobility of newly hatched tobacco budworms (Lepidoptera: Noctuidae). *Journal of Economic Entomology* 77(3), 619–621.

Reeves, R.G. and Valle, C.C. (1932) Anatomy and microchemistry of the cotton seed. *Botanical Gazette* 93(3), 259–277.

Ruan, Y.L. (2005) Recent advances in understanding cotton fibre and seed development. *Seed Science Research* 15(4), 269–280.

Salnikov, V.V., Agreea, M.V., Yumashev, V.N. and Lozovaya, V.V. (1993) The ultrastructure of bast fibres. *Soviet Plant Physiology* 40(3), 399–404.

Seagull, R.W. (1995) Cotton fibre growth and development: evidence for tip synthesis and intercalary growth in young fibres. *Plant Physiology (Life Sci. Adv.)* 14, 27–38.

Seagull, R.W., Grimson, M.J., Muehring, T.C. and Haigler, C.H. (1998) Analysis of cotton fibre ultrastructure during secondary wall deposition using ultra-rapid freezing and freeze substitution. In: *Proceedings of the Beltwide Cotton Conference*, San Diego, California, pp. 5–9.

Seagull, R.W., Oliveri, V., Murphy, K., Binder, A. and Kothari, S. (2000) Cotton fibre growth and development. Changes in cell diameter and wall birefringence. *Journal of Cotton Science* 4, 97–104.

Sengupta, G. and Palit, P. (2004) Characterization of a lignified secondary phloem fibre-deficient mutant of jute (*Corchorus capsularis*). *Annals of Botany* 93, 211–220.

Shaikh, M.A.Q., Ahmed, Z.U., Khan, A.I. and Majid, M.A. (1980) An anatomical screening approach to selection of high yielding mutants of jute (*Corchorus capsularis* L.). *Environmental and Experimental Botany* 20(3), 287–296.

Snider, J. (2010) Effects of high temperature stress on the anatomy and biochemistry of pollen-pistil interactions in cotton. Doctoral thesis/dissertation, 188 pp.

Steer, M.W. and Steer, J.M. (1989) Pollen tube tip growth. *New Phytologist* 111, 323–358.

Stewart, J.M. (1975) Fibre initiation on the cotton ovule (*G. hirsutum*). *American Journal of Botany* 62, 723–730.

Thanki, Y.J., Garasia, K.K. and Shah, K. (2000) Studies on the nodal anatomy of the seedlings of some Tiliaceae. *Journal of Phytological Research* 13(2), 179–181.

Tiwari, S.C. and Wilkins, T.A. (1995) Cotton (*Gossypium hirsutum*) seed trichomes expand via diffuse growing mechanism. *Canadian Journal of Botany* 73, 746–757.

Vavilov, N.I. (1951) The origin, variation, immunity and breeding of cultivated plants. *Chron. Bot.* 13, 21–26.

Wise, R.R., Sassenrath-Cole, G.F. and Percy, R.G. (2000) A comparison of leaf anatomy in field-grown *Gossypium hirsutum* and *G. barbadense*. *Annals of Botany* 86(4), 731–738.

Wullschleger, S.D. and Oosterhuis, D.M. (1987) Electron microscope study of cuticular abrasion on cotton leaves in relation to water potential measurements. *Journal of Experimental Botany* 38(4), 660–667.

9

Vegetable Crops

Vegetables form our daily dietary supplements as they are a rich source of vitamins, minerals and fibres. Most vegetables are used for nutrition but some even have therapeutic properties used in medicine. There are diverse forms of vegetables in nature, some commonly found in all geographical regions but a few restricted to specific regions. Such species, termed endemic species, have a small range of adaptability, hence are found growing in small pockets of climatic regions. Thus vegetable crops develop different adaptations, either morphologically and anatomically in order to survive under different habitat conditions. Each crop variety has its own distinct range of adaptations that give it a characteristic identity.

Large variability exists among vegetable crop species in their characters. Morphological and anatomical characters differ from one species to another. We can identify the species based on these characters. Anatomy gives the details of internal structure and functions of different plant parts and how they are adapted to different environmental conditions.

9.1 Tomato

9.1.1 Introduction

The tomato (*Lycopersicon esculentum*) is an herbaceous and profusely branched annual herb widely cultivated for its edible fruit. Savoury in flavour; the fruit of most varieties ripens to a distinctive red colour. It belongs to family Solanaceae. Diploid chromosome number is 24 ($2n = 2x = 24$).

9.1.2 Origin

The tomato is native to South America. Genetic evidence shows the progenitors of tomatoes were herbaceous green plants with small green fruit and a centre of diversity in the highlands of Peru. One species, *Solanum lycopersicum*, was transported to Mexico where it was grown and consumed by Mesoamerican civilizations. The exact date of domestication is not known. The first domesticated tomato may have been a little yellow fruit, similar in size to a cherry tomato, grown by the Aztecs of central Mexico. Aztec writings mention tomatoes were prepared with peppers, corn and salt. The word 'tomato' comes from the Aztec (Nahuatl) *tomatl*, literally 'the swelling fruit'.

9.1.3 Utilization

Tomato is frequently used as a vegetable for curries all over the world. Tomatoes are now eaten freely throughout the world, and

©R. Maiti, P. Satya, D. Rajkumar and A. Ramaswamy 2012. *Crop Plant Anatomy*
(R. Maiti *et al.*)

their consumption is believed to benefit the heart among other things. They contain lycopene, one of the most powerful natural antioxidants. In some studies lycopene, especially in cooked tomatoes, has been found to help prevent prostate cancer but other research contradicts this claim. Lycopene has also been shown to improve the skin's ability to protect against harmful UV rays. Tomato varieties are available with double the normal vitamin C, 40 times normal vitamin A, high levels of anthocyanin and two to four times the normal amount of lycopene (numerous available cultivars with the high crimson gene).

Tomato consumption has been associated with decreased risk of breast cancer, head and neck cancers and might be strongly protective against neurodegenerative diseases.

Tomatoes are used extensively in Mediterranean cuisine, especially Italian and Middle Eastern cuisines. The tomato is acidic; this acidity makes tomatoes especially easy to preserve in home canning whole, in pieces, as tomato sauce, or paste. Tomato juice is often canned and sold as a beverage; unripe green tomatoes can also be breaded and fried, used to make salsa, or pickled. The fruit is also preserved by drying, often by the sun, and sold either in bags or in jars in oil.

9.1.4 Morphological description

Vegetative characters

The plant is a mesophytic herb, with a profusely branched taproot system. The stem is aerial, erect, branched, and pubescent and greenish. Leaves are pinnately lobed, ovate with a wavy margin and rounded apex, stipulate, petiolate (small), alternate and pubescent with reticulate venation (Fig. 9.1).

Floral characters

The inflorescence is auxiliary with a scorpioid cyme. Flowers are small and yellow in colour, bracteate, pedicellate, complete, bisexual, pentamerous and hypogynous. There are five gamosepalous, persistent sepals and five gamopetalous petals. The

Fig. 9.1. Tomato plant showing pinnately lobed, ovate, wavy margin, pubescent leaves and small yellow flowers.

male part of the flower has five epipetalous stamens, and the anthers are dithecous and basifixed. The gynoecium is a bicarpellary, syncarpus superior ovary with axile placentation. The berry retains a marcescent calyx. The seed contains small hairs, is light weight, brown in colour, ovate to kidney shaped with an average size of 2–3 mm.

9.1.5 Anatomical description

Root anatomy

Transverse section of the root shows three main regions, the epidermis, cortex and stele (Fig. 9.2).

The epidermis is uniseriate, having a single layer of compactly arranged thin-walled living cells. The epidermis provides protection and absorption of water and minerals. Epidermal cells produce unicellular root hairs, which are useful in increasing the surface area for absorption of water and help in obtaining water from soil.

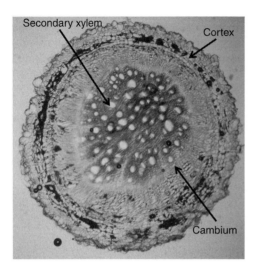

Fig. 9.2. Transverse section of root showing secondary xylem and compressed cortex due to secondary growth achieved by the division of cambium in the stele.

The cortex is multiseriate and relatively homogeneous and consists of general cortex and endodermis.

The general cortex consists of loosely arranged thin-walled parenchyma cells with prominent intercellular spaces. The cells are colourless and store starch and are usually round or oval.

The endodermis is single layered with compactly arranged barrel-shaped cells. Endodermal cells are characterized by the presences of Casparian strips. The endodermis in the root acts as a barrier between the cortex and stele.

The stele forms the central part of the root and consists of pericycle, vascular strands and conjunctive tissue.

The pericycle is the outermost region of the stele, and is uniseriate and parenchymatous. Cells of the pericycle retain meristematic activity. Lateral roots arise endogenously from the pericycle.

Vascular strands are radial and the xylem is exarch, closed type. Secondary xylem vessels are regularly arranged.

The parenchymatous conjunctive tissue occurs between the xylem and phloem strands. Pith is absent.

When the epidermis ruptures, the cortex undergoes compression and primary structures become disturbed due to pressure created by the secondary tissues by the action of cambial rings (secondary growth) formed in the stele (Fig. 9.2).

Stem anatomy

The primary stem shows three regions, the epidermis, cortex and stele (Fig. 9.3).

The epidermis is the outermost region surrounding the stem. It is uniseriate, having a single layer of compactly arranged barrel-shaped living cells. The outer walls of the epidermal cells are cutinized. Cuticle reduces the transpiration. Only a few stomata are present in the epidermis for exchange of gases and transpiration. All the epidermal cells are colourless except the guard cells. Multicellular hairs are present on the epidermis.

The cortex is the middle region present between the epidermis and stele, and is smaller than the stele. It shows hypodermis, middle cortex and endodermis. Hypodermis forms the outermost region of the cortex and is multilayered (two to six) and collenchymatic, which gives considerable strength, flexibility and elasticity to young stems. Having chloroplasts, it may carry out photosynthesis. The middle cortex is multilayered and parenchymatic. The cells may be round or oval with prominent intercellular spaces. This region also stores food materials temporarily. Endodermis forms the innermost layer of the cortex, and is usually distinct in stems without Casparian bands.

The stele is the central region of the stem and is thicker than the cortex. It consists of pericycle, vascular bundles, pith rays and pith. Pericycle is the outermost non-vascular region of the stele and is multiseriate and sclerenchymatic. Xylem and phloem are organized into vascular bundles and 15–20 vascular bundles are arranged in one ring called the eustele. Each vascular bundle is wedge-shaped, conjoint, collateral, open and endarch. The vascular bundles are separated from each other by radial rows of parenchyma cells known as pith rays. The pith ray cells are usually elongated in a radial direction and serve primarily for the

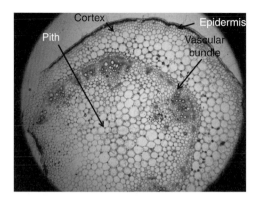

Fig. 9.3. Transverse section of stem showing primary structure consisting of epidermis, cortex, vascular bundle and prominent pith at the centre of the stem.

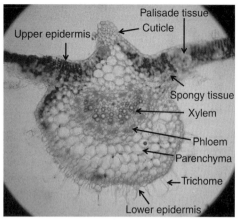

Fig. 9.4. Transverse section of leaf (100×) showing various parts. Note: palisade tissue is not very compactly arranged and trichomes are observed in the lower epidermis.

conduction of food and water radially in the stem and for the storage of food. The centre of the stem is composed of thin-walled parenchyma cells and known as pith. The cells have distinct intercellular spaces.

Leaf anatomy

The leaf shows a dorsi-ventral nature, i.e. the leaf blade consists of distinct dorsal and ventral surfaces. In transverse section, the leaf shows three distinct regions, the epidermis, mesophyll and vascular system (Fig. 9.4).

Epidermis is present on both surfaces of the leaf blade, i.e. upper epidermis and lower epidermis, and is mainly meant for protection. The epidermis is single layered with compactly arranged barrel-shaped (tabular) cells. The outer walls of the epidermal cells are cutinized and the epidermis on its outer surface is covered by a continuous layer of cutin called cuticle. The cutinized outer walls and cuticle reduce the loss of water due to transpiration. Anisocytic type of stomata are present in both the epidermal layers but are usually more frequent in the lower epidermis. The stomatal index of upper epidermis is 13.4 and lower epidermis is 22.5. Stomata facilitate the exchange of gases between the leaf and environment. All the epidermal cells except guard cells are colourless. Epidermis contains unicellular glandular trichomes; glandular hairs provide resistance to insects.

The ground tissue of the leaf, present between the upper and the lower epidermal layers, is called mesophyll. The mesophyll is chlorenchymatic and is concerned mainly with photosynthesis. The mesophyll is differentiated into an upper palisade tissue and lower spongy tissue. Palisade tissue is present below the upper epidermis. The palisade cells are thin walled, cylindrical and contain numerous chloroplasts. The cells are compactly arranged in one layer, perpendicular to the upper epidermis. Narrow intercellular spaces are present in the palisade tissue. Palisade tissue is the most highly specialized type of photosynthetic tissue. The upper surface of the leaf is dark green in colour due to palisade tissue. Spongy tissue forms the lower part of the mesophyll present towards the lower epidermis. Cells of the spongy tissue are thin-walled, irregular in shape, and arranged loosely with large continuous intercellular spaces. Stomata open into large intercellular spaces called sub-stomatal chambers. The cells of spongy parenchyma contain lower numbers of chloroplasts, hence the lower surface of the leaf is light green in colour.

The vascular tissue occurs in the form of discrete bundles called veins. Veins are interconnected to form reticulate venation.

Collateral and closed vascular bundles (veins) occur between the palisade and spongy tissue (median in position). In the vascular bundle, xylem is present towards the upper epidermis and phloem towards the lower epidermis. The vascular bundle is surrounded by parenchymatous bundle sheath.

Role of leaf trichome in insect resistance

Intensive research has been undertaken on the role of glandular trichomes in insect resistance.

Tomato varieties used for commercial production in Dutch glasshouses possess a high density of glandular trichomes on the stem, but a very low density on the leaves. The two-spotted spider mite *Tetranychus urticae* Koch, and the predatory mite, *Phytoseiulus persimilis* Athias-Henriot disperse from leaf to leaf via the stem and face a high risk of death caused by exudates of the glandular trichomes. These risks have been quantified on the tomato cv. Turbo and an accession of *Lycopersicon peruvianum* that is almost free of glandular trichomes (Van Haren *et al.*, 1987).

The tomato plant glandular trichome breaks to release a yellow chemical substance that produces the characteristic tomato plant smell. Current pest management techniques depend on pesticides but trichome-based host-plant resistance may reduce pesticide use. Most studies are focused on Lepidoptera and Hemiptera, although a few studies have been undertaken on Coleoptera, Diptera and Acarina. Both antibiotic and anti-xenotic effects have been demonstrated. It is emphasized that trichome-based host-plant resistance could be utilized as a pest management tool; trichomes of wild species need to be introgressed into the cultivated tomato. Hybrids between the cultivated tomato and the wild species *Lycopersicon hirsutum* f. *glabratum, Lycopersicon pennellii* and *Lycopersicon cheesmanii* f. *minor* have been produced, which demonstrate useful levels of resistance to Acarina, Diptera and Hemiptera pests, although these effects may be tempered by effects on natural enemies (Simmons and Gurr, 2005).

It has been reported that the foliar tetracellular glandular trichomes (tetrads) of the tomato plant, *Lycopersicon esculentum* Mill., contribute significantly to the antibiotic effect of the leaf against the fruitworm *Heliothis zea* (Boddie), as measured by reduction in larval growth. This effect is attributable to phenolic compounds localized within the tetrads (Duffey and Isman, 1980).

Removal of the glandular trichome exudate from leaflets of the wild tomato *L. hirsutum* by swabbing with ethanol resulted in loss of resistance to larvae of the tomato fruitworm (*H. zea*). *H. zea* larvae were killed by fumes from the surface extract and from pure 2-tridecanone. The air surrounding leaflets was found to be rich in 2-tridecanone vapours (Dimock and Kennedy, 1983).

The active compounds in the *L. pennellii* rinsates were identified as 2,3,4-tri-O-acylglucoses bearing short to medium chain length fatty acids. These compounds are localized in the glandular exudate of the type IV trichomes and may accumulate to levels in excess of 400 g/cm^2 (Goffreda *et al.*, 1989).

2-Tridecanone/glandular trichome-mediated resistance to the tobacco hornworm *Manduca sexta* (L.) and the Colorado potato beetle *Leptinotarsa decemlineata* (Say) in a wild tomato, *L. hirsutum* f. *glabratum* C. H. Mull, PI 134417, adversely affects several species of parasitoids and predators of the tomato fruitworm *H. zea* (Boddie) (Kennedy *et al.*, 1991).

L. hirsutum proved to be the least affected by the leaf curl virus because it did not support whitefly infestation to any extent. *L. hirsutum* differed from other species by the presence of glandular trichomes designated type VIc. Whiteflies became entrapped in the exudate of type VIc trichome glands before they could transmit the virus. Thus, it may be possible to control tomato leaf curl virus transmission by breeding plants with certain types of trichomes, especially trichome type VIc (Channarayappa *et al.*, 1992).

Cuticle

Fruit anatomy and ultra-structure associated with cuticle cracking (CC) were compared for five cultivars of tomato. Cultigens resistant to CC had combined epidermal and cuticle (epicarp) layers significantly thicker (10.38–11.37 µm) than susceptible cultigens (6.45–7.76 µm) (Emmons and Scott, 1998).

Three tomato mutants, cutin deficient 1 (cd1), cd2 and cd3, have been identified, the fruit cuticles of which have a dramatic (95–98%) reduction in cutin content and substantially altered, but distinctly different, architectures. This cutin deficiency resulted in an increase in cuticle surface stiffness. Cutin plays an important role in protecting tissues from microbial infection. The three cd mutations were mapped to different loci, and the cloning of cd2 revealed it to encode a homoeodomain protein, which may serve as a key regulator of cutin biosynthesis in tomato fruit (Isaacson *et al.*, 2009).

Disease resistance

Natural openings such as stomata, broken trichomes and cuticular cracks were counted. It was found that susceptibility increased in proportion to the number of these potential penetration sites. However, their role is probably limited because even in the highly resistant cultivars there are enough natural openings to enable successful bacterial penetration (Bashan *et al.*, 1985).

The silverleaf whitefly is one of most important pests of tomato (*Lycopersicon* spp.). In order to determine the oviposition for non-preference of *Bemisia tabaci* biotype B in different tomato plants, a correlation has been undertaken between oviposition preference and pubescence in the genotypes. Four wild tomato genotypes, and two commercial genotypes, were evaluated for number of eggs/cm^2 in free and no-choice oviposition tests, using randomized block design and completely randomized design, respectively. The number of trichomes was measured in 4 mm^2 and coefficients of correlation between number of trichomes and number of eggs were calculated. The wild genotypes were less preferred showing oviposition non-preference resistance type in both tests (Toscano *et al.*, 2002).

Drought resistance

Morpho-anatomical traits contribute to drought resistance in tomato.

Total petiole thickness, compact parenchyma and greater width of phloem vessels, and a higher number of xylem vessels were important characters of resistant tomato genotypes. Leaves of resistant types had longer palisade mesophyll cells with compact arrangement, and a thin spongy mesophyll layer leading to higher tissue ratio (palisade mesophyll:spongy mesophyll).

Thickness of leaves was invariably higher in drought-resistant genotypes (500–550 µm) as compared to susceptible ones (400–450µm). A remarkably reduced number of stomata, a larger size and more distance between them was the highlighting character of drought-resistant genotypes as compared to abundant, smaller size and closely placed stomata in susceptible genotypes (Kulakarni and Deshpande, 2006). This coincides with the findings obtained in Vibha Seeds, Hyderabad, in the case of drought-resistant cotton, sunflower and castor genotypes (Maiti, not published).

Significant variations in anatomy

The following are possible avenues for future research and breeding. In tomato, the presence of a thick cuticle with a higher number of collenchyma layers possibly contributes to drought resistance. Having a greater number of glandular trichomes gives disease resistance to insects. Presence of a higher number of trichomes in leaf anatomy reduces the transpiration under drought condition. In fruit, a thick epicarp will increase storage times and fruit borer resistance. The presence of silica may increase resistance to powdery mildew.

Petiole anatomy

The chief anatomical characters of the petiole as shown by transverse section are the epidermis, hypodermis, ground tissue and vascular tissue (Fig. 9.5).

The epidermis is uniseriate with a single layer of compactly arranged barrel-shaped living cells. The outer walls of the epidermal cells are cutinized. Cuticle reduces the transpiration. Multicellular hairs are present on the epidermis.

Multilayered (two to six layers) hypodermis of collenchyma cells is found immediately beneath the epidermis. Collenchyma is living mechanical tissue. The hypodermis gives considerable strength, flexibility and elasticity to young petioles. As it contains chloroplasts, it may carry out photosynthesis.

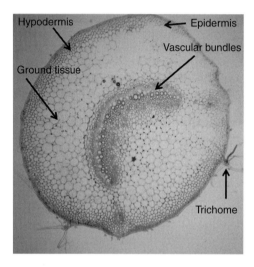

Fig. 9.5. Transverse section of petiole showing epidermis, trichomes and vascular system. Note: vascular bundles arranged in the form of a 'C' shape.

Just beneath the hypodermis ground tissue is found, consisting of thin-walled parenchymatous cells with well defined intercellular spaces among them. Vascular bundles are arranged in a half ring scattered in ground tissue.

The vascular bundles are of various sizes in the same petiole. Each vascular bundle is wedge-shaped. In the petiole, the xylem is always found towards the upper side whereas phloem occurs towards the lower side (as in the leaf).

Seed anatomy

In full-grown tomato seeds the embryo is surrounded by a large quantity of endosperm cells and by the testa or seed coat. The embryos are bent and flattened. The seeds are discoid in nature and a micropylar cap-like organization having endosperm and testa covers the radicle tip. This micropylar cap is the place of radicle protrusion, but there is no observable distinction between testa rupture and endosperm rupture. Profuse hairs (Fig. 9.6) are present on the thin testa, which help in absorption of water, hold the water and make the water easily accessible for the embryo. Macroscereids are round in shape and not very compactly arranged, so water easily enters into

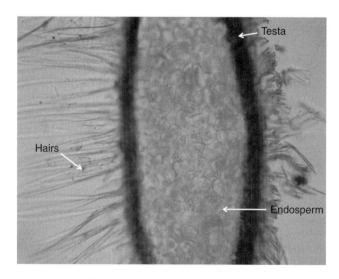

Fig. 9.6. Transverse section of seed (100×) showing abundant hairs, testa and plentiful endosperm.

the embryo, reducing the germination time of the seed; this is 36 h under laboratory conditions (Maiti, unpublished).

Research needs

In the context of the literature review mentioned earlier it is assessed that anatomical traits play an important role in improving tomato for tolerance to various insects, such as the presence and abundance of glandular trichomes which exude toxic materials inhibiting the attack of insects and some disease resistance specially related to cuticle thickness. Tomato production is highly affected by these biotic stresses, which need to be addressed effectively. There exists a large variability among tomato cultivars in the intensity of these traits, thereby offering good scope for selection of cultivars for tolerance to these biotic stresses. Therefore, concerted research activities need to be directed in this direction for genetic improvement of tomato for biotic stress factors, including virus that is transmitted by whitefly.

It has been reported that the presence of thick cuticle and compact palisade is related to drought resistance. In addition, the presence of thick cuticle and thick collenchyma contributes to stiffness of the petiole as well as drought resistance in tomato cultivars. More research emphasis needs to be directed towards these aspects, and there is a requirement to select genotypes possessing these traits.

9.2 Chilli

9.2.1 Introduction

Capsicum peppers are widespread and the fruit contribute to burning and stinging of hands, mouth and eyes in conjunction with food preparation and ingestion (Fett, 2003). Capsicum pepper, popularly known as chilli (*Capsicum annuum* L.), has a great demand throughout the world being a very important condiment of high commercial value. It also has medicinal values and contains antioxidants, anticancerous and many other properties. Extensive research activities have been directed for improvement of yield and quality as well as for controlling biotic and abiotic factors affecting the crop growth and productivity. Diploid chromosome number is 24 $(2n=2x=24)$.

9.2.2 Origin

Chilli peppers originated in the Americas. After the Columbian Exchange, many cultivars of chilli pepper spread across the world, used in both food and medicine.

9.2.3 Utilization

Chilli peppers are used around the world to make a countless variety of sauces, known as hot sauce, chili sauce, or pepper sauce. The fruit is eaten raw or cooked for its fiery hot flavour, concentrated along the top of the pod. The stem end of the pod has most of the glands that produce the capsaicin. The white flesh surrounding the seeds contains the highest concentration of capsaicin, therefore removing the inner membranes is effective at reducing the heat of a pod.

Chilli is sold fresh, dried and powdered.

Keitel *et al.* (2001) recommended the usefulness of *Capsicum* pain plaster in chronic non-specific low back pain. In patients with osteoarthritis or fibromyalgia, it was observed that a *Capsicum* plaster preparation can also be used to advantage in chronic non-specific back pain.

Takashi *et al.* (2001) investigated the cancer chemopreventive activity of carotenoids in the fruits of red paprika *C. annuum* L. and observed the presence of capsanthin and related carotenoids.

Gupta *et al.* (2002) reported a hypocholesterolaemic effect of the oleoresin of *C. annuum* L. in gerbils (*Meriones hurrianae* Jerdon). It prevented the accumulation of cholesterol and triglycerides in the liver and aorta. The faecal excretion of cholesterol and triglycerides were significantly increased in oleoresin-fed gerbils.

9.2.4 Morphological description

Vegetative characters

The plant is a mesophytic perennial herb with a taproot system, and fine lateral roots spreading superficially (not deep penetrating). The stem is aerial, erect, branched, quadrangular and greenish (Fig. 9.7). The leaves are simple, opposite and pubescent, obviate, with a wavy margin, acute apex, stipulate, petiolate (small), with reticulate venation.

Floral characters

The flowers are small and white, bracteate, pedicellate, complete, bisexual, pentamerous and hypogynous. There are five sepals, which are gamosepalous and persistent, and there are five gamopetalous petals. There are five stamens, epipetalous, and the anthers are dithecous and basifixed. The gynoecium is a bicarpellary syncarpus superior ovary with exile placentation. The fruit is a berry with a marcescent calyx. Seeds contain capsaicin, and are rounded in shape, flat, with a brown colour, smooth surface, and sized 1.5–2.3 mm.

9.2.5 Anatomical description

Root anatomy

The root anatomy in transverse section consists of distinct epidermis, cortex and vascular tissue (Fig. 9.8).

The epidermis is uniseriate with a single layer of compactly arranged thin-walled living cells. The epidermal cells produce root hairs, which are useful for absorption of water.

The cortex is multiseriate and relatively homogeneous. It consists of general cortex and endodermis.

The general cortex consists of loosely arranged thin-walled parenchyma cells with prominent intercellular spaces. The cells are colourless and store starch and are usually round or oval.

The endodermis is single layered with compactly arranged barrel-shaped cells. Endodermal cells are characterized by the presence of Casparian strips.

The stele forms the central part of the root and consists of pericycle, vascular strands and conjunctive tissue. Pericycle is the outermost region of the stele and is uniseriate and parenchymatous. Cells of the pericycle retain meristematic activity.

Fig. 9.7. Chilli plant morphology: quadrangular stem, leaves are simple, obviate, with a wavy margin and acute apex, and flowers are small and white.

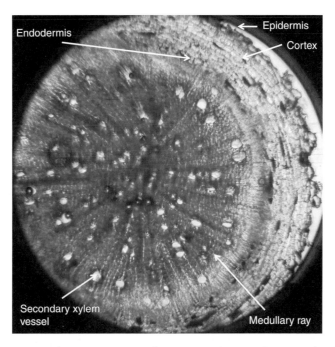

Fig. 9.8. Transverse section of mature root (100×) showing secondary growth. Ruptured epidermis, compressed cortex, secondary phloem, secondary xylem, medullary rays and medulla are clearly visible.

Later roots arise endogenously from the pericycle. Vascular strands are radial, the xylem is exarch and closed type. The secondary xylem of chilli root is composed of mostly isolated xylem vessels accompanied with medium thick-walled sclerenchyma giving strength to the root system. Conjunctive tissue is the non-vascular tissue present between the xylem and phloem strands. It consists of medullary rays or pith rays, which help in the lateral conduction of water and store food material.

Stem anatomy

A transverse section of stem shows three regions, the epidermis, cortex and stele (Fig. 9.9).

The epidermis is uniseriate consisting of compactly arranged barrel-shaped living cells. The outer walls of the epidermal cells are cutinized; the cuticle reduces the transpiration. Few stomata are present in the epidermis for the exchange of gases and transpiration. All the epidermal cells are colourless except the guard cells. Trichomes are absent on the epidermis.

The cortex consists of hypodermis, middle cortex and endodermis. Hypodermis is outermost region of the cortex and is multilayered (two to six layers) and collenchymatic. Chlorenchyma is of the angular type and is living mechanical tissue imparting flexibility to the stem. The hypodermis gives considerable strength, flexibility and elasticity to young stems, and, having chloroplasts, it may carry out photosynthesis. The presence of a thick layer of collenchyma offers strength, rigidity and drought resistance. The middle cortex is multilayered and parenchymatic. The cells may be round or oval with prominent intercellular spaces. This region also stores food materials temporarily. Endodermis forms the innermost layer of the cortex, and is usually distinct and without Casparian bands.

The stele is the central region of the stem and is thicker than the cortex. It consists of pericycle, vascular bundles and medulla. Medullary rays are prominent structures in mature stems. Pericycle is the outermost non-vascular region of the stele and is multiseriate and sclerenchymatic. The eustele is formed from 15–20 vascular bundles arranged in

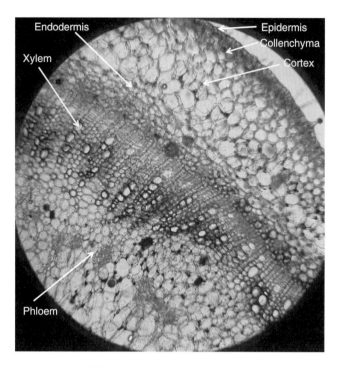

Fig. 9.9. Transverse section of stem (100×) showing various parts.

one ring. Each vascular bundle is wedge-shaped, conjoint, collateral, open and endarch. Medulla is distinct, present at the centre of the stem and is usually parenchymatous. The cells are round or oval with intercellular spaces. The medulla stores food materials and assists in lateral conduction.

It can be observed that in the secondary xylem vessels are arranged in radial rings interspersed with medullary rays and medium thick-walled sclerenchyma, offering strength to the stem.

It was reported that cross-sectional areas of pepper stems, thickness of secondary xylem, numbers of intra-xylary phloem bundles in the periphery of stem pith tissues, leaf thickness, numbers of chloroplasts per palisade mesophyll cell, and thickness of palisade and spongy mesophyll tissues were greatest in peppers grown under MH (metal halide) lamps, intermediate in plants grown under the 660/blue LED array, and lowest in peppers grown under the 660 or 660/735 LED arrays. Most anatomical features of pepper stems and leaves were similar among plants grown under 660 or 660/735 LED arrays. The effects of spectral quality on anatomical changes in stem and leaf tissues of peppers generally correlate to the amount of blue light present in the primary light source (Schuerger *et al.*, 1997).

Leaf anatomy

Leaf anatomy in transverse section shows epidermis, mesophyll and vascular tissue (Fig. 9.10).

Epidermis is present on both surfaces of the leaf blade, i.e. upper epidermis and lower epidermis. The epidermis is single layered with compactly arranged barrel-shaped (tabular) cells. The outer walls of the epidermal cells are cutinized. The outer surface of the epidermis is covered by a continuous layer of cutin called cuticle. The cutinized outer walls and cuticle reduce the loss of water due to transpiration. Anisocytic type of stomata are present in both the epidermal layers, but the stomata are usually more frequent in the lower epidermis. Stomata facilitate

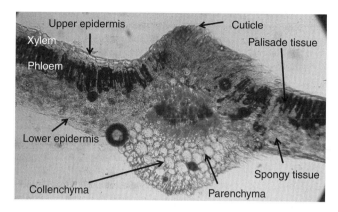

Fig. 9.10. Transverse section of leaf (100×) showing arrangement of tissues.

the exchange of gases between the leaf and environment. The stomatal index of upper epidermis ranges from 11 to 13 and lower epidermis from 26 to 29. All the epidermal cells except guard cells are colourless.

The ground tissue of the leaf, present between the upper and the lower epidermal layers, is called mesophyll. The mesophyll is chlorenchymatic and is concerned mainly with photosynthesis. The mesophyll is differentiated into an upper palisade tissue and lower spongy tissue. Palisade tissue is present below the upper epidermis. The palisade cells are thin walled, cylindrical and contain numerous chloroplasts. The cells are arranged compactly in one layer, perpendicular to the upper epidermis. Narrow intercellular spaces are present in the palisade tissue. Palisade tissue is the most highly specialized type of photosynthetic tissue. The upper surface of the leaf is dark green in colour due to palisade tissue. Spongy tissue forms the lower part of the mesophyll present towards the lower epidermis. Cells of the spongy tissue are thin-walled, irregular in shape, arranged loosely with large continuous intercellular spaces. Stomata open into large intercellular spaces called sub-stomatal chambers. The cells of spongy parenchyma contain a lower number of chloroplasts, hence the lower surface of the leaf is light green in colour.

The vascular tissue occurs in the form of discrete bundles called veins. Veins are interconnected to form reticulate venation. Vascular bundles (veins) occur between the palisade and spongy tissue (median in position), and are collateral and closed. In the vascular bundle, xylem is present towards the upper epidermis and phloem towards the lower epidermis. The vascular bundle is surrounded by parenchymatous bundle sheath.

Trichome in relation to leaf curl virus

Studies revealed that negative association of trichome density and length with thrips and mites scored at flowering and maturity and corresponding positive association with dry chilli yield in all the populations, indicating the role of trichomes in conferring resistance to the leaf curl virus complex (Yadwad *et al.*, 2008).

Petiole anatomy

The chief internal tissues of the petiole can be seen as four main divisions: the epidermis, hypodermis, ground tissue and vascular bundles (Fig. 9.11).

The epidermis is uniseriate with a single layer of compactly arranged barrel-shaped living cells. The outer walls of the epidermal cells are cutinized. Cuticle reduces the transpiration.

Multilayered (two to six layers) hypodermis of collenchyma cells is found immediately beneath the epidermis. Collenchyma is living mechanical tissue. It is annular collenchyma. Hypodermis gives considerable strength, flexibility and elasticity to young petioles and, having chloroplasts, it may carry out photosynthesis.

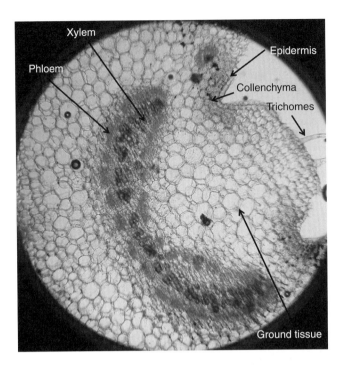

Fig. 9.11. Transverse section of petiole (400×) showing different parts.

Just beneath the hypodermis ground tissue is found. This consists of thin-walled parenchymatous cells with well-defined intercellular spaces among them.

Vascular bundles are arranged in a half ring scattered in ground tissue. The vascular bundles are wedge-shaped and of various sizes in the same petiole. In the petiole, the xylem is always found towards the upper side whereas phloem is found towards the lower side (as in the leaf).

Seed coat anatomy

The seed coat of chilli has thick testa (Fig. 9.12). The testa and tegman are not clearly differentiated; the rate of imbibition is slow due to the thick seed coat so time taken for germination is 4–5 days under 30°C temperature in laboratory conditions. It can be observed from the seed coat anatomy of chilli that there is a thick cuticle giving a glistening appearance to the seed surface and the presence of a highly compact layer of rectangular macrosclereids,

which delays the imbibition of water thereby delaying time of germination.

Significance of variations in anatomy

Future research may be directed in the following directions. The possession of a higher number of trichomes gives disease resistance to a plant. The presence of thick epicarp will help in long storage and fruit borer resistance. A high amount of capsaicin at the placental region has more commercial value in the market. Insignificant research has been directed on the use of anatomical characters for stress resistance in chilli, unlike in other vegetable crops.

9.3   Aubergine, Eggplant or Brinjal

9.3.1   Introduction

Aubergine (also called eggplant or brinjal) (*Solanum melongena* L.) belongs to the family Solanaceae and is extensively distributed

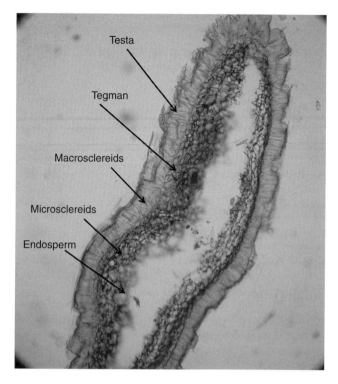

Fig. 9.12. Transverse section of seed (400×) showing different parts.

throughout the temperate and tropical regions. The fruit is a berry and the seeds have a large endosperm and are grown mainly for food and medicinal purposes. The aubergine is usually called brinjal in India, and is an edible vegetable fruit of high commercial value and demand in various countries and well adapted. Aubergine (*S. melongena*) is consumed extensively in Brazil and on the Indian subcontinent. It has various medicinal values, but aubergine pollen is reported to cause allergic reactions (Gill *et al.*, 2002). The diploid chromosome number is 24 ($2n = 2x = 24$).

9.3.2 Origin

Aubergine is native to India and Sri Lanka and has been widely cultivated in all temperate regions of the world. They have also grown in southern and eastern Asian countries since time immemorial. In the 16th century, Arabic traders introduced this vegetable in the West. China currently leads the world in aubergine production, followed by India, Japan, Turkey and Egypt.

9.3.3 Utilization

Fruits of aubergine are used as vegetables for cooking different dishes all over the world. Several studies report the reduction of cholesterol with the extract or powdered infusion of aubergine fruit. There are also reports on its effects on cholesterol metabolism and its possible hypocholesterolaemic effect. Flavonoids extracted from the fruits of *S. melongena* orally administered at a dose of 1 mg/100 g BW/day showed significant hypolipidaemic action in normal and cholesterol fed rats. HMG CoA reductase activity was found to be enhanced, while activities of glucose-6-phosphate dehydrogenase and malate dehydrogenase reduced significantly. Silva *et al.* (1999) studied the effect of aubergine extract on serum and hepatic cholesterol and triglyceride levels in adult rats. The results indicated that the aubergine extract increased serum but decreased hepatic cholesterol and had little or no effect on both serum and hepatic triglycerides.

9.3.4 Morphological description

Vegetative characters

The plant is a mesophytic perennial herb with a taproot system with deep roots and extending inclined lateral roots. The stem is aerial, erect, branched, pubescent, greenish and woody. Leaves are simple, ovate, with a wavy margin, acute apex, stipulate, petiolate (small size), with reticulate venation, opposite and pubescent (Fig. 9.13).

Floral characters

The inflorescence is an auxiliary, scorpioid cyme. Flowers are mostly pink in colour, bracteate, pedicellate, complete, bisexual, pentamerous and hypogynous, obliquely zygomorphic. There are five gamosepalous (united) flowers, persistent in the fruit, and five gamopetalous petals. The flower has five epipetalous stamens, the anthers are dithecous (four-chambered) and basifixed, and show porous dehiscence (anthers open by means of apical pores). The gynoecium is bicarpellary syncarpous superior ovary, with axile placentation. The fruit is a berry, i.e. fleshy with marcescent calyx (no further growth of calyx after fertilization). Seed is rounded in shape, brown, with a smooth and shining surface.

9.3.5 Anatomical description

Root anatomy

A transverse section of root shows three main parts, the epidermis, cortex and stele (Fig. 9.14).

The epidermis of the root is also called the rhizodermis, epiblema, or siliferous layer, and is the outermost region of the root. It is a uniseriate with a single layer of compactly arranged thin-walled living cells. The epidermis gives protection and absorption of water and minerals. Epidermal cells produce unicellular root hairs, which are useful in increasing the surface area for absorption of water and help in obtaining water from soil. The epidermis ruptures after secondary growth.

Fig. 9.13. Aubergine morphology: leaves are simple, ovate, stipulate, with a wavy margin and acute apex. The inflorescence is auxiliary with a scorpioid cyme. The flowers are mostly pink in colour.

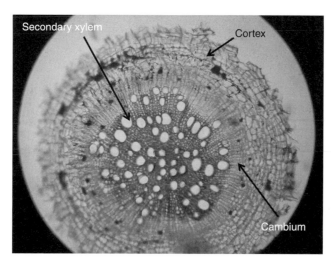

Fig. 9.14. Transverse section of root showing secondary xylem and compressed endodermis and cortex due to secondary growth achieved by the division of cambium in the stele.

Cortex forms the middle region, lying between the epidermis and the stele. Cortex is well developed and is thicker than the stele. It is multiseriate and relatively homogeneous and is differentiated into general cortex and endodermis. General cortex is extensively developed in the root and consists of loosely arranged thin-walled parenchyma cells with prominent intercellular spaces and flavonoid deposits. The cells are usually round or oval, are colourless and store starch. This region helps in lateral conduction of water and salts. After secondary growth the general cortex compresses. The endodermis is a very distinct single layer with compactly arranged barrel-shaped cells. Endodermal cells are characterized by the presences of Casparian strips or Casparian bands made of suberin. In transverse section, the Casparian bands appear as lens-shaped thickening on the radial walls of the endodermis. Endodermis is also called the starch sheet layer, as starch grains are present in the endodermal cells. The endodermis of the root acts as a barrier between the cortex and stele, and ruptures due to secondary growth.

The stele consist of pericycle, vascular strands and pith. Pericycle forms the outermost region of the stele and is uniseriate and parenchymatous. Cells of the pericycle retain meristematic activity. Lateral roots arise endogenously from the pericycle, which help in absorption of water. During secondary growth, pericycle forms phellogen or cork cambium and a small amount of vascular cambium. Xylem and phloem are the vascular tissues. These are arranged in the form of separate strands on different radii, alternating with each other; hence the vascular strands are described as separate and radial. Xylem is exarch, having protoxylem towards the pericycle and metaxylem towards the centre and scattered secondary xylem vessels. Cambium is absent and hence vascular bundles are described as closed type.

Secondary growth in root

During secondary growth, pericycle becomes meristematic and divides, giving rise to cork cambium or phellogen. It produces a few brownish layers of cork cells or phellem towards outside, and the phelloderm on the inside. Secondary xylem and secondary phloem are formed due to the activity of the cambium ring (Fig. 9.14). After secondary growth the secondary vascular tissues form a continuous cylinder and usually the primary xylem becomes embedded in it. Primary xylem located at the centre and primary phloem elements are generally seen in a crushed condition. It is observed that the secondary xylem of root contains isolated scattered larger xylem vessels. Pith is absent. The non-vascular tissue present between xylem and phloem strands is called conjunctive tissue, and it may be totally parenchymatous or a part of it may be converted into sclerenchyma. During secondary growth the cambial cells that originate from the pericycle lying against the group of protoxylem function as ray initials and produce medullary rays. These rays are transversed in the xylem and phloem through cambium.

Stem anatomy

Transverse section of the primary stem shows epidermis, cortex and stele.

The epidermis is the outermost region surrounding the stem. It is uniseriate with a single layer of compactly arranged barrel-shaped living cells. The outer walls of the epidermal cells are cutinized and the epidermis is covered by a separate layer of cutin called cuticle on its outer surface (cuticularized), which reduces transpiration. Few stomata are present in the epidermis for exchange of gases and transpiration. All the epidermal cells are colourless except the guard cells. Multicellular hairs are present on the epidermis. The epidermis ruptures after secondary growth and bark occupies the place of epidermis.

Cortex is the middle region present between the epidermis and stele, and is thinner than the stele. It shows hypodermis, middle cortex and endodermis. The hypodermis is the outermost region of the cortex and is multilayered (two to six layers) and collenchymatic. Hypodermis gives considerable strength, flexibility and elasticity to young stems. The middle cortex is also multilayered and parenchymatic. The cells may be round or oval with prominent intercellular spaces. This region also stores food materials temporarily and flavonoids are deposited. The endodermis forms the innermost layer of the cortex and is usually distinct in stems without Casparian bands. In young stems the innermost layer of the cortex contains abundant starch, hence this layer is called the starch sheath. The starch sheath layer is homologous to endodermis, but morphologically unspecialized, hence such layer may be called endodermoid. After secondary growth the endodermis ruptures due to pressure created by inner secondary tissues.

The stele is the central region of the stem and is thicker than the cortex. It consists of pericycle, vascular bundles and medulla and medullary rays. Pericycle (perivascular region) forms the outermost non-vascular region of the stele; it is multiseriate and sclerenchymatic. Sclerenchyma is mechanical tissue and gives strength and rigidity to the stem. Xylem and phloem are organized into wedge-shaped vascular bundles and from 10 to 20 vascular bundles are arranged in one ring consisting of xylem towards the centre of the axis and phloem towards the periphery. In the vascular bundle xylem and phloem are present together (conjoint) on the same radius (collateral), with a strip of cambium (fascicular cambium) between xylem and phloem (open), hence the vascular bundle is described as conjoint, collateral and open type. Xylem is endarch with protoxylem present towards the centre. Medulla and medullary rays are distinct, present at the centre of the stem and are usually parenchymatous. The cells are round or oval with intercellular spaces. The extensions of parenchymatous pith between vascular bundles are called primary medullary rays, which help in lateral conduction and may give rise to secondary meristem (interfascicular cambium). The medulla stores food materials.

Secondary growth in stem

The ever increasing shoot system requires the supply of plentiful water and mineral salts and the root system requires much food material. Primary vascular tissues are inadequate to supply them. The vascular cambium becomes active and produces secondary vascular tissue (Fig. 9.15). During the secondary growth cambial strips are formed from medullary rays; these are called interfascicular cambium. Fascicular cambium and inter-fascicular cambium combines and forms a cambial ring. The meristematic cells

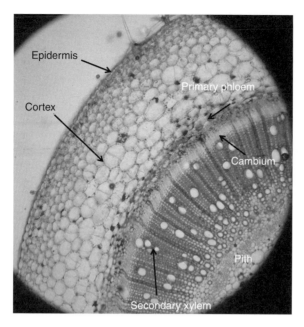

Fig. 9.15. Transverse section of stem (100×) showing different parts. Secondary growth has been initiated and cambium is forming secondary xylem.

of the cambial ring form secondary xylem towards the inner side and a smaller amount of secondary phloem towards the outside. Phellogen or cork cambium is formed from the cells of hypodermis or middle cortex. Cork cambium forms secondary cortex towards the inner side and cork tissue towards the outer side. A greater amount of cork tissue is produced than the secondary cortex.

It can be observed from the transverse section of stem that the presence of thick cuticle on the stem surface and several layers of collenchyma give strength to the stem. The secondary xylem contains radial rows of vessels interspersed with medullary rays and sclerenchyma.

Shoot and fruit borer

The presence of a thick cuticle, broad and thick collenchymatous area (hypodermis), compact parenchyma cells in the cortical tissue, small area in the cortical tissues, more vascular bundles with narrower spaces in the inter-fascicular region, and compact arrangement of vascular tissue with lignified cells and small pith were the main characters of tolerant varieties. On the other hand, a thinner cuticle and collenchymatous area (hypodermis), loose parenchyma cells in the cortical region, larger spaces between vascular bundles, a smaller number of trichomes, soft parenchymatous cells in the inter-fascicular region, might be responsible for the susceptibility to brinjal shoot and fruit borer (Hossain *et al.*, 2002).

Leaf anatomy

Epidermis is mainly intended for protection and is present on both surfaces of the leaf blade. Epidermis covering the upper surface of the leaf is called upper epidermis or ventral epidermis or adaxial epidermis. The epidermis present towards the lower side of the leaf is called lower epidermis or dorsal epidermis or abaxial epidermis. The epidermis is single layered with compactly arranged barrel-shaped (tabular) cells. The outer walls of the epidermal cells are cutinized and the epidermis on its outer surface is covered by a continuous layer of cutin called cuticle.

The cutinized outer walls and cuticle reduce the loss of water due to transpiration. Anisocytic type of stomata are present in both of the epidermal layers, but stomata are usually more frequent in the lower epidermis. The stomatal index of the upper epidermis is 21.3 and the lower epidermis is 29.6. Stomata facilitate the exchange of gases between the leaf and environment. All the epidermal cells except guard cells are colourless. Epidermis contains multicellular hairs.

The ground tissue of the leaf, present between the upper and the lower epidermal layers, is called mesophyll. The mesophyll is chlorenchymatic and is concerned mainly with photosynthesis. The mesophyll is differentiated into an upper palisade tissue and lower spongy tissue. The palisade tissue is present below the upper epidermis (Fig. 9.16) and the cells are thin walled, cylindrical and contain numerous chloroplasts. The cells are arranged compactly in one layer, perpendicular to the upper epidermis. Narrow intercellular spaces are present in the palisade tissue. Palisade tissue is the most highly specialized type of photosynthetic tissue. The upper surface of the leaf is dark green in colour due to palisade tissue. Compactly arranged palisade tissue reduces the high transpiration loss of water and offers drought

resistance. The lower part of the mesophyll present towards the lower epidermis is the spongy tissue. These cells are thin-walled, irregular in shape, and arranged loosely with large continuous intercellular spaces. Stomata open into large intercellular spaces called sub-stomatal chambers. The cells of spongy parenchyma contain a smaller number of chloroplasts, hence the lower surface of the leaf is light green in colour.

The leaf receives the vascular supply from the stem. The vascular tissue occurs in the form of discrete bundles called veins, which are interconnected to form reticulate venation. Vascular bundles (veins) occur between the palisade and spongy tissue (median in position). They are collateral and closed. In the vascular bundle, xylem is present towards the upper epidermis and phloem towards the lower epidermis. The vascular bundle is surrounded by parenchymatous bundle sheath, the cells of which are called border parenchyma. The larger vascular bundles are supported by hypodermal parenchymatous strands. These strands are considered to be the sheath extension.

The presence of thick cuticle and compactly arranged palisade cells definitely reduces transpiration loss, which needs to be assessed in aubergine in relation to drought-resistant cultivars. No studies are available in the literature.

The leaf area, leaf thickness, trichome density and chlorophyll content were correlated with leafhopper and whitefly population. The data indicate that leaf area, leaf thickness and chlorophyll content exerted no effect on leafhopper population, while trichome density had a negative correlation (Naqvi *et al.*, 2009).

Effect of chromium on leaf anatomy

Morphological changes are observed on applying chromium at different concentration levels in aubergine. Chromium caused reduction in root length, shoot length and decreased number of branches. The number of stomata and epidermal cells decreased with increasing level of chromium in aubergine, whereas the number of trichomes increased and

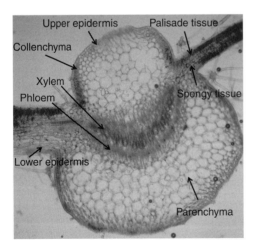

Fig. 9.16. Transverse section of leaf (100×) showing arrangement of tissues. Note: below the epidermis collenchyma consisting of chlorophyll is involved in synthesis of starch by photosynthesis.

the size of trichomes and stomata decreased (Purohit *et al.*, 2003).

Insect resistance – fruit anatomy

Length of fruits had a significant negative correlation with the fruit borer infestation. The fruit diameter had significant positive correlation with the fruit borer infestation. Significant positive correlation was also found between percentage infestation of fruit and thickness of pericarp and meso-carp of aubergine varieties screened. The increase in size of pericarp and mesocarp enhanced the infestation of fruit borer. The varieties having compact seed rings with closely arranged seeds in mesocarp had a low degree of fruit infestation, whereas highly susceptible varieties neither had compact seed rings nor closely arranged seeds in the mesocarp (Naqvi *et al.*, 2009).

Petiole anatomy

The epidermis is uniseriate with a single layer of compactly arranged barrel-shaped living cells. The outer walls of the epidermal cells are cutinized, which, with the cuticle, reduces the transpiration. Multicellular hairs are present on the epidermis.

A multilayered (two to six layers) hypodermis of angular collenchyma cells is found immediately beneath the epidermis. Collenchyma is living mechanical tissue. The hypodermis gives considerable strength, flexibility and elasticity to young petioles and, having chloroplasts, it may carry out photosynthesis. Due to the presence of the thick layer of collenchyma the petiole is strong, stiff.

Just beneath the hypodermis ground tissue is found. This consists of thin-walled parenchymatous cells with well-defined intercellular spaces among them. Vascular bundles are arranged in a half ring scattered in ground tissue.

The vascular bundles are of various sizes in the same petiole. Each vascular bundle is wedge-shaped. In the petiole, the xylem is always found towards upper side whereas phloem is found towards the lower side (as in the leaf).

Seed coat anatomy

Integuments of fertilized ovule become seed coat. The seed coat covers the embryo and provides protection for the embryo. The seed coat consists of two layers, the testa and tegman (Fig. 9.17).

The testa is the outermost layer of the seed coat and is formed from the outer integument of the ovule. Generally it is hard and impermeable to water and sometimes shows ornamentations. The testa contains macro-and microsclereids. Macrosclereids are arranged very compactly and are dumbbell/boat/bone shaped (osteosclereids). Microsclereids are small in size with a square shape and dentate edges. The hardness of the seed coat depends on thickness of testa and the arrangement of sclereids. Rate of absorption of water (imbibition) depends on the thickness of testa. Very hard and thick testa causes seed dormancy in some plants.

The tegman is the inner layer of the seed coat and is formed from the outer integument. It is thin and soft, and covers the embryo. The thin seed coat and the fact that the macro-and microsclereids are not very compactly arranged, means that water enters easily into the seed and on imbibition it ruptures easily. Time of germination is about 30–36 h.

Significance of variations in anatomy

Future research may be directed in the following directions. The possession of a thick cuticle and multilayered collenchyma will help in severe drought conditions. In aubergine, a higher number of trichomes gives disease resistance to insects. The epicarp and pericarp is thick and so is important in long storage and fruit borer resistance. Due to the wax coating, it may be resistant to fungal disease.

9.4 Okra

9.4.1 Introduction

The name 'okra' (*Abelmoschus esculentus* L.) is most often used in the USA and is of East African origin. Okra is frequently known as lady's fingers outside the USA. The okra

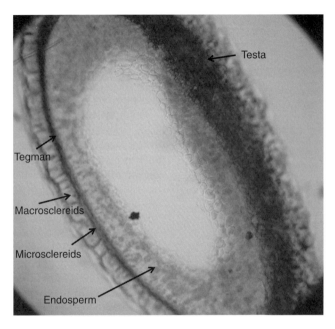

Fig. 9.17. Transverse section of seed (400×) showing different parts. Note: microsclereids are rectangular in shape.

pods are usually harvested at 2–4 day intervals. The pods are harvested at certain intervals depending on marketable size and consumer's preference. In general the pods should be harvested when the pods are 4.5–7 cm in length.

9.4.2   Origin and Distribution

The species apparently originated in the Ethiopian highlands, though the manner of distribution from there is undocumented. The Egyptians and Moors of the 12th and 13th centuries used the Arab word for the plant, suggesting that it had come from the east. The plant may thus have been taken across the Red Sea or the Bab-el-Mandeb strait to the Arabian Peninsula, rather than north across the Sahara. One of the earliest accounts is by a Spanish Moor who visited Egypt in 1216, who described the plant under cultivation by the locals who ate the tender, young pods with meals.

From Arabia, the plant spread around the shores of the Mediterranean Sea and eastward. The lack of a word for okra in the ancient languages of India suggests that it arrived there in the Common Era. The plant was introduced to the Americas by ships plying the Atlantic slave trade by 1658, when its presence was recorded in Brazil. It was further documented in Suriname in 1686.

Okra may have been introduced to south-eastern North America in the early 18th century. It was being grown as far north as Philadelphia by 1748.

9.4.3   Utilization

Okra is a popular and important food worldwide. Its tender fruits are used in making many dishes. Okra gum is used industrially. The yield in oil, the quality of its proteins and the use of the stem in paper-making reveal that okra has economic potential for cultivation. Okra fruit rhamno-galacturonans increased cell proliferation.

In Asian medicine the fruit of the okra plant *A. esculentus* (L.) Moench is used as a mucilaginous food additive against gastric irritative and inflammatory diseases.

9.4.4 Morphological description

Vegetative characters

The plant is a mesophytic herb with a tap-root system with inclined deep lateral roots. The stem is aerial, erect, branched, and stellate hairs are present. The leaves are alternate, pubescent, simple, heart shaped, with a dentate margin and obtuse apex, stipulate, petiolate and have palmately reticulate venation (Fig. 9.18).

Floral characters

Flowers are born on solitary inflorescence and are bracteate, pedicellate, with the presence of an epicalyx, complete, bisexual, pentamerous and hypogynous, actinomorphic. There are five gamosepalous (united

Fig. 9.18. Okra plant morphology: leaves are simple, heart shaped with a dentate margin. Flowers are born on solitary inflorescence and fruit is loculicidal capsule.

sepals) sepals, which are bowl shaped with trichiferous nectar present at the base of calyx. There are five petals, polypetalous, large and attractive. The stamens are numerous and monodelphous and the kidney-shaped anthers are monothecous. The gynoecium is multicarpellary, syncarpus, with a superior ovary and axile placentation. The number of stigma is equal to number of locules in the ovule. The fruit is a loculicidal capsule and seeds are round, grey-ash colour with a rough surface, and the seed size is from 2.5 mm to 3 mm.

9.4.5 Anatomical description

Root anatomy

The epidermis is uniseriate with a single layer of compactly arranged thin-walled living cells. Epidermal cells produce root hairs, which are useful for absorption of water.

The cortex is multiseriate, relatively homogeneous and consists of general cortex and endodermis. The endodermis is single layered with compactly arranged barrel-shaped cells. Endodermal cells are characterized by the presences of Casparian strips.

Stele is the central part of the root and consists of pericycle, vascular strands and conjunctive tissue. The pericycle is the outermost region of the stele and is uniseriate and parenchymatous. Cells of the pericycle retain meristematic activity. Later roots arise endogenously from the pericycle. The vascular strands are radial and the xylem is exarch, closed type. The non-vascular tissue present between xylem and phloem strands is called conjunctive tissue. Pith is absent. Secondary tissues are formed by the activity of cambium causing the primary structure of the root to lose its initial appearance in transverse section (Fig. 9.19).

Stem anatomy

The epidermis is the outermost region surrounding the stem. It is uniseriate with a single layer of compactly arranged barrel-shaped living cells. The outer walls of the epidermal cells are cutinized and the cuticle

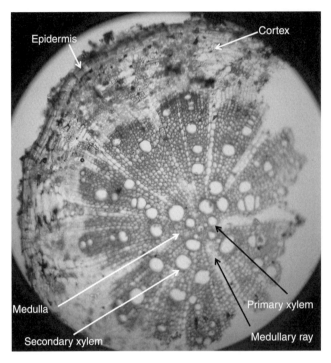

Fig. 9.19. Transverse section of root (100×) showing secondary growth and various parts including epidermis, cortex, secondary xylem and medullary rays.

reduces the transpiration. Few stomata are present in the epidermis for exchange of gases and transpiration. All the epidermal cells are colourless except the guard cells. Stellate, tufted trichomes are present on the epidermis, which give insect resistance.

The hypodermis is the outermost region of the cortex. It is multilayered (two to six layers) and collenchymatic. Collenchyma is living mechanical tissue. The hypodermis give considerable strength, flexibility and elasticity to young stems. As it contains chloroplasts, it may carry out photosynthesis. The endodermis is innermost layer of the cortex. It is usually distinct in stems without Casparian bands.

Pericycle is the outermost non-vascular region of the stele, and is multiseriate and sclerenchymatic. Each vascular bundle is wedge-shaped, conjoint, collateral, open and endarch. Medulla and medullary rays are distinct, present at the centre of the stem, and are usually parenchymatous. The cells are round or oval with intercellular spaces.

Medulla stores food materials and helps in lateral conduction.

During the secondary growth primary phloem, endodermis and cortex are crushed due to pressure created by inner secondary tissues (Fig. 9.20).

Leaf anatomy

The epidermis is present on both surfaces of the leaf blade, i.e. upper epidermis and lower epidermis. Epidermis is mainly meant as a protective layer. The epidermis is single layered with compactly arranged barrel-shaped (tabular) cells. The outer walls of the epidermal cells are cutinized. The cutinized outer walls and cuticle reduce the loss of water due to transpiration. The stomatal index of the upper epidermis is 25.9 and of the lower epidermis is 31.3. Anisocytic stomata (Fig. 9.21) are present in both of the epidermal layers. Stellate, tuft trichomes are present on the epidermis. All the epidermal cells except guard cells are colourless.

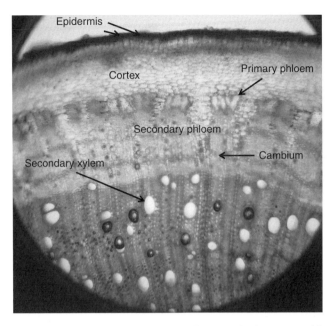

Fig. 9.20. Transverse section of stem (100×) showing secondary growth. The primary phloem and endodermis is crushed due to pressure created by inner secondary tissues.

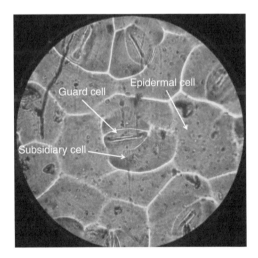

Fig. 9.21. Leaf upper epidermal layer (1000×) showing epidermal cells, guard cell covering stomatal opening (pore) and subsidiary cells constituting the stomatal complex (anisocytic type).

The mesophyll is chlorenchymatic and is differentiated into an upper palisade tissue and lower spongy tissue (Fig. 9.22). The palisade tissue is present below the upper epidermis. The palisade cells are thin walled, cylindrical, contain numerous chloroplasts and are arranged compactly in one layer, perpendicular to the upper epidermis. Narrow intercellular spaces are present in the palisade tissue. Palisade tissue is the most highly specialized type of photosynthetic tissue. The upper surface of the leaf is dark green in colour due to palisade tissue. The lower part of the mesophyll present towards the lower epidermis is the spongy tissue. Cells of the spongy tissue are thinwalled, irregular in shape, and arranged loosely with large continuous intercellular spaces. The cells of spongy parenchyma contain fewer chloroplasts, hence the lower surface of the leaf is light green in colour.

The vascular tissue occurs in the form of discrete bundles called veins. Veins are interconnected to form reticulate venation. Vascular bundles (veins) occur between the palisade and spongy tissue (median in position), and are collateral and closed. In the vascular bundle, xylem is present towards the upper epidermis and phloem towards the lower epidermis. The vascular bundle is surrounded by parenchymatous bundle sheath.

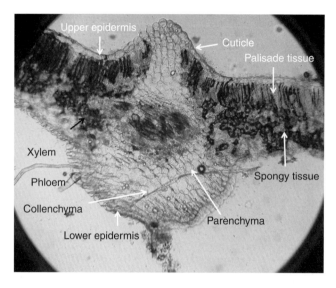

Fig. 9.22. Transverse section of leaf (100×) showing different parts.

Trichome density

The trichome density on the main vein and lateral veins in relation to oviposition was negative and significant. Similarly, trichome length on lateral veins and subveins also showed a negative and significant correlation with oviposition of leafhopper. The number of eggs laid per leaf was less in the resistant genotypes than that in susceptible ones in okra (Lokesh and Singh, 2005).

Petiole anatomy

The epidermis is uniseriate with a single layer of compactly arranged barrel-shaped living cells. The outer walls of the epidermal cells are cutinized. Cuticle reduces the transpiration. Multicellular hairs are present on the epidermis.

Multilayered (two to six layers) hypodermis of angular collenchyma cells is found immediately beneath the epidermis (Fig. 9.23). Collenchyma is living mechanical tissue. The hypodermis gives considerable strength, flexibility and elasticity to young petioles and it may carry out photosynthesis as it contains chloroplasts.

Just beneath the hypodermis ground tissue is found. It consists of thin-walled parenchymatous cells with well-defined intercellular spaces among them. Vascular bundles are arranged in a half ring scattered in ground tissue.

The vascular bundles are of various sizes in the same petiole. Each vascular bundle is wedge-shaped. In the petiole, the xylem is always found towards the upper side whereas phloem is found towards the lower side (as in the leaf).

Fruit dehiscence – role of anatomy

During development of the mature pericarp in okra the epicarpic cells become enlarged, vacuolated and thick walled and resemble collenchyma. The mesocarp is derived from the inner zone of ground parenchyma of the ovary wall and is parenchymatous. The large and vacuolated parenchyma of the developing mesocarp appears disorganized at maturation of the fruit. The dehiscence of the ripe capsule of *A. esculentus* is the result of differentiation of mechanically weak cells in the median plane of each carpel, as well as in the central column and porous endocarp (Inamdar *et al.*, 1989).

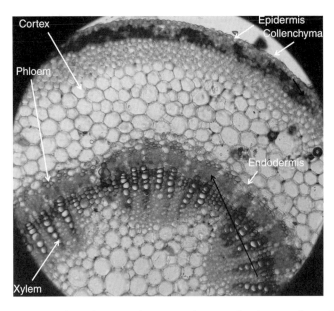

Fig. 9.23. Transverse section of petiole (100×) showing epidermis, collenchyma and vascular bundles.

Seed coat anatomy

The seed coat of okra is thin and shows clear distinction between layers of testa (Fig. 9.24). The interval between imbibition and germination is less. Time for germination is 20 h.

Significance variations in anatomy

Future research may be directed in the following directions. Less space present in the palisade and sponge tissue, and the ratio of palisade and sponge tissues will help in severe drought condition. The presence of stellate tuft hairs helps in resistance to insect-borne diseases. Compact arranged palisade tissue with a strong stereome is somewhat resistant to yellow vein disease. Early development of sclerenchyma in fruit is not desirable for edibility but is useful for fibre. The epicarp and pericarp is thick, so is helpful in long storage and fruit borer resistance.

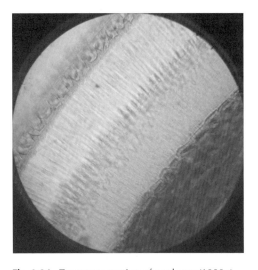

Fig. 9.24. Transverse section of seed coat (1000×) showing different layers.

9.5 Watermelon

9.5.1 Introduction

Watermelon (*Citrullus vulgaris* L.) is an important vegetable crop of the world belonging to the family Cucurbitaceae. Botanically, the fruit of watermelon is described as a pepo and it is a highly nutritious food. The seeds of watermelon, on the other hand, are a good source of essential minerals and amino acids, except lysine and methionine (Swamy *et al.*, 2007). In India it is a popular vegetable crop grown during the summer season and good market demand is noted throughout the country.

9.5.2   Origin

Although it is disputed, it is widely belie-ved that watermelon originated in tropical to subtropical Africa and two important centres of diversity for the cultivated forms are found in tropical to subtropical Africa and India (Swamy and Sadashiva, 2007). The crop was introduced to India in the prehistoric period. The early Vedic and post-Vedic Sanskrit literature describes that various wild and cultivated plants belonging to the family Cucurbitaceae played an important role as food and medi-cines in the lives of Indo Aryans in north-ern India during 2000 to 200 BC. These literatures also inform that the Indo Aryans, after entering the Indian subcontinent, learned the cultivation of at least 11 more species belonging to the Cucurbitaceae family from the local dwellers of the region, who were the speakers of the Dravidian or Munda language (Decker-Walters, 1999).

9.5.3   Utilization

Fresh fruit of watermelon is consumed all over the world. Watermelon rinds are also edible, and sometimes used as a vegetable. In China, they are stir-fried, stewed or more often pickled. Pickled watermelon rind is also generally consumed in the southern USA. Watermelon juice can also be made into wine and is also slightly di-uretic. Watermelons have large amounts of beta-carotene. Watermelon seeds also serve as a good source of vegetable oil and in its centre of origin specific local races are found that are purposely cultivated for seed-oil pressing.

Scientific reports are found regarding the tremendous possibility of exploiting the antioxidant properties of watermelon lyco-pene against oxidative stress-related prob-lems in human beings.

9.5.4   Morphological description

Vegetative characters

The plant is a mesophytic climbing (with tendrils) herb, pubescent and succulent. It has a taproot system with fewer penetrating lateral roots into the soil. The hollow stem is aerial, weak, succulent, angular, branched, pubescent and greenish, with tendrils devel-oped from the leaf axial. The leaves are oppo-site and pubescent, simple, heart shaped with a dentate margin and acuminate apex, stipulate and petiolate (small), with reticu-late venation (Fig. 9.25).

Floral characters

The flower is an auxiliary inflorescence, with alternate male and female flowers formed in a 9:1 ratio. Flowers are mostly white and yellow in colour, ebracteate, unisexual,

Fig. 9.25. Watermelon plant morphology: the plant is a climbing herb. The stem is aerial, weak, succulent, angular and pubescent. The leaves are simple, heart shaped, with a dentate margin. Auxiliary inflorescence with the male and female flowers arranged alternately.

pentamerous and hypogynous. There are five gamosepalous pubescent sepals: in the male flower these are bell shaped, in female flower the calyx tube is connate with the ovary. There are five free, yellow and pubescent petals. The male flower has five syandrous stamens, the anthers are dithecous and a pistillode is present.

The female flower has from three to five staminodes present on the calyx edges. The gynoecium is unilocular, tricarpellary syncarpous and with parietal placentation. The fruit is a pepo with red coloured endocarp.

9.5.5   Anatomical description

Root anatomy

The epidermis is uniseriate with a single layer of compactly arranged thin-walled living cells. Epidermal cells produce root hairs, which are useful for absorption of water.

The cortex is multiseriate and relatively homogeneous, and consists of general cortex and endodermis. General cortex has loosely arranged thin-walled oval parenchyma cells with prominent intercellular spaces. the endodermis is single layered with compactly arranged barrel-shaped cells. Endodermal cells are characterized by the presence of a Casparian strip.

The pericycle is uniseriate and parenchymatous. Cells of the pericycle retain meristematic activity. Later roots arise endogenously from the pericycle. The vascular strands are radial; xylem is exarch, closed type. Pith is absent. The non-vascular tissue present between xylem and phloem strands is called conjunctive tissue.

Stem anatomy

A transverse section of the stem shows ridges and grooves in outline and shows epidermis, cortex and stele (Fig. 9.26).

The epidermis is uniseriate, with a single layer of compactly arranged barrel-shaped living cells. The outer walls of the epidermal cells are cutinized. Cuticle reduces the transpiration. Few stomata are present in the epidermis for the exchange of

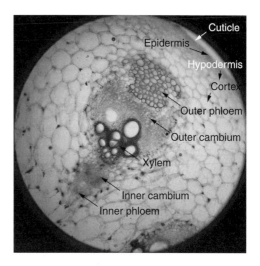

Fig. 9.26. Transverse section of stem (1000×) showing different parts. Note: bicollateral vascular bundles consisting of outer phloem, outer cambium, xylem, inner cambium and inner phloem.

gases and transpiration. Multicellular hairs are present on the epidermis.

Below the ridges, three to four layers of angular collenchyma (hypodermis) are present. The hypodermis gives considerable strength, flexibility and elasticity to young stems. It may carry out photosynthesis as it possesses chloroplasts. The cortex is multilayered. Endodermis is the innermost layer of the cortex and is distinct without Casparian bands.

The pericycle is multiseriate and sclerenchymatic. Vascular bundles are bicollateral and arranged in two rows. Each vascular bundle is wedge-shaped, conjoint, open and endarch. Medulla and medullary rays are absent and the centre part of stem is hollow; this is an important characteristic of Cucurbitaceae members.

Leaf anatomy

The epidermis is single layered, with compactly arranged barrel-shaped (tabular) cells. The outer walls of the epidermal cells are cutinized and covered by a continuous layer of cuticle. Anamocytic stomata present in both of the epidermal layers. The stomatal index of the upper epidermis is 18.47 and that of the lower epidermis is 28.98. The foot cell of the trichome and subsidiary cell surrounding the foot cell contains cystolith. Trichomes are of the glandular type.

The mesophyll is differentiated into an upper palisade tissue and lower spongy tissue. The palisade tissue is present below the upper epidermis. The palisade cells are thin walled, cylindrical and contain numerous chloroplasts. The cells are compactly arranged in one layer, perpendicular to the upper epidermis. Narrow intercellular spaces are present in the palisade tissue. Palisade tissue is the most highly specialized type of photosynthetic tissue. The upper surface of the leaf is dark green in colour due to palisade tissue. Spongy tissue forms the lower part of the mesophyll present towards the lower epidermis. Cells of the spongy tissue are thin-walled, irregular in shape and arranged loosely with large continuous intercellular spaces. The cells of spongy parenchyma contain fewer chloroplasts, hence the lower surface of the leaf is light green in colour.

The vascular tissue occurs in the form of discrete bundles called veins. Veins are interconnected to form reticulate venation. Vascular bundles occur between the palisade and spongy tissue (median in position). They are collateral and closed. In the vascular bundle, xylem is present towards the upper epidermis and phloem towards the lower epidermis. The vascular bundle is surrounded by a parenchymatous bundle sheath.

Petiole anatomy

The epidermis is uniseriate, with a single layer of compactly arranged barrel-shaped living cells. The outer walls of the epidermal cells are cutinized. Multicellular hairs are present on the epidermis. A multilayered (two to six layers) hypodermis of angular collenchyma cells is found immediately beneath the epidermis (Fig. 9.27). Collenchyma is living mechanical tissue. The hypodermis gives considerable strength, flexibility and elasticity to young petioles. As it contains chloroplasts, it may carry out photosynthesis.

Just beneath the hypodermis ground tissue is found. This consists of thin-walled parenchymatous cells with well-defined intercellular spaces among them. Vascular

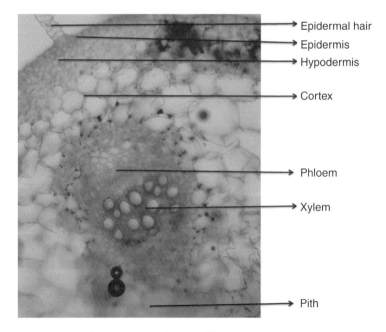

Fig. 9.27. Transverse section of petiole (100×) showing different parts.

bundles are arranged in a half ring scattered in ground tissue. Bicollateral vascular bundles are arranged in two rows.

The central part of petiole is hollow, which is an important character of Cucurbitaceae members.

Anatomical changes in stigma after pollination

The following changes were observed on stigma. The wall thickenings of the papilla transfer cells contained callose and their presence prior to pollination was confirmed using EM-autoradiography, freeze-fracture and fixation. No further callose thickenings were produced following pollination. Pollination resulted in a rapid increase in aqueous stigma secretion and localized disruption of the cuticle, which appeared to remain on the surface of the secretion. The reaction was localized to the papilla cells adjacent to the pollen tube only. Both pollen-grain wall and stigma secretion contained proteins, carbohydrates, acidic polysaccharides, lipids and phenolics (Sedgley, 1982).

Anatomical changes during fruit development

Parthenocarpic fruits were larger and had higher fresh weight and percentage water than pollinated fruits at day 1 but the positions were reversed by day 9. Unpollinated fruits did not increase in size after day 3. Pericarp cells were small, of regular shape and showed no obvious change with either time or treatment. Cell number increased in the pollinated and parthenocarpic but not in the unpollinated fruits. Cells divided in the flesh of the parthenocarpic but not of the pollinated fruits, which increased in size by cell enlargement only. Starch, present in the cells of the flesh and placenta at day 0, was absent from the unpollinated fruits at day 6. Ovules grew in both pollinated and parthenocarpic fruits largely due to cell division in the nucellus and integuments; the pollinated ovules were larger than the parthenocarpic throughout. Embryo and endosperm development occurred in the pollinated but not in the parthenocarpic ovules. Starch was present throughout the 9-day period in the integuments of the pollinated and

parthenocarpic ovules but was lost from the integuments of the unpollinated ovules by day 6. Pollinated and parthenocarpic ovules contributed increasingly to fruit dry weight over the 9-day period.

The ovule tissues, in particular the nucellus and integument, may exert control over early development in both pollinated and parthenocarpic fruits (Sedgley *et al.*, 1977).

Anatomical changes under flooding stress

There are morpho-anatomical changes in flooding lines comparing with control. The obtained results revealed that growing watermelon on land flooded with a 2-week period was the best. Parameters of plants in this treatment exceeded those of unflooded plants in plant length (47.3%), plant stem diameter (26.9%), leaf area (43.7%), leaf area index (82.8%), fruit number per plant (30%), yield (48.8%), number of stomata/mm^2 (18.3%), number of cells/mm^2 (24.3%) and width of second trichome cell (64.4%), narrowest vessel diameter (28.75%). Moreover, this treatment revealed superiority over a 1-week flooding period in stem dry matter percentage (26.9%), narrowest width of vessel (27.84%), highest vessel width (12.12%). A 2-week flooding period also overwhelmed that of a 3-week period in fruit number per plant (26.2%), width of narrowest vessel (12.84%) and width of hair second cell (43.7%)(Abdel and Bamerni, 2011).

Anatomy of embryo sac

In this study embryo sacs from watermelon were observed over a 13-day period following flowering with: (i) normal pollination; (ii) non-pollination; and (iii) induction of parthenocarpic fruit development with naphthalene acetic acid. With normal pollination, approximately 2 days after fertilization the embryo sacs completed development and consisted of two synergids with prominent filiform apparatus, an egg cell, a central cell with two polar nuclei and three antipodal cells. Sperm nuclei were observed within the embryo sac at 2 days and by 4 days the endosperm was proliferating. In the non-pollination treatment the embryo sac was still intact after 4 days although the antipodal nuclei were becoming hard to distinguish. By 7 days only the two synergids and the egg cell were still well defined, the polar nuclei appeared in some preparations to be fused, and the antipodals had degenerated. By 10 days the embryo sac was a structure-less watery mass. In parthenocarpic fruit the fate of the embryo sac was similar to that in non-pollinated fruit except that final breakdown was delayed past 10 days (Buttrose and Sedgley, 1979).

Significance variations in anatomy

Future research may be directed in the following directions. Having a higher number of trichomes gives disease resistance to insects. The epicarp and pericarp is thick, and so is helpful in long storage and fruit borer resistance. The presence of glandular trichomes, somewhat resistant to fungal diseases, is still to be studied.

9.6 Bottle Gourd

9.6.1 Introduction

The calabash or bottle gourd (*Lagenaria leucantha* Rusby) or opo squash is a vine grown for its fruit, which can either be harvested young and used as a vegetable or harvested mature, dried, and used as a bottle, utensil, or pipe. The rounder varieties are called calabash gourds whereas the longer and slimmer kinds are usually known as bottle gourds. The fresh fruit has a light green smooth skin and a white flesh.

9.6.2 Utilization

The calabash, as a vegetable, is frequently used in southern Chinese cuisine either in a stir-fry or in a soup. In Japan, the species is known as hyōtan, large fruiting varieties are used mostly for making containers or other handicrafts and the smaller-fruiting varieties are more edible. In parts of India, the dried, unpunctured gourd is used as a float to learn swimming in rural areas. The

dried and cored thick outer skin has traditionally been used to make musical instruments like the tanpura. The baul singers of Bengal (India) and Bangladesh have their musical instruments made out of it. The practice is also common among Buddhist and Jain sages. In Vietnam, it is used in a variety of dishes: boiled, stir-fried, soup dishes and as a medicine. The shoots, tendrils and leaves of the plant may also be eaten as greens. The juice of bottle gourd is considered to have many medicinal properties and is very good for health.

9.6.3 Morphological description

Vegetative characters

The plant is a mesophytic herb, climbing (with tendrils), pubescent and succulent. It has a taproot system with less penetrating lateral roots. The stem is aerial, hollow, weak, succulent, angular, branched, pubescent and greenish, with tendrils developed from the leaf axile. The leaves are simple, heart shaped, opposite and pubescent, with a dentate margin, acuminate apex, stipulate and petiolate (small) (Fig. 9.28).

Floral characters

The inflorescence is auxiliary, and male and female flowers are arranged alternately. The flower is mostly white and yellow in colour, ebracteate, unisexual, pentamerous and hypogynous. There are five gamosepalous sepals: in the male flower these are bell shaped, in the female flower the calyx tube is connate with the ovary and pubescent. There are five free, yellow and pubescent petals. The male flower has five syandrous stamens and the anthers are dithecous. A pistillode is present. The female flower has from three to five staminodes present on the calyx edges. The gynoecium is unilocular, tricarpellary syncarpous and with parietal placentation. The fruit is a pepo with white coloured endocarp. The seeds are cylindrical in shape, brown coloured with a smooth surface and an average length of 15 mm and average breadth of 6 mm.

Fig. 9.28. Bottle gourd plant morphology: plant is a climbing (with tendrils) herb and pubescent. The stem is aerial, weak, succulent, angular and branched. The leaves are simple, heart shaped, with a dentate margin. Auxiliary inflorescence with the male and female flowers arranged alternately.

9.6.4 Anatomical description

Root anatomy

The epidermis is uniseriate with a single layer of compactly arranged thin-walled living cells. Epidermal cells produce root hairs, which are useful for the absorption of water (Fig. 9.29).

The cortex is multiseriate, relatively homogeneous and consists of general cortex and endodermis. The endodermis is single layered with compactly arranged barrel-shaped cells. Endodermal cells are characterized by the presences of Casparian strips. The cortex is multilayered and parenchymatic in nature.

The pericycle is the outermost region of the stele and is uniseriate and parenchymatous. Cells of the pericycle retain meristematic activity. Later roots arise endogenously from the pericycle. Vascular strands are radial, xylem is exarch and closed type. Pith is absent. The non-vascular tissue present between xylem and phloem strands is called conjunctive tissue.

Stem anatomy

The single outermost layer of the epidermis consists of compact barrel-shaped cells having no intercellular spaces (Fig. 9.30). The epidermis remains covered with a thin cuticle, which reduces transpiration. Multicellular epidermal hairs are present on the epidermis.

The cortex region consists of external collenchyma, chlorenchyma (photosynthetic tissue) and endodermis. The collenchyma lies immediately beneath the epidermis and consists of many layers of cells in the ridges, whereas in the furrows it is only two or three layered. Just below the collenchyma two layers of parenchyma containing chloroplasts are present, which help in the process of assimilation of photosynthates. The endodermis is the innermost layer of the cortex, lying immediately outside the sclerenchymatous zone of pericycle. This layer is wavy and contains many starch grains. Just beneath the endodermis there is a multilayered zone of sclerenchymatous pericycle. The cells are lignified and appear polygonal in cross section. The vascular bundles are found lying embedded in the thin-walled parenchyma cells of ground tissue. The ground tissue extends from just below the sclerenchymatous pericycle to the central medullary cavity.

Ten vascular bundles are arranged in two rows, those of the outer row corresponding to the ridges and those of the inner to the furrows. The vascular bundles are bicollateral, each consisting of xylem, two strips (inner and outer) of cambium and two strands of phloem (inner and outer). Bicollateral vascular bundles are important character of Cucurbitaceae family members. Medulla and medullary rays absent. The central part of stem contains a hollow region called the medullary cavity or pith cavity.

Leaf anatomy

The epidermis is single layered with compactly arranged barrel-shaped (tabular)

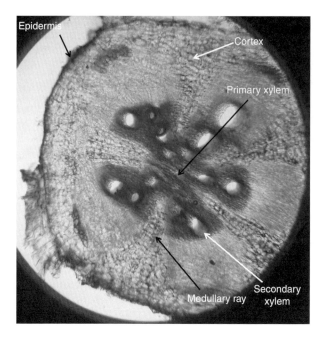

Fig. 9.29. Transverse section of root (100×) showing epidermis, cortex, primary xylem, secondary xylem and medullary rays.

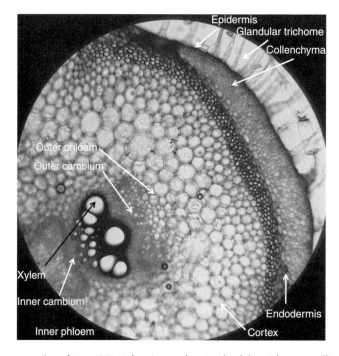

Fig. 9.30. Transverse section of stem (100×) showing epidermis, glandular trichomes, collenchyma, cortex and bicollateral vascular bundles.

cells. The outer walls of the epidermal cells are cutinized and covered by a continuous layer of cuticle. Anamocytic stomata are present in both of the epidermal layers. The stomatal index of the upper epidermis is 11.76 and of the lower epidermis is 19.39. The foot cell of the trichome and the subsidiary cell surrounding it contain cystolith. Trichomes are of the glandular type (Fig. 9.31).

The mesophyll is differentiated into an upper palisade tissue and lower spongy tissue. The palisade tissue is present below the upper epidermis; the cells are thin walled, cylindrical and contain numerous chloroplasts. The cells are arranged compactly in one layer, perpendicular to the upper epidermis. Narrow intercellular spaces are present in the palisade tissue. Palisade tissue is the most highly specialized type of photosynthetic tissue. The upper surface of the leaf is dark green in colour due to palisade tissue. The lower part of the mesophyll present towards the lower epidermis is the spongy tissue. Cells of the spongy tissue are thin walled, irregular in shape, and arranged loosely with large continuous intercellular spaces. The cells of spongy parenchyma contain fewer chloroplasts.

The vascular tissue occurs in the form of discrete bundles called veins, which are interconnected to form reticulate venation. Vascular bundles (veins) occur between the palisade and spongy tissue (median in position) and are collateral and closed. In the vascular bundle, xylem is present towards the upper epidermis and phloem towards the lower epidermis. The vascular bundle is surrounded by parenchymatous bundle sheath.

Petiole anatomy

The epidermis is uniseriate with a single layer of compactly arranged barrel-shaped living cells. The outer walls of the epidermal cells are cutinized. Cuticle reduces transpiration. Multicellular hairs are present on the epidermis.

Multilayered (from two to six layers) hypodermis of collenchyma cells is found immediately beneath the epidermis (Fig. 9.32). Collenchyma is living mechanical tissue and is of the angular type. Hypodermis gives considerable strength, flexibility and elasticity to young petioles and, having chloroplasts, it may carry out photosynthesis.

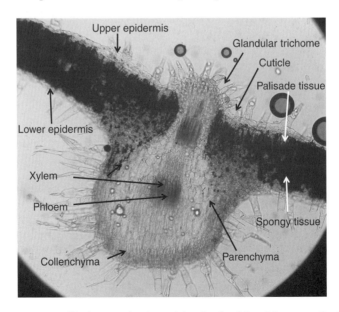

Fig. 9.31. Transverse section of leaf (100×) showing epidermis, glandular trichomes, palisade and spongy tissue and vascular bundles.

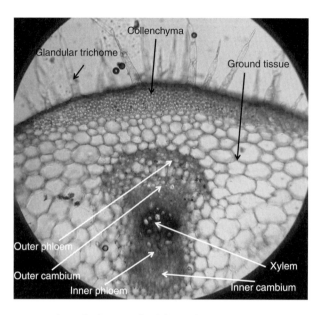

Fig. 9.32. Transverse section of petiole showing glandular trichomes on epidermis, collenchyma and bicollateral vascular bundles.

Just beneath the hypodermis ground tissue is found. It consists of thin-walled parenchymatous cells with well defined intercellular spaces among them. Vascular bundles are arranged in a half ring scattered in ground tissue.

Bicollateral vascular bundles are arranged in two rows. The central part of the petiole contains a hollow region, which is an important character of Cucurbitaceae members.

Seed coat anatomy

The seed coat is moderately thick. Testa is thick and macrosclereids are arranged compactly (Fig. 9.33).

Significance variations in anatomy

Future research may be directed in the following directions. Having a greater number of multicellular trichomes gives disease resistance to insects. The epicarp and pericarp is thick, so maintaining a long storage life and fruit borer resistance. The presence of glandular trichomes that are somewhat resistant to fungal diseases needs to be studied.

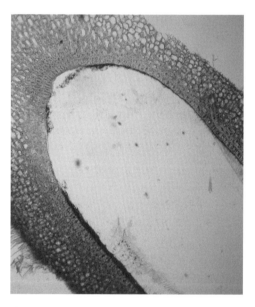

Fig. 9.33. Transverse section of seed coat showing micro- and macrosclereids.

9.7 Bitter Gourd

9.7.1 Introduction

Bitter gourd or bitter melon (*Momordica charantia* L.) is considered as a minor cucurbitaceous vegetable in spite of its considerable nutritional and medicinal properties.

Monoecious bitter gourd (*M. charantia* L. var. *minima* and *maxima* Williams & Ng), a cucurbit of major economic importance, is widely cultivated in India, China, Africa and South America. The morphology (i.e. growth habit and fruit shape, size, colour and surface texture) of Indian bitter gourd is diverse where gynoecious sexual forms exist.

Momordica charantia is a tropical and subtropical species belonging to the family Cucurbitaceae, and is widely grown for its edible fruit, which is among the most bitter of all fruits. Various names exist for the plant and its fruit, including bitter melon, bitter gourd, goya from the Japanese or Karela/Karella, ampalayá from Tagalog, and cerasee (Caribbean and South America; also spelled cerasse).

9.7.2 Origin

The original home of the species is not known, other than that it is a native of the tropics. It is widely grown in India and other parts of the Indian subcontinent, South-east Asia, China, Africa and the Caribbean. Bitter gourd contains a bitter compound called momordicin that is said to have a stomachic effect.

9.7.3 Utilization

Fruits of bitter gourd are used in the preparation of numerous dishes in India and some Asian countries. Bitter gourd has been used in various Asian traditional medicine systems for a long time. Like most bitter-tasting foods, bitter gourd stimulates digestion. While this can be helpful in people with sluggish digestion, dyspepsia and constipation, it can sometimes make heartburn and ulcers worse. The fact that bitter melon is also a demulcent and at least a mild inflammation modulator, however, means that it rarely does have these negative effects, based on clinical experience and traditional reports. Leaves of the plant are brewed in hot water to create a tea to treat malaria and diabetes. The leaves are allowed to steep in hot water before being strained thoroughly so that only the remaining liquid is used for the tea.

9.7.4 Morphological description

Vegetative characters

The plant is a mesophytic herb, climbing, (with tendrils) highly pubescent and succulent. It has a taproot system with superficially spreading and not profuse roots. The stem is aerial, weak, succulent, angular, branched, highly pubescent and greenish, with tendrils developed from the leaf axial. The leaves are alternate, petiolate, (pubescent), stipulate, simple, palmately lobed (seven lobes) and heart shaped, with a serrate margin, acuminate apex and reticulate venation (Fig. 9.34).

Fig. 9.34. Bitter gourd plant morphology: plant is a climbing (with tendrils) herb and pubescent. The stem is aerial, weak, succulent, angular and branched. The leaves are simple, heart shaped, with a dentate margin. Auxiliary inflorescence with the male and female flowers arranged in racemose.

Floral characters

The inflorescence is auxiliary, with male and female flowers alternately arranged in racemose. The flower is mostly yellow in colour, ebracteate, unisexual, pentamerous and hypogynous. There are five sepals: in the male flower they are bell shaped; in the female flower the calyx tube is connate with the ovary, pubescent. There are five free, yellow and pubescent petals. The male flower has five stamens, syandrous, and the anthers are dithecous. A pistillode is present. The female flower has from three to five staminodes present on the calyx edges. The gynoecium is unilocular, tricarpellary syncarpous and with parietal placentation. The fruit is dehiscent and fleshy. Seeds are ovate in shape, brown, with a rough surface and average length of 15 mm and breadth of 8 mm.

9.7.5 Anatomical description

Root anatomy

The epidermis is a uniseriate with single layer of compactly arranged thin-walled living cells. Epidermal cells produce root hairs, which are useful for absorption of water.

The cortex is multiseriate and relatively homogeneous. It consists of general cortex and endodermis (Fig. 9.35). It consists of loosely arranged thin walled parenchyma cells with prominent intercellular spaces. The cells are usually round or oval, colourless and store starch. Endodermis is single layered with compactly arranged barrel-shaped cells. Endodermal cells are characterized by the presences of Casparian strips.

Stele forms the central part of the root and consists of pericycle, vascular strands and conjunctive tissue. Pericycle is the outermost region of the stele and is uniseriate and parenchymatous. Cells of the pericycle retain meristematic activity. Later roots arise endogenously from the pericycle. The vascular strands are radial and the xylem is exarch, closed type. Pith is absent. Conjunctive tissue is present between xylem and phloem strands.

Stem anatomy

The epidermis is uniseriate with a single layer of compactly arranged barrel-shaped living cells. The outer walls of the epidermal cells are cutinized. Cuticle reduces transpiration. Few stomata are present in the epidermis for the exchange of gases and transpiration. All the epidermal cells are colourless except

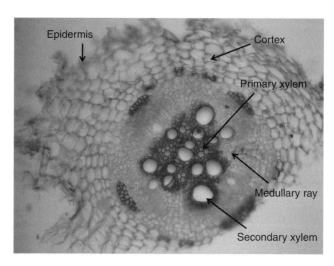

Fig. 9.35. Transverse section of root (100×) showing epidermis, cortex, primary xylem, secondary xylem and medullary ray.

the guard cells. Multicellular hairs are present on the epidermis.

Cortex is the middle region present between the epidermis and stele, and is thinner than the stele. A section shows hypodermis, middle cortex and endodermis.

There are three to four layers of angular collenchyma present below the ridges. Collenchyma is living mechanical tissue. Hypodermis gives considerable strength, flexibility and elasticity to young stems. It may carry out photosynthesis, as chloroplasts are present. Endodermis forms the innermost layer of the cortex, and is usually distinct in stems without Casparian bands.

Stele is the central region of the stem and is thicker than the cortex. It consists of pericycle, vascular bundles, medulla and medullary rays. Pericycle is the outermost non-vascular region of the stele and is multiseriate and sclerenchymatic.

Bicollateral vascular bundles are arranged in two rows (Fig. 9.36). Each vascular bundle is wedge-shaped, conjoint, open and endarch. The central part of stem contains a hollow region, which is an important character of Cucurbitaceae family members.

Leaf anatomy

Epidermis is present on both surfaces of the leaf blade, i.e. the upper epidermis and lower epidermis. The epidermis is single layered with compactly arranged barrel-shaped (tabular) cells. The outer walls of the epidermal cells are cutinized and covered by a continuous layer of cuticle. Anamocytic stomata is present in both of the epidermal layers. The foot cell of the trichome and the subsidiary cell surrounding it contain cystolith. Trichomes are of the glandular type.

The ground tissue of the leaf, present between the upper and the lower epidermal layers, is called mesophyll. The mesophyll is chlorenchymatic and is concerned mainly with photosynthesis. The mesophyll is differentiated into an upper palisade tissue and lower spongy tissue (Fig. 9.37). Palisade tissue is present below the upper epidermis. The palisade cells are thin walled, cylindrical and contain numerous chloroplasts. The cells are arranged compactly in one layer, perpendicular to the upper epidermis. Narrow intercellular spaces are present in the palisade tissue.

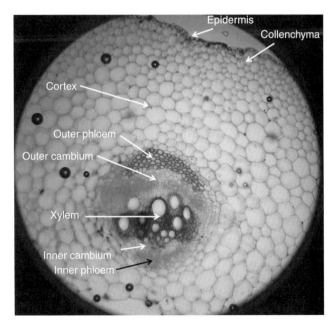

Fig. 9.36. Transverse section of stem (100×) showing epidermis, collenchyma, cortex and bicollateral vascular bundles.

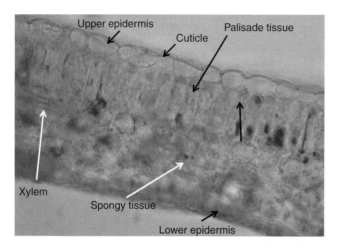

Fig. 9.37. Transverse section of leaf (100×) showing different parts.

Palisade tissue is the most highly specialized type of photosynthetic tissue. The upper surface of the leaf is dark green in colour due to palisade tissue. The lower part of the mesophyll present towards the lower epidermis is the spongy tissue. Cells of the spongy tissue are thin walled, irregular in shape, arranged loosely with large continuous intercellular spaces. Stomata open into large intercellular spaces called sub-stomatal chambers. The cells of spongy parenchyma contain a lower number of chloroplasts, hence the lower surface of the leaf is light green in colour.

The vascular tissue occurs in the form of discrete bundles called veins. Veins are interconnected to form reticulate venation. Vascular bundles (veins) occur between the palisade and spongy tissue (median in position) and are collateral and closed. In the vascular bundle, xylem is present towards the upper epidermis and phloem towards the lower epidermis. The vascular bundle is surrounded by parenchymatous bundle sheath.

Petiole anatomy

The epidermis is uniseriate with a single layer of compactly arranged barrel-shaped living cells. The outer walls of the epidermal cells are cutinized. Cuticle reduces transpiration. Multicellular hairs are present on the epidermis. Multilayered (two to six layers) hypodermis of angular collenchyma cells is found immediately beneath the epidermis. Collenchyma is living mechanical tissue. The hypodermis gives considerable strength, flexibility and elasticity to young petioles, and, having chloroplasts, it may carry out photosynthesis.

Just beneath the hypodermis ground tissue is found (Fig. 9.38). It consists of thin walled parenchymatous cells with well-defined intercellular spaces among them. Vascular bundles are arranged in a half ring scattered in ground tissue.

Bicollateral vascular bundles are arranged in two rows. The centre part of petiole is hollow, which is an important character of Cucurbitaceae family members.

Seed coat anatomy

The seed coat is thick and consists of hairs. Macrosclereids are rectangular in shape and very compactly arranged.

9.7.6 Fruit development

Fruit development in bitter gourd depends on temperature, and early development is observed in summer compared to winter. It took 12–16 days for harvest, with fruits turning yellow in 16 days in Taiwan, whereas it took 20–22 days for harvest under lower temperatures in the cold growth season of

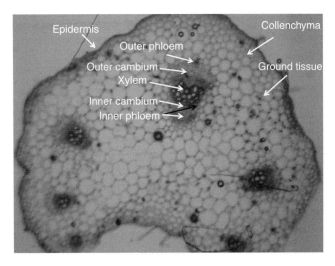

Fig. 9.38. Transverse section of petiole showing different parts.

1992, with fruits turning yellow 24 days after pollination. Seeds also developed rapidly at high temperatures. About 50% of seeds developed into third grade with sizes near maturity at 9–10 days after pollination. Fruit growth, seed and embryo development were delayed in the cold growth season of 1992 (Chang *et al.*, 2000).

9.7.7 Significance variations in anatomy

Future research may be directed in the following directions. Having a higher number of multicellular trichomes gave resistance to insect borne diseases. The presence of glandular trichomes makes the plant somewhat resistant to fungal diseases. the epicarp is thick, so is helpful in long storage and fruit borer resistance.

9.8 Cucumber

9.8.1 Introduction

Cucumber (*Cucumis sativa* L.) is a very important vegetable fruit used as salad for its high nutritive value. The genus includes some economically important and widely grown vegetables such as cucumbers and melons.

9.8.2 Origin

Cucumbers originated in India large where a large amount of genetic variety of cucumber has been observed. It has been cultivated for at least 3000 years in western Asia, and was probably introduced to parts of Europe by the Romans. Records of cucumber cultivation appear in France in the ninth century, England in the 14th century and in North America by the mid-16th century.

9.8.3 Morphological description

Vegetative characters

The plant is a mesophytic herb, climbing (with tendrils), highly pubescent and succulent. It has a taproot system with superficially spreading lateral roots. The stem is aerial, weak, succulent, angular, branched, highly pubescent and greenish, with tendrils developed from the leaf axial. Leaves are simple, heart shaped, with a serrate margin, acuminate apex and reticulate venation, alternate and pubescent, stipulate and petiolate (pubescent) (Fig. 9.39).

Floral characters

The inflorescence is auxiliary, male and female flowers are arranged in racemose.

Fig. 9.39. Cucumber plant morphology: plant is a climbing (with tendrils) herb and pubescent. The stem is aerial, weak, succulent, angular and branched. The leaves are simple, heart shaped, with a serrate margin. Auxiliary inflorescence with the male and female flowers arranged in racemose.

The flower is mostly yellow in colour, ebracteate, unisexual, pentamerous and hypogynous. There are five gamosepalous sepals; in the male flower these are bell shaped, in the female flower the calyx tube is connate with the ovary and pubescent. There are five free, yellow and pubescent petals. The male flower has five syandrous stamens and the anthers are dithecous (four-chambered); the presence of a pistillode is notable. The female flower has three to five staminodes present on the calyx edges. The gynoecium is unilacular, tricarpellary syncarpous and with parietal placentation. The fruit is a pepo, green to light yellow in colour and oblong to round.

9.8.4 Anatomical description

Root anatomy

Epidermis is the outermost region of the root and is uniseriate, with a single layer of compactly arranged thin-walled living cells consisting of root hairs. The epidermis ruptures due to secondary growth.

The cortex is the middle region, lying between the epidermis and the stele. Cortex is well developed, multiseriate and relatively homogeneous and is thicker than the stele. It consists of general cortex and endodermis. The general cortex consists of loosely arranged thin-walled parenchyma cells with prominent intercellular spaces (Fig. 9.40), which is compressed due to secondary growth. The cells are usually round or oval, colourless and store starch. This region helps in the lateral conduction of water and salts. Endodermis forms the innermost layer of the cortex and is very distinct in roots. It is single layered, with compactly arranged barrel-shaped cells. Endodermal cells are characterized by the presence of Casparian strips or Casparian bands. After secondary growth the general endodermis ruptures. Endodermis of the root acts as barrier between the cortex and stele.

The stele consists of pericycle, vascular strands and pith. Xylem and phloem form the vascular tissues. Xylem is meant for conduction of water and salts and phloem for the transport of food materials. The xylem is exarch, with protoxylem towards the pericycle and metaxylem towards the centre and scattered secondary xylem vessels. Cambium is absent and hence vascular bundles are described as closed type. Pith is absent. The non-vascular tissue present between xylem and phloem strands is called conjunctive tissue and may be totally parenchymatous or a part of it may be converted into sclerenchyma. During secondary growth the cambial cells that originate from the pericycle

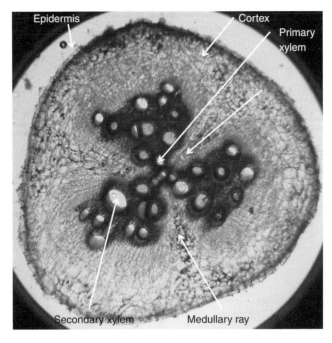

Fig. 9.40. Transverse section of root (100×) illustrating various parts.

lying against the group of protoxylem function as ray initials and produce medullary rays. These rays are transversed in the xylem and phloem through cambium; this is a characteristic feature of the roots.

Stem anatomy

The epidermis is the outermost region surrounding the stem and is uniseriate, with a single layer of compactly arranged barrel-shaped living cells. The outer walls of the epidermal cells are cutinized. Cuticle reduces the transpiration. Few stomata are present in the epidermis for the exchange of gases and transpiration. All the epidermal cells are colourless except the guard cells. Multicellular hairs are present on the epidermis.

The cortex is the middle region present between the epidermis and stele and is thinner than the stele. In section, it shows hypodermis, middle cortex and endodermis. Below the ridges from three to four layers of angular collenchyma are present (Fig. 9.41). Collenchyma is living mechanical tissue. The hypodermis gives considerable strength, flexibility and elasticity to young stems and,

having chloroplasts, it may carry out photosynthesis. The middle cortex is also multi-layered and parenchymatic. The cells may be round or oval with prominent intercellular spaces. This region also stores food materials temporarily. Endodermis is innermost layer of the cortex and is usually distinct in stems without Casparian bands.

The stele is the central region of the stem and is thicker than the cortex. It consists of pericycle, vascular bundles, and medulla and medullary rays. Pericycle is the outermost non-vascular region of the stele and is multiseriate and sclerenchymatic. Bicollateral vascular bundles are arranged in two rows. Each vascular bundle is wedge-shaped, conjoint, open and endarch. The central part of the stem contains a hollow region, which is an important character of Cucurbitaceae family members.

Leaf anatomy

Epidermis is present on both surfaces of the leaf blade, i.e. upper epidermis and lower epidermis. The epidermis is single layered, with compactly arranged barrel-shaped (tabular)

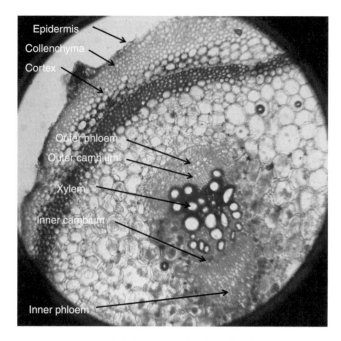

Fig. 9.41. Transverse section of stem (100×) showing epidermis, collenchyma, cortex and bicollateral vascular bundles.

cells. The outer walls of the epidermal cells are cutinized and covered by a continuous layer of cuticle. Anamocytic stomata are present in both the epidermal layers. The foot cell of the trichome and the subsidiary cell that surrounds the foot cell of the trichome contains a cystolith. Trichomes are of the glandular type. The stomatal index of the upper epidermis ranges from 18 to 23 and that of the lower epidermis from 26 to 28.

The ground tissue of the leaf, present between the upper and the lower epidermal layers, is called mesophyll. The mesophyll is chlorenchymatic and is concerned mainly with photosynthesis. The mesophyll is differentiated into upper palisade tissue and lower spongy tissue.

The palisade tissue is present below the upper epidermis. The palisade cells are thin walled, cylindrical, contain numerous chloroplasts and are compactly arranged in one layer, perpendicular to the upper epidermis. Narrow intercellular spaces are present in the palisade tissue. Palisade tissue is the most highly specialized type of photosynthetic tissue. The upper surface of the leaf is dark green in colour due to palisade tissue.

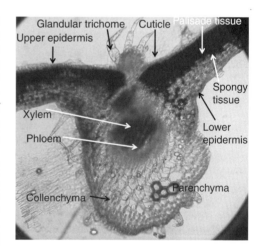

Fig. 9.42. Transverse section of leaf (100×) showing epidermis, palisade and spongy tissue and vascular bundle.

The lower part of the mesophyll present towards the lower epidermis is the spongy tissue (Fig. 9.42). Cells of the spongy tissue are thin-walled, irregular in shape and arranged loosely with large continuous intercellular spaces. Stomata open into large intercellular spaces called sub-stomatal chambers. The cells of spongy parenchyma contain fewer chloroplasts, hence the lower surface of the leaf is light green in colour.

The vascular tissue occurs in the form of discrete bundles called veins. Veins are interconnected to form reticulate venation. Vascular bundles (veins) occur between the palisade and spongy tissue (median in position). They are collateral and closed. In the vascular bundle, xylem is present towards the upper epidermis and phloem towards the lower epidermis. The vascular bundle is surrounded by parenchymatous bundle sheath.

Glandular trichomes – insect resistance

The leaf epidermis and their effects on behaviour of *Aphis gossypii* Glover were evaluated in two *Cucumis melo* L. genotypes (aphid susceptible, aphid resistant). No differential effects of epicuticular waxes on aphid behaviour were observed. The type, distribution and number of trichomes on melon leaves were also studied. Pubescence in melon, measured as the number of non-glandular trichomes per square centimetre, was not sufficient to prevent aphid settling. There was a high density of type I glandular trichomes on leaves of the aphid-resistant genotype. These results indicate that a high density of glandular trichomes and chemicals secreted by them deter *A. gossypii* and disturb aphid settling on aphid-resistant varieties (Sarria *et al.*, 2010).

Leaf senescence

Detached cucumber (*Cucumis sativa* L.) leaves show senescence and rejuvenation after cultivation in nutrient solution for 4 weeks. Rooting of the petiole elicited a combination of different morphological, anatomical and physiological changes in the lamina. Extensive growth in area and thickness, changes of chloroplast structure and activity, as well as the pattern of Chl-protein complexes were observed and compared either to the corresponding parameters of young detached leaves. The hormones showed mutuality in their effects, benzyladenine being responsible for the growth of cells, while indolylacetic acid and kinetin promoted an increase in chlorophyll content (Kovács *et al.*, 2007).

Petiole anatomy

The epidermis is uniseriate, with a single layer of compactly arranged barrel-shaped living cells (Fig. 9.43). The outer walls of

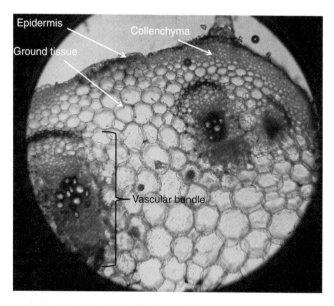

Fig. 9.43. Transverse section of petiole (100×) showing epidermis, collenchyma, ground tissue and bicollateral vascular bundles.

the epidermal cells are cutinized. Cuticle reduces the transpiration. Multicellular hairs are present on the epidermis.

Multilayered (two to five layers) hypodermis of angular collenchyma cells is found immediately beneath the epidermis. Collenchyma is living mechanical tissue. The hypodermis (collenchyma) gives considerable strength, flexibility and elasticity to young petioles and, having chloroplasts, it may carry out photosynthesis. Just beneath the hypodermis ground tissue is found, which consists of thin-walled parenchymatous cells having well-defined intercellular spaces among them. Vascular bundles are arranged in a half ring scattered in ground tissue. Bicollateral vascular bundles are arranged in two rows. The centre part of the petiole contains a hollow region, which is an important character of Cucurbitaceae members.

Development and anatomy of the staminate flower

All organs arise in a low spiral arrangement; the staminate flower produces three stamens. The morphological nature of these is uncertain. Heimlich (1927) inclines to the view that two of them are normal, complete stamens, and the odd one a stamen in which one theca fails to develop. One large bundle enters the base of each complete stamen. This is a double bundle in the sense that an original bundle coming from the vascular plate in the receptacle forks and then very soon reunites its two branches. The large bundle sends off a right and a left branch in the broad connective. Each branch forks sharply to give off one ascending and one descending branch. The main bundle continues above the right and left branches and ultimately forks again, sending a branch into each of the two stamen prolongations. The odd stamen is similar to the complete stamens except that the whole vascular distribution is exactly one-half that of the complete stamen. This difference is apparently due to the failure of the respective original

bundle to fork. A large schizo-lysigenous cavity develops in the perianth tube, extending into the connective and stamen prolongations. Stomata are found on the outer surfaces of the perianth tube and the calyx lobes. The corolla seems regularly free from stomata except at the extreme tips of the lobes. The mesophyll of the corolla is undifferentiated (Heimlich, 1927).

Seed coat anatomy

The seed coat is very thick; the testa is thick but the macrosclereids are not very compactly arranged, so water enters into the seeds resulting in a low germination time, i.e. 24 h (Fig. 9.44).

Significance variations in anatomy

Future research may be directed in the following directions. The possession of a higher number of multicellular trichomes gives disease resistance to insects. The epicarp and pericarp is thick, so is helpful in long storage and fruit borer resistance. The presence of silica may provide resistance to powdery mildew.

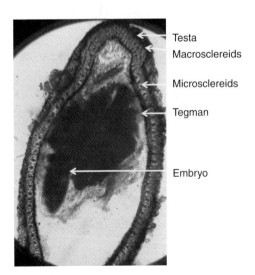

Fig. 9.44. Transverse section of cucumber seed showing inner embryo covered by outer testa and inner tegman. Macrosclereids are almost rectangular in shape, microsclereids are small dot-like structures.

9.9   Ridge Gourd

9.9.1   Introduction

The luffa, loran, or luau are tropical and sub-tropical vines comprising the genus *Luffa*, the only genus of the subtribe Luffinae. The fruit of at least two species, *Luffa acutangula* and *Luffa aegyptiaca* (*L. cylindrica*), is grown, harvested before maturity, and eaten as a vegetable in Asia and Africa.

9.9.2   Morphological description

Vegetative characters

The plant is mesophytic climbing (with tendrils) herb, pubescent and succulent. It has a taproot system and lateral roots do not penetrate very deep. The stem is hollow, aerial, weak, succulent, angular, branched, pubescent and greenish, with tendrils developed from the leaf axial. Leaves are simple, alternate and pubescent, heart shaped, with a wavy margin, acute apex, stipulate, petiolate and with reticulate venation (Fig. 9.45).

Fig. 9.45. Ridge gourd plant morphology: plant is a climbing (with tendrils) herb and pubescent. The stem is aerial, weak, succulent, angular and branched. Leaves are simple, heart shaped, with a serrate margin. The inflorescence is auxiliary, with male and female flowers arranged on the same axial of the leaf.

Floral characters

The inflorescence is auxiliary, with female flower and male flowers arranged on separate peduncles arising from same axial of the leaf. The flowers are mostly yellow in colour, ebracteate, unisexual, pentamerous and hypogynous. There are five gamosepalous sepals; in the male flower they are bell shaped, in the female flower the calyx tube is connate with the ovary, pubescent. There are five free, yellow and pubescent petals. The male flower has five stamens; free anthers are dithecous and a pistillode is present. The female flower has three to five staminodes present on the calyx edges and the gynoecium is unilacular, tricarpellary syncarpous, and with parietal placentation. The fruit is a pepo, green in colour, with ridges and grooves. Seeds are ovate, black, with shining band-like structures on the rough surface of the seed and an average length of 12 mm and average breadth of 6 mm.

9.9.3   Anatomical description

Root anatomy

The epidermis is uniseriate, with a single layer of compactly arranged thin-walled living cells. Epidermal cells produce root hairs, which are useful for absorption of water.

The cortex is multiseriate, relatively homogeneous and consists of general cortex and endodermis. Endodermis is single layered, with compactly arranged barrel-shaped cells. Endodermal cells are characterized by the presence of Casparian strips.

Stele forms the central part of the root. The stele consists of pericycle, vascular strands and conjunctive tissue. Pericycle is the outermost region of the stele and is uniseriate and parenchymatous. Cells of the pericycle retain meristematic activity. Later roots arise endogenously from the pericycle. Vascular strands are radial; xylem is exarch, closed type. The non-vascular tissue present between xylem and phloem strands is called conjunctive tissue.

Stem anatomy

The stem shows ridges and grooves. Primary stem consists of epidermis, cortex and stele.

The epidermis is the outermost region surrounding the stem. It is uniseriate, with a single layer of compactly arranged barrel-shaped living cells. The outer walls of the epidermal cells are cutinized. Cuticle reduces the transpiration. Few stomata are present in the epidermis for exchange of gases and transpiration. All the epidermal cells are colourless except the guard cells. Multicellular hairs are present on the epidermis.

Below the ridges from three to four layers of angular collenchyma is present. Collenchyma is living mechanical tissue. The hypodermis gives considerable strength, flexibility and elasticity to young stems and, having chloroplasts, it may carry out photosynthesis. The middle cortex is also multi-layered and parenchymatic. The cells may be round or oval with prominent intercellular spaces. This region also stores food materials temporarily. Endodermis is usually distinct in stems without Casparian bands.

Pericycle is the outermost non-vascular region of the stele and is multiseriate and sclerenchymatic. Bicollateral vascular bundles are arranged in two rows. Each vascular bundle is wedge-shaped, conjoint, open and endarch. The central part of stem contains a hollow region.

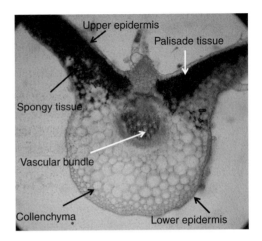

Fig. 9.46. Transverse section of leaf (100×) showing epidermis, collenchyma and vascular bundles.

Leaf anatomy

The epidermis is single layered, with compactly arranged barrel-shaped (tabular) cells. The outer walls of the epidermal cells are of cutinized and covered by a continuous layer of cuticle. Anamocytic stomata are present in both the epidermal layers. The stomatal index of the upper epidermis is 18.26 and of the lower epidermis is 26.5. The foot cell of the trichome and the subsidiary cell surrounding the foot cell contains cystoliths. Trichomes are glandular type.

The ground tissue of the leaf, present between the upper and the lower epidermal layers, is called mesophyll (Fig. 9.46). The mesophyll is chlorenchymatic and is concerned mainly with photosynthesis. The mesophyll is differentiated into an upper palisade tissue and lower spongy tissue. The palisade cells are thin walled, cylindrical and contain numerous chloroplasts. The cells are compactly arranged in one layer, perpendicular to the upper epidermis. Narrow intercellular spaces are present in the palisade tissue. Palisade tissue is the most highly specialized type of photosynthetic tissue. The upper surface of the leaf is dark green in colour due to palisade tissue. Cells of the spongy tissue are thin-walled, irregular in shape and arranged loosely with large continuous intercellular spaces. Stomata open into large intercellular spaces called sub-stomatal chambers.

The vascular tissue occurs in the form of discrete bundles called veins. Veins are interconnected to form reticulate venation. Vascular bundles (veins) occur between the palisade and spongy tissue (median in position) and are collateral and closed. In the vascular bundle, xylem is present towards the upper epidermis and phloem towards the lower epidermis. The vascular bundle is surrounded by parenchymatous bundle sheath.

Petiole anatomy

The epidermis is uniseriate, with a single layer of compactly arranged barrel-shaped living cells. The outer walls of the epidermal cells are cutinized. Cuticle reduces the transpiration. Multicellular hairs are present on the epidermis.

Multilayered (two to five layers) hypodermis of angular collenchyma cells is found immediately beneath the epidermis.

Collenchyma is living mechanical tissue. The hypodermis gives considerable strength, flexibility and elasticity to young petioles and, having chloroplasts, it may carry out photosynthesis.

Just beneath the hypodermis, ground tissue is found (Fig. 9.47), consisting of thin-walled parenchymatous cells having well defined intercellular spaces among them. Vascular bundles are arranged in a half ring scattered in ground tissue.

Bicollateral vascular bundles are arranged in two rows. The central part of the petiole contains a hollow region, which is an important character of Cucurbitaceae members.

Seed coat anatomy

During the germination of the seed, mucilage is produced, which enhances the absorption of water. The testa contains long macrosclereids with gaps between them, allowing water to enter into the seed. Time for initial germination is 36 h (Fig. 9.48).

Significance variations in anatomy

Future research may be directed in the following directions. A higher number of multicellular trichomes gives resistance to insect borne diseases. The epicarp is thick, so helpful in long storage and fruit borer resistance. The presence of silica may be involved with resistance to powdery mildew.

9.10 Sponge Gourd

9.10.1 Introduction

Sponge gourd belongs to the family Cucurbitaceae. It is a minor vegetable and predominantly grown in south Asian countries such as India, Pakistan etc. The fruit is used in preparing dishes.

9.10.2 Morphological description

Vegetative characters

The plant is mesophytic, climbing (with tendrils) herb, pubescent and succulent, and has a taproot system with shallow rooting. The stem is hollow, aerial, weak, succulent, angular, branched, pubescent and greenish, with tendrils developed from the leaf axil. The leaves are simple, alternate, pubescent, larger than in ridge gourd, heart

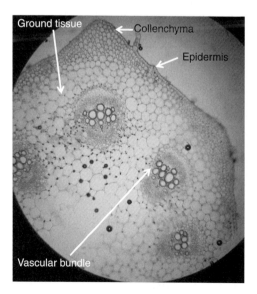

Fig. 9.47. Transverse section of petiole (100×) showing epidermis, collenchyma, ground tissue and vascular bundles.

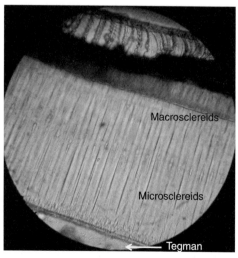

Fig. 9.48. Transverse section of seed coat showing micro- and macrosclereids.

shaped, with a dentate margin, reticulate venation and acute apex, stipulate and petiolate (small) (Fig. 9.49).

Floral characters

The inflorescence is auxiliary, male and female flowers are arranged alternately. The flowers are mostly yellow in colour, ebracteate, unisexual, pentamerous and hypogynous. There are five gamosepalous sepals; in the male flower these are shaped, in the female flower the calyx tube is connate with the ovary, pubescent. There are five petals, free, yellow and pubescent. In the male flower there are five free stamens; the anthers are dithecous and a pistillode is present. The female flower has from three to five staminodes present on the calyx edges. The gynoecium is unilacular, tricarpellary syncarpous and with parietal placentation. The fruit is a pepo, and the seeds are ovate, black, smooth, with a shiny surface and average length of 12 mm and width of 6 mm.

Fig. 9.49. Sponge gourd plant morphology: plant is a climbing (with tendrils) herb and pubescent. The stem is aerial, weak, succulent, angular and branched. The leaves are simple, larger than ridge gourd, heart shaped, with a dentate margin and acute apex. Auxiliary inflorescence with the male and female flowers arranged alternately.

9.10.3 Anatomical description

Root anatomy

The epidermis is uniseriate, with a single layer of compactly arranged thin-walled living cells. Epidermal cells produce root hairs, which are useful for absorption of water. The cortex is multiseriate, relatively homogeneous (Fig. 9.50) and consists of general cortex and endodermis. The general cortex consists of loosely arranged thin-walled parenchyma cells with prominent intercellular spaces. The cells are usually round or oval.

The endodermis is single layered, with compactly arranged barrel-shaped cells. The endodermal cells are characterized by the presence of Casparian strips.

Stele forms the central part of the root, and consists of pericycle, vascular strands and conjunctive tissue. Pericycle is the outermost region of the stele and is uniseriate and parenchymatous. Cells of the pericycle retain meristematic activity. Later roots arise endogenously from the pericycle. The vascular strands are radial, xylem is exarch, closed type. Pith is absent. The non-vascular tissue present between xylem and phloem strands is called conjunctive tissue.

Stem

The epidermis is the outermost region surrounding the stem. It is uniseriate, with a single layer of compactly arranged barrel-shaped living cells (Fig. 9.51). The outer walls of the epidermal cells are cutinized. Cuticle reduces the transpiration. Few stomata are present in the epidermis for exchange of gases and transpiration. All the epidermal cells are colourless except the guard cells. Multicellular hairs are present on the epidermis.

The cortex is the middle region present between the epidermis and stele and is thinner than the stele. A section of the stem shows hypodermis, middle cortex and endodermis.

Below the ridges, from three to four layers of angular collenchyma are present in the hypodermis. Collenchyma is living mechanical tissue. The hypodermis gives considerable strength, flexibility and elasticity to young stems and, having chloroplasts, it may carry out photosynthesis.

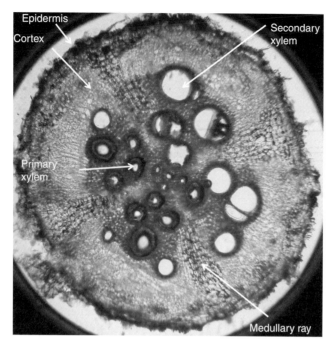

Fig. 9.50. Transverse section of root (100×) showing epidermis, cortex, primary and secondary xylem and medullary rays.

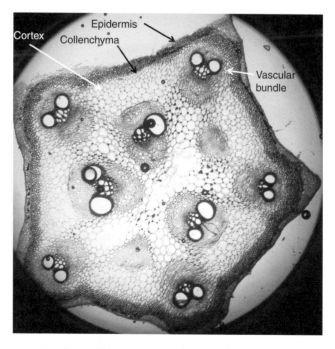

Fig. 9.51. Transverse section of stem (100×) showing epidermis, collenchyma and vascular bundles.

The middle cortex is also multilayered and parenchymatic. The cells may be round or oval with prominent intercellular spaces. This region also stores food materials temporarily. Endodermis is innermost layer of the cortex and is usually distinct in stems without Casparian bands.

Stele is the central region of the stem and is thicker than the cortex. It consists of pericycle, vascular bundles and medulla and medullary rays. Pericycle is the outermost non-vascular region of the stele, and is multiseriate and sclerenchymatic. Bicollateral vascular bundles are arranged in two rows (Fig. 9.51). Each vascular bundle is wedge-shaped, conjoint, open and endarch. Medulla and medullary rays are absent, and the central part of stem contains a hollow region, which is an important character of Cucurbitaceae members.

Leaf anatomy

Epidermis is present on both surfaces of the leaf blade, i.e. upper epidermis and lower epidermis. The epidermis is single layered, with compactly arranged barrel-shaped (tabular)

cells. The outer walls of the epidermal cells are cutinized and covered by a continuous layer of cuticle. The stomatal index of upper epidermis is 10.52 and that of the lower epidermis is 14.6. Anamocytic stomata are present in both of the epidermal layers. The foot cell of the trichome and the subsidiary cell surrounding it contain cystoliths. The trichomes are of the glandular type.

The ground tissue of the leaf, present between the upper and the lower epidermal layers, is called mesophyll. The mesophyll is chlorenchymatic and is concerned mainly with photosynthesis. The mesophyll is differentiated into an upper palisade tissue and lower spongy tissue (Fig. 9.52).

The palisade tissue is present below the upper epidermis. The palisade cells are thin walled, cylindrical and contain numerous chloroplasts, and are compactly arranged in one layer, perpendicular to the upper epidermis. Narrow intercellular spaces are present in the palisade tissue. Palisade tissue is the most highly specialized type of photosynthetic tissue. The upper surface of the leaf is dark green in colour due to palisade tissue.

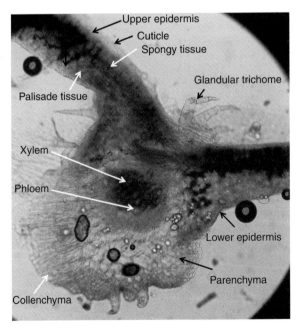

Fig. 9.52. Transverse section of leaf (100×) showing epidermis, collenchyma, ground tissue and vascular bundles.

The lower part of the mesophyll present towards the lower epidermis is the spongy tissue. Cells of the spongy tissue are thin walled, irregular in shape, and arranged loosely with large continuous intercellular spaces. Stomata open into large intercellular spaces called sub-stomatal chambers. The cells of spongy parenchyma contain fewer chloroplasts, hence the lower surface of the leaf is light green in colour.

The vascular tissue occurs in the form of discrete bundles called veins. Veins are interconnected to form reticulate venation. Vascular bundles (veins) occur between the palisade and spongy tissue (median in position) and are collateral and closed. In the vascular bundle, xylem is present towards the upper epidermis and phloem towards the lower epidermis. The vascular bundle is surrounded by parenchymatous bundle sheath.

Petiole anatomy

The epidermis is uniseriate, with a single layer of compactly arranged barrel-shaped living cells. The outer walls of the epidermal cells are cutinized. Cuticle reduces the transpiration. Multicellular hairs are present on the epidermis.

Multilayered (from two to six layers) hypodermis of angular collenchyma cells is found immediately beneath the epidermis (Fig. 9.53). Collenchyma is living mechanical tissue. The hypodermis gives considerable strength, flexibility and elasticity to young petioles and, having chloroplasts, it may carry out photosynthesis.

Just beneath the hypodermis, ground tissue is found. It consists of thin-walled parenchymatous cells having well-defined intercellular spaces among them. Vascular bundles are arranged in a half ring scattered in ground tissue.

Bicollateral vascular bundles are arranged in two rows. The central part of the petiole contain a hollow region, which is an important character of Cucurbitaceae members.

Seed coat anatomy

The seed coat is thin and macrosclereids are not very compactly arranged, meaning that the rate of imbibition and speed of germination requires less time (Fig. 9.54).

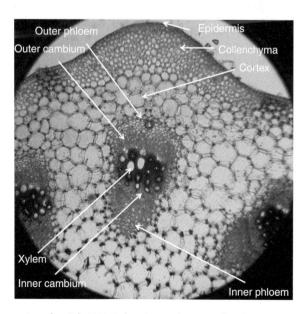

Fig. 9.53. Transverse section of petiole (100×) showing epidermis, collenchyma, ground tissue and vascular bundles.

Fig. 9.54. Transverse section of seed (100×) showing testa and tegman.

Significance variations in anatomy

Future research may be directed in the follow-ing directions. The possession of a higher number of multicellular trichomes gives disease resistance to insects. The epicarp is thick, so is helpful in long storage and fruit borer resistance, and the presence of silica may increase resistance to powdery mildew.

9.11   Cabbage

9.11.1   Introduction

The cabbage is a popular cultivar of the spe-cies *Brassica oleracea* Linne (Capitata group) of the family Brassicaceae (or Cruciferae), and is used as a leafy green vegetable. It is an herbaceous, biennial, dicotyledonous flow-ering plant distinguished by a short stem upon which is crowded a mass of leaves, usually green but in some varieties red or purplish, which while immature form a characteristic compact, globular cluster (cab-bagehead). The plant is also called head cab-bage or heading cabbage, and in Scotland a bowkail, from its rounded shape. The Scots call the stalk a castock, and the British occa-sionally call its head a loaf. It is in the same genus as the turnip, *Brassica rapa* L.

Cabbage leaves often have a delicate, powdery, waxy coating called bloom. The occasionally sharp or bitter taste of cabbage is due to glucosinolate(s). Cabbages are also a good source of riboflavin.

9.11.2   Origin

The cultivated cabbage is derived from a leafy plant called the wild mustard plant, native to the Mediterranean region, where it is common along the seacoast. Also called sea cabbage and wild cabbage, it was known to the ancient Greeks and Romans; Cato the Elder praised this vegetable for its medici-nal properties, declaring that 'It is the cab-bage that surpasses all other vegetables.' The English name derives from the Normanno-Picard *caboche* (head), perhaps from *boche* (swelling, bump). Cabbage was developed by ongoing artificial selection for suppression of the internode's length.

9.11.3   Utilization

The only part of the plant that is normally eaten is the leafy head; more precisely, the spherical cluster of immature leaves, excluding the partially unfolded outer leaves. Cabbage is used in a variety of dishes for its naturally spicy flavour. The so-called 'cabbage head' is widely consumed raw, cooked, or preserved in a great variety of dishes. It is the principal ingredient in coleslaw.

Cabbage is an excellent source of vitamin C. It also contains significant amounts of glutamine, an amino acid that has anti-inflammatory properties. Cabbage can also be included in dieting programmes, as it is a low calorie food. In European folk medicine, cabbage leaves are used to treat acute inflammation: a paste of raw cabbage may be placed in a cabbage leaf and wrapped around the affected area to reduce discomfort. Some claim that it is effective in relieving painfully engorged breasts in breastfeeding women. Fresh cabbage juice has been shown to promote rapid healing of peptic ulcers.

9.11.4 Morphological description

Vegetative characters

The plant is a mesophytic herb with a taproot system with bushy lateral roots. The stem is stunted, aerial, erect and greenish. The stem is the largest vegetative bud. Leaves are opposite, thick, simple, ovate, with a wavy margin, reticulate venation, rounded apex and waxy coating, exstipulate and sessile, and food material is preserved in the leaves (vegetative bud) (Fig. 9.55).

Floral characters

The inflorescence is auxiliary, racemose type. The flower is ebracteate, pedicellate, complete, bisexual, tetramerous and hypogynous. There are four polysepalous sepals and four petals, which are polypetalous, in a cruciform arrangement. There are six tetradynamous stamens and the anthers are dithecous and basifixed. The gynoecium has a bicarpellary syncarpous superior ovary, with parietal placentation. The fruit is a siliqua and the seed is round (approximately 1 mm), pink to black with a smooth surface.

9.11.5 Anatomical description

Root anatomy

The epidermis is uniseriate, with a single layer of compactly arranged thin-walled living cells. The cortex is multiseriate and relatively homogenous. The general cortex consists of loosely arranged thin-walled parenchyma cells (Fig. 9.56) with prominent intercellular spaces. The cells are usually round or oval.

Fig. 9.55. Cabbage plant morphology: leaves are thick, opposite, sessile, exstipulate, simple, ovate, with a wavy margin and rounded apex, reticulate venation and waxy coating, and food material preserved in leaves (vegetative bud).

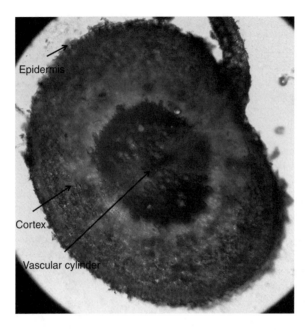

Fig. 9.56. Transverse section of root (100×) showing epidermis, cortex and central vascular cylinder.

The endodermis is single layered, with compactly arranged barrel-shaped cells. Endodermal cells are characterized by the presences of Casparian strips.

Pericycle is the outermost region of the stele and is uniseriate and parenchymatous. Cells of the pericycle retain meristematic activity. Later roots arise endogenously from the pericycle. Vascular strands are radial, xylem is exarch, closed type. Pith is absent. The non-vascular tissue present between xylem and phloem strands is called conjunctive tissue.

Stem anatomy

Epidermis is the outermost region surrounding the stem and is uniseriate, with a single layer of compactly arranged barrel-shaped living cells. The outer walls of the epidermal cells are cutinized. Cuticle reduces the transpiration. Few stomata are present in the epidermis for exchange of gases and transpiration. All the epidermal cells are colourless except the guard cells. Hairs are absent on the epidermis.

Cortex is the middle region, present between the epidermis and stele (Fig. 9.57).

Cortex is thinner than the stele and shows hypodermis, middle cortex and endodermis. Hypodermis is the outermost region of the cortex. It is multilayered (from two to six layers) and collenchymatic. Collenchyma is living mechanical tissue. The hypodermis gives considerable strength, flexibility and elasticity to young stems and, having chloroplasts, it may carry out photosynthesis. The middle cortex is also multilayered and parenchymatic. The cells may be round or oval with prominent intercellular spaces. This region also stores food materials temporarily. The endodermis forms the innermost layer of the cortex. It is usually distinct in stems without Casparian bands.

Stele is the central region of the stem and is thicker than the cortex. It consists of pericycle, vascular bundles and medulla and medullary rays. Pericycle is the outermost non-vascular region of the stele and is multiseriate and sclerenchymatic. Xylem and phloem are organized into vascular bundles. Each vascular bundle is wedge-shaped, conjoint, collateral, open and endarch. Medulla and medullary rays are distinct, present at the centre of the stem and usually parenchymatous. The cells are round or oval with

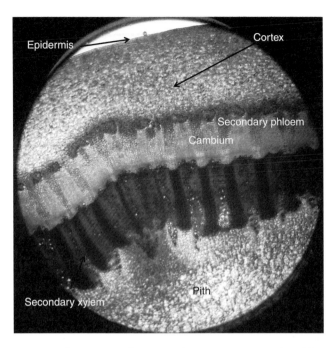

Fig. 9.57. Transverse section of stem (100×) showing epidermis, cortex, secondary phloem, secondary xylem and pith.

intercellular spaces. The medulla stores food materials and helps in lateral conduction.

Anatomic changes during shoot development

Zhu *et al.* analysed shoot apical anatomic changes during the development of *Brassica* plants Chinese cabbage (*Brassica campestris* spp. *pekinensis*) and cabbage (*Brassica oleracea* L.). It was shown that all of their apical meristems changed from the original tunica-corpus structure to the intergrade sub-area, to typical five-subarea structure and to four-subarea structure when they went into reproductive development. The bilateral cells of sub-tunica of Chinese cabbage and cabbage divided into apical leafy primordial, in which lateral inflorescence meristem arose (Zhu *et al.*, 2006).

Leaf anatomy

Epidermis is present on both surfaces of the leaf blade, i.e. upper epidermis and lower epidermis. Epidermis is mainly for protec-

tion. The epidermis is single layered, with compactly arranged barrel shaped (tabular) cells. The outer walls of the epidermal cells are cutinized. The epidermis is also covered on its outer surface by a continuous layer of cutin called cuticle. The cutinized outer walls and cuticle reduce the loss of water due to transpiration. Anisocytic stomata are present in both of the epidermal layers, but the stomata are usually more frequent in the lower epidermis. The stomatal index of the upper epidermis is 19.1 and of the lower epidermis is 26.8. The stomata facilitate the exchange of gases between the leaf and environment. All the epidermal cells except guard cells are colourless. Hairs are absent on the epidermis. The presence of epicuticular wax on both sides of the epidermis is an important character.

The ground tissue of the leaf, present between the upper and the lower epidermal layers, is called mesophyll. The mesophyll is chlorenchymatic and is concerned mainly with photosynthesis. The mesophyll is not differentiated into palisade tissue and spongy tissue (Fig. 9.58).

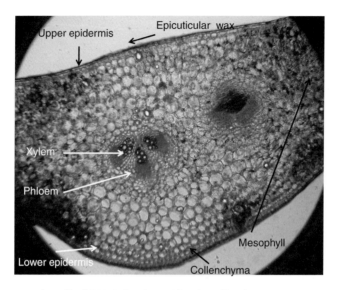

Fig. 9.58. Transverse section of leaf (100×) showing epidermis, collenchyma, spongy tissue and vascular bundles. Note: thick epicuticular wax and the mesophyll is not differentiated into palisade and spongy tissues – these are anatomical features of cabbage.

The vascular tissue occurs in the form of discrete bundles called veins. Veins are interconnected to form reticulate venation. Vascular bundles (veins) occur between the palisade and spongy tissue (median in position) and are collateral and closed. In the vascular bundle, xylem is present towards the upper epidermis and phloem towards the lower epidermis. The vascular bundle is surrounded by parenchymatous bundle sheath.

Effect of nickel on leaf anatomy

There are morpho-anatomical changes in cabbage when grown on nickel (different concentrations of 0, 5, 10 and 20 g/m^3 (Ni$_0$, Ni$_5$, Ni$_{10}$, Ni$_{20}$)) medium agar. Reduction of leaf blade area, of succulence and of leaf density and growth of specific leaf area were noticed in plants treated with all concentrations of Ni. In Ni-treated plants the total number of stomata and open stomata decreased, and the number of defective stomata in both adaxial and abaxial sides of leaves was higher. In all Ni-treated samples the volume of spongy and palisade mesophyll cells was smaller in comparison to control. In comparison to control, the intercellular spaces of mesophyll tissue decreased in Ni$_{10}$ and Ni$_{20}$ plants and increased in Ni$_5$

plants. In Ni$_5$ plants the number of chloroplasts in mesophyll cells was higher than in the Ni$_0$ control (Molas, 1997).

Cuticle

A scanning electron microscope was used to observe variations in wax formations on leaves of cabbage. The size and shape of the wax formations on adaxial and abaxial surfaces of the same leaf differed in all genera. Individual wax crystals on the first-formed leaves were two to three times larger than those on the sixth-formed leaves (Davis, 1971).

Flea beetles (*Phyllotreta* spp.), cabbage stink bugs (*Eurydema ventrale*) and onion thrips (*Thrips tabaci*) cause significant economic problems to cabbage growers in Slovenia. The aim of this study was to assess the potential effect of the epicuticular wax on leaves as defence mechanism against these three cabbage pests. These insect pests showed weak preference on cabbage heads with high epicuticular wax. There was a strong negative relationship between epicuticular wax content and the level of plants infested. High epicuticular wax was resistant to insects (Žnidarčič *et al.*, 2008).

9.12 Coriander

9.12.1 Introduction

Coriander (*Corundum sativa*) is an annual herb in the family Apiaceae (Umbelliferae) and is native to southern Europe and North Africa to south-western Asia. It is a soft, hairless plant growing to 50 cm (20 in) tall. The leaves are variable in shape, broadly lobed at the base of the plant, and slender and feathery higher on the flowering stems. The flowers are borne in small umbels, white or very pale pink, asymmetrical, with the petals pointing away from the centre of the umbel longer (5–6 mm) than those pointing towards it (only 1–3 mm long). The fruit is a globular dry schizocarp 3–5 mm diameter. Coriander is common in Middle Eastern, central Asian, Mediterranean, Indian, south Asian, Mexican, Texan, Latin American, Chinese and African countries.

9.12.2 Utilization

All parts of the plant are edible, but the fresh leaves and the dried seeds are the parts most commonly used in cooking. Coriander, like many spices, contains antioxidants, which can delay or prevent the spoilage of food seasoned with this spice. A study found both the leaves and seed to contain antioxidants, but the leaves were found to have a stronger effect.

Chemicals derived from coriander leaves were found to have antibacterial activity against *Salmonella choleraesuis*, and this activity was found to be caused in part by these chemicals acting as non-ionic surfactants.

Coriander has been used as a folk medicine for the relief of anxiety and insomnia in Iran. Experiments in mice support its use as an anxiolytic. Coriander seeds are used in traditional Indian medicine as a diuretic by boiling equal amounts of coriander seeds and cumin seeds, then cooling and consuming the resulting liquid. In holistic and traditional medicine, it is used as a carminative and as a digestive aid.

Coriander seeds were found in a study on rats to have a significant hypolipidaemic effect, resulting in lowering of levels of total cholesterol and triglycerides, and increasing levels of high-density lipoprotein. This effect appeared to be caused by increasing synthesis of bile by the liver and increasing the breakdown of cholesterol into other compounds.

Coriander juice (mixed with turmeric powder or mint juice) is used as a treatment for acne, applied to the face in the manner of toner.

9.12.3 Morphological description

Vegetative characters

The plant is a mesophytic annual herb with a taproot system with less branching and superficial lateral roots, so it is adapted for moist areas only. The stem is aerial, erect, branched, palisade shape, smooth. The decompound leaves are opposite, spathulate, exstipulate, petiolate, with a serrate margin and reticulate venation, and the leaf base is amplexicaul (Fig. 9.59).

Floral characters

The inflorescence is a compound umbel, with the presence of involucres and involucel. The flower is mostly white, bracteate, pedicellate, complete, bisexual, pentamerous and zygomorphic. There are five gamosepalous sepals, with valvate aestivation. There are five polypetalous petals, the lateral flowers in the umbel consisting of unequal petals. There are five free stamens and the anthers are dithecous and basifixed. The gynoecium is bicarpellary syncarpous, with an inferior ovary, axile placentation and the presence of a stylopodium. Fruit is called a cremocarp and breaks up vertically into two mericarps.

9.12.4 Anatomical description

All parts of the plant show resin-filled schizogenous cavities. This is an important character.

Fig. 9.59. Coriander plant morphology: decompound leaves, spathulate, with a serrate margin. The inflorescence has a compound umbel, and involucres and involucel are present.

Root anatomy

Transverse section of the root shows mainly three parts, the epidermis, cortex and stele.

The epidermis is uniseriate, with a single layer of compactly arranged thin-walled living cells. Epidermal cells produce root hairs, which are useful for absorption of water.

The cortex is multiseriate and relatively homogeneous (Fig. 9.60). It consists of general cortex and endodermis. The general cortex consists of loosely arranged thin-walled parenchyma cells with prominent intercellular spaces. The cells are usually round or oval. The endodermis is single layered, with compactly arranged barrel-shaped cells, which are characterized by the presences of Casparian strips.

The stele is the central part of the root and consists of pericycle, vascular strands and conjunctive tissue. Pericycle is uniseriate and parenchymatous and its cells retain meristematic activity. Later roots arise endogenously from the pericycle. Vascular strands are radial, xylem is exarch, closed type. The non-vascular tissue present between xylem and phloem strands is called conjunctive tissue.

Stem anatomy

The epidermis is the outermost region surrounding the stem, and is uniseriate with a single layer of compactly arranged barrel-shaped living cells. The outer walls of the epidermal cells are cutinized. Cuticle reduces the transpiration. Few stomata are present in the epidermis for exchange of gases and transpiration. All the epidermal cells are colourless except the guard cells. Trichomes are absent on the epidermis.

Hypodermis is the outermost region of the cortex. It is multilayered, with from two to six layers, and collenchymatic. Collenchyma is living mechanical tissue. The hypodermis gives considerable strength, flexibility and elasticity to young stems and, having chloroplasts, it may carry out photosynthesis. The middle cortex is also multilayered and parenchymatic. The cells may be round or oval with prominent intercellular spaces. This region also stores food materials temporarily and contains resin-filled schizogenous cavities (Fig. 9.61). Endodermis is the innermost layer of the cortex and is usually distinct in stems without Casparian bands.

Pericycle is the outermost non-vascular region of the stele and is multiseriate and sclerenchymatic. Xylem and phloem are the vascular tissues, organized into vascular bundles. Each vascular bundle is wedge-shaped, conjoint, collateral, open and endarch. The vascular bundles are arranged in one ring called a eustele. Medulla and medullary rays are distinct, present at the centre of the stem, and usually parenchymatous. The cells are round or oval with intercellular spaces. The medulla stores food materials and medullary rays help in lateral conduction.

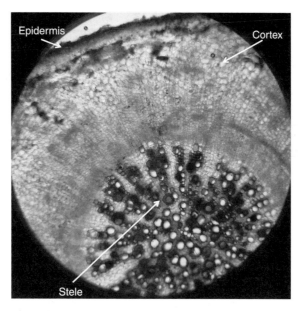

Fig. 9.60. Transverse section of root (100×) showing epidermis, cortex and stele (vascular cylinder).

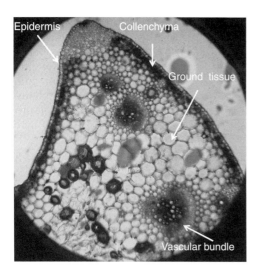

Fig. 9.61. Transverse section of stem (100×) showing epidermis, collenchyma, ground tissue and vascular bundles. Note: schizogenous cavities are distributed in the cortex.

Leaf anatomy

Epidermis is present on both surfaces of the leaf blade, i.e. upper epidermis and lower epidermis, and is mainly for protection. The epidermis is single layered, with compactly arranged barrel-shaped (tabular) cells. The outer walls of the epidermal cells are cutinized. The epidermis is also covered on its outer surface by a continuous layer of cutin called cuticle. The cutinized outer walls and cuticle reduce the loss of water due to transpiration. Anisocytic stomata are present in both of the epidermal layers, but the stomata are usually more frequent in the lower epidermis. The stomatal index of upper epidermis ranges from 23 to 25 and that of the lower epidermis from 31 to 34. The stomata facilitate the exchange of gases between the leaf and environment. All the epidermal cells except guard cells are colourless. Epidermis contains multicellular hairs.

The ground tissue of the leaf, present between the upper and the lower epidermal layers, is called mesophyll and is concerned mainly with photosynthesis. The mesophyll is differentiated into an upper palisade tissue and lower spongy tissue. The palisade tissue is present below the upper epidermis. The palisade cells are thin walled, cylindrical and contain numerous chloroplasts. The cells are compactly arranged in one layer, perpendicular to the upper epidermis. Narrow intercellular spaces are present in the palisade tissue. Palisade tissue is the most highly specialized type of photosynthetic tissue. The upper surface of the leaf is dark green in colour due to palisade tissue. The lower part of the mesophyll present towards the lower epidermis is the spongy tissue. Cells of the spongy tissue are thin walled, irregular in shape, and arranged loosely with large continuous intercellular spaces. The cells of spongy parenchyma contain a smaller number of chloroplasts, hence the lower surface of the leaf is light green in colour.

The vascular tissue occurs in the form of discrete bundles called veins. Veins are interconnected to form reticulate venation. Vascular bundles (veins) occur between the palisade and spongy tissue (median in position) and are collateral and closed. In the vascular bundle, xylem is present towards the upper epidermis and phloem towards the lower epidermis. The vascular bundle is surrounded by parenchymatous bundle sheath.

Petiole anatomy

The epidermis is uniseriate, with a single layer of compactly arranged barrel-shaped living cells. The outer walls of the epidermal cells are cutinized. Cuticle reduces the transpiration. Multicellular hairs are present on the epidermis.

Multilayered (two to three layers) hypodermis of collenchyma cells is found immediately beneath the epidermis.

Just beneath the hypodermis ground tissue is found, which consists of thin-walled parenchymatous cells with well-defined intercellular spaces among them. Vascular bundles are arranged in a half ring

scattered in ground tissue. The vascular bundles are of various sizes in the same petiole. Each vascular bundle is wedge-shaped. In the petiole, the xylem is always found towards upper side whereas phloem is found towards the lower side (as in the leaf).

Seed coat anatomy

The seed contain deposits of aromatic volatile compounds as storage food material. The testa is thin and there is no clear difference between testa and tegman (Fig. 9.62). Rate of imbibition is slow.

Significance variations in anatomy

Future research may be directed in the following directions. Possessing a greater number of multicellular trichomes gives disease resistance to insects. The presence of volatile compounds is helpful in fungal resistance.

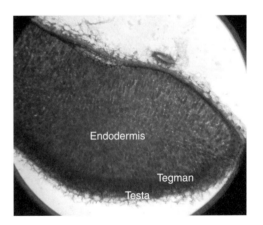

Fig. 9.62. Transverse section of seed (100×) showing testa, tegman and endosperm.

References

Abdel, C.G. and Bamerni, K.A.A. (2011) Effect of pre-planting land flooding durations on growth, yield and anatomical parameters of three watermelon [*Citrulluslanatus* (Thunb.) Matsum.] cultivars. *American Journal of Experimental Agriculture* 1(4), 187–213.

Bashan, Y., Okon, Y. and Henis, Y. (1985) Morphology of leaf surfaces of tomato cultivars in relation to possible invasion into the leaf by *Pseudomonas syringae* pv. Tomato. *Annals of Botany* 55(6), 803–809.

Buttrose, M.S. and Sedgley, M. (1979) Anatomy of watermelon embryo sacs following pollination, non-pollination or parthenocarpic induction of fruit development. *Annals of Botany* 43(2), 141–146.

Chang, Y.M., Cheng, Y.H., Hsu, W.S. and Huang, P.L. (2000) Observation of fruit anatomy and development of bitter gourd: II. Development of embryo, seed and fruit of bitter gourd. *Journal of Agricultural Research of China* 49(2), 49–60.

Channarayappa, C., Shivashankar, G., Muniyappa, V. and Frist, R.H. (1992) Resistance of *Lycopersicon* species to *Bemisia tabaci*, a tomato leaf curl virus vector. *Canadian Journal of Botany* 70(11), 2184–2192.

Davis, D.G. (1971) Scanning electron microscopic studies of wax formations on leaves of higher plants. *Canadian Journal of Botany* 49(4), 543–546.

Decker-Walters, D.S. (1999) Cucurbits, Sanskrit and the Indo-Aryans. *Economic Botany* 53(1), 98–112.

Dimock, M.B. and Kennedy, G.G. (1983) The role of glandular trichomes in the resistance of *Lycopersicon hirsutum* F. *Glabratum* to *Heliothis zea*. *Entomologia Experimentalis et Applicata* 33(3), 263–268.

Duffey, S.S. and Isman, M.B. (1980) Inhibition of insect larval growth by phenolics in glandular trichomes of tomato leaves. *Cellular and Molecular Life Sciences* 37(6), 574–576.

Emmons, C.L.W. and Scott, J.W. (1998) Ultra structural and anatomical factors associated with resistance to cuticle cracking in tomato (*Lycopersicon esculentum* Mill.). *International Journal of Plant Sciences* 159(1), 14–22.

Fett, D.D. (2003) Botanical briefs: Capsicum peppers. *CUTIS* 72, 21–23.

Gill, M., Hogendijk, S. and Hauser, C. (2002) Allergy to eggplant flower pollen. *Allergy* 57, 652.

Goffreda, J.C., Mutschler, M.A., Avé, D.A., Tingey, W.M. and Steffens, J.C. (1989) Aphid deterrence by glucose esters in glandular trichome exudate of the wild tomato, *Lycopersicon pennellii*. *Journal of Chemical Ecology* 15(7), 2135–2147.

Gupta, R.S., Dixit, V.P. and Dobhal, M.P. (2002) Hypocholesterolaemic effect of the oleoresin of *Capsicum annum* L. in gerbils (*Meriones hurrianae* Jerdon). *Phytotherapy Research* 16, 273–275.

Heimlich, L.F. (1927) The development and anatomy of the staminate flower of the cucumber. *American Journal of Botany* 14(5), 227–237.

Hossain, M.M., Shahjahan, M., Prodhan, A.K.M.A., Islam, M.S. and Begum, M.A. (2002) Study of anatomical characters in relation to resistance against brinjal shoot and fruit borer. *Pakistan Journal of Biological Sciences* 5(6), 672–678.

Inamdar, J.A., Ramana Rao, T.V. and Dave, Y. (1989) Histological structure of the pericarp of *Abelmoschus esculentus* L. (Moench) in relation to growth and dehiscence. Proceedings. *Plant Sciences* 99(6), 547–556.

Isaacson, T., Kosma, D.K., Matas, A.J., Buda, G.J., He, Y.H., Yu, B.W., Pravitasari, A., Batteas, J.D., Stark, R.E., Jenks, M.A. and Rose, J.K.C. (2009) Cutin deficiency in the tomato fruit cuticle consistently affects resistance to microbial infection and biomechanical properties, but not transpirational water loss. *Plant Journal* 60(2), 363–377.

Keitel, W.G., Frerick, H., Kuhn, U., Schmidt, U., Kuhlmann, M. and Bredehorst, A. (2001) Capsicum pain plaster in chronic non-specific low back pain. *Arzneimittel-Forschhung Drug Research* 51, 896–903.

Kennedy, G.G., Farrar, R.R. Jr and Kashyap, R.K. (1991) 2-Tridecanone-glandular trichome-mediated insect resistance in tomato: effect on parasitoids and predators of *Heliothiszea*. In: Heddin, P.A. (ed.) *Naturally Occurring Pest Bioregulators*. ACS Symposium Series 449, American Chemical Society, Washington, DC, pp. 150–165.

Kovács, E., Sárvári, E., Nyitrai, P., Darók, J., Cseh, E., Láng, F. and Keresztes, A. (2007) Structural-functional changes in detached cucumber leaves, and modelling these by hormone-treated leaf discs. *Plant Biology* 9(1), 85–92.

Kulakarni, M. and Deshpande, U. (2006) Anatomical breeding for altered leaf parameters in tomato genotypes imparting drought resistance using leaf strength index. *Asian Journal of Plant Sciences* 5(3), 414–420.

Lokesh and Singh, R. (2005) Influence of leaf vein morphology in okra genotypes (Malvaceae) on the oviposition of the leafhopper species *Amrasca biguttula* (Hemiptera: Cicadellidae). *Entomologia Generalis* 28(2), 103–114.

Molas, J. (1997) Changes in morphological and anatomical structure of cabbage (*Brassica oleracea* L.) outer leaves and in ultrastructure of their chloroplasts caused by an *in vitro* excess of nickel. *Photosynthetica* 34(4), 513–522.

Naqvi, A.R., Pareek, B.L., Nanda, U.S. and Mitharwal, B.S. (2009) Biophysical characters of brinjal plant governing resistance to shoot and fruit borer, *Leucinodes orbonalis*. *Indian Journal of Plant Protection* 37(1/2), 1–6.

Purohit, S., Varghese, T.M. and umari, M. (2003) Effect of chromium on morphological features of tomato and brinjal. *Indian Journal of Plant Physiology* 8(1), 17–22.

Sarria, E., Palomares-Rius, F.J., López-Sesé, A.I., Heredia, A. and Gómez-Guillamón, M.L. (2010) Role of leaf glandular trichomes of melon plants in deterrence of *Aphis gossypii* Glover. *Plant Biology* 12(3), 503–511.

Schuerger, A.C., Brown, C.S. and Stryjewski, E.C. (1997) Anatomical features of pepper plants (*Capsicum annuum* L.) grown under red light-emitting diodes supplemented with blue or far-red light. *Annals of Botany* 79(3), 273–282.

Sedgley, M. (1982) Anatomy of the unpollinated and pollinated watermelon stigma. *Journal of Cell Science* 54, 341–355.

Sedgley, M., Newbury, H.J. and Possingham, J.V. (1977) Early fruit development in the watermelon: anatomical comparison of pollinated, auxin-induced parthenocarpic and unpollinated fruits. *Annals of Botany* 41(6), 1345–1355.

Silva, M.E., Santos, R.C., Leary, O. and Santos, R.S. (1999) Effect of aubergine (*S. melongena*) on serum and hepatic cholestrol and triglycerides in rats. *Brazilian Archives of Biology and Technology* 42, 339–342.

Simmons, A.T. and Gurr, G.M. (2005) Trichomes of *Lycopersicon* species and their hybrids: effects on pests and natural enemies. *Agricultural and Forest Entomology* 7, 265–276.

Swamy, K.R.M. and Sadashiva, A.T. (2007) Tropical vegetable crops. In: Peter, K.V. and Abraham, Z. (eds) *Biodiversity in Horticulture*, vol. 1. Daya Publishing House, New Delhi, pp. 109–147.

Takashi, M., Mochida, K., Kozuka, M., Ito, Y., Fujiwara, Y., Hashimoto, K., Enjo, F., Ogata, M., Nobukuni, Y., Tokuda, H. and Nishino, H. (2001) Cancer chemopreventive activity of carotenoids in the fruits of red paprika *Capsicum annuum* L. *Cancer Letters* 172, 103–109.

Toscano, L.C., Boiça, A.L. Jr and Maruyama, W.I. (2002) Nonpreference of whitefly for oviposition in tomato genotypes. *Scientia Agricola (Piracicaba, Brazil)* 59(4), 677–681.

Van Haren, R.J.F., Steenhuis, M.M., Sabelis, M.W. and De Ponti, O.M.B. (1987) Tomato stem trichomes and dispersal success of *Phytoseiulus persimilis* relative to its prey *Tetranychus urticae. Experimental and Applied Acarology* 3(2), 115–121.

Yadwad, A., Sridevi, O. and Salimath, P.M. (2008) Leaf trichomes-based resistance in chilli (*Capsicum annuum*) to thrips and mites. *Indian Journal of Agricultural Science* 78(6), 518–521.

Zhu, J.X., Li, X.F., Liu, P.L. and He, Y.K. (2006) Aanalysis of shoot apical anatomic changes during the development of brassica plants. *Fen Zi Xi Bao Sheng Wu XueBao (Journal of Molecular Cell Biology)* 39(3), 199–207.

Žnidarčič, D., Valič, N. and Trdan, S. (2008) Epicuticular wax content in the leaves of cabbage (*Brassica oleracea* L. var. *capitata*) as a mechanical barrier against three insect pests. *Acta Agriculturae Slovenica* 91(2), 361–370.

10

Anatomical Adaptation to Defence Against Biotic Stresses

10.1 Introduction

To a crop plant, biotic stress implies all the negative impacts it faces from any living organism. The definition of a pest of crop plants refers specifically to a harmful organism for crops, including animals plus all other harmful plants (crop weed, fungi, parasitic plants, bacteria) and viruses. Broadly, pests of crop plants include primarily insect and non-insect pests, pathogens, parasites, weeds and sometimes also grazing animals. An organism in one environment may be harmful to a crop plant by being a pest; however, in another environment it may not pose a threat to the crop. The crop is generally referred as host (when it harbours the pest, the size of the pest being smaller than the crop). In other cases, the pest may be a predator being larger in size and the crop plant is the prey. A disease on the other hand is a symptom or syndrome expressed by a plant, which negatively affects the normal growth and development. Disease may be caused by biotic agents such as pathogens and parasites or abiotic agents such as hypoxia, nutritional deficiency, pollution, heavy metals and other abiotic agents.

Crop plants are damaged by more than ten thousand insects and one million pathogens (Dhaliwal *et al.*, 2007). About 10% of these insects and diseases are considered as major pests of crops. Global estimates reveal that the damage caused by the insect pests and diseases cause pre-harvest loss of about 42% in the major food and cash crops (Oerke *et al.*, 1994). Along with weeds, insect pests and diseases are the principal enemies of crop plants. Apart from these, insects as well as vertebrate pests (rodents, birds) causing loss of stored foods are also classified under the major pests of crop plants.

The history of emergence of pests of crop plants is as old as agriculture, and often dates back to the pre-domestication phase of co-evolution of host and pest/ pathogen. Many of today's major pests have evolved through the intensification of agriculture, incessant and indiscriminate use of pest-controlling chemicals and drastic reduction in genetic variability of host species under cultivation. Plants have evolved a number of structural defence systems against the various forms of biotic stresses, which operate best in the natural population. However, crop species are more vulnerable, since the genetic base of varieties of major crop plants grown around the world is very narrow. These constitutive defence properties are thus found in a higher number in the landraces and wild relatives of crop plants rather than the advanced varieties.

10.2   Structural Modifications for Defence Against Insect Pests and Pathogens

Insect pests damage the crop by phytophagy, through chewing the tissues or sucking saps using specialized structures. In the battle against insect pests, plants rely on two mechanisms – structural barriers that prohibit or inhibit insects from phytophagy and chemical warfare through producing chemicals to kill, trap or stop insects from feeding on the host plant. In both mechanisms, the structural anatomy of plants plays a very important role in mechanical defence or delivery of the chemicals through pores or glands.

10.2.1   Modifications of outer protective layers

Cuticle formation

The epidermis, which constitutes the outermost layer of plant tissues, is the first line of defence against insect pest or pathogen attack (Riederer, 2006). It serves as a protective tissue system in leaves, floral parts, fruits, seeds, stems and roots of plants. The epidermal cells of aerial plant parts are often covered in a waxy cuticle that not only prevents water loss from the plant, but also prevents microbial pathogens from coming into direct contact with epidermal cells and thereby limits infection. The cuticle can be relatively thin as in aquatic plants, or extremely thick as in cacti.

Cuticles are formed from the outer epidermal cell layers by deposition of long chain lipids on the surface of epidermal cells. These lipids are of two types – cutin (C_{16} and C_{18} fatty acids and glycerol) and waxes (comprising of very long chain fatty acids). These fatty acids are synthesized in the plastids and endoplasmic reticulum of epidermal cells by fatty acid elongases and are mobilized to the exterior of the epidermal cells by lipid transporters (Kunst and Samuels, 2009).

Wax and lipids serve three major purposes in defence against insect pests and diseases.

1. Wax and lipids form a barrier layer, which inhibits penetration of fugal hypha, bacterial spores or sucking insects. The deposition of cuticular waxes increases with plant age. In many cases, increased cuticle thickness increases the resistance towards the pathogen, as observed in the case of charcoal rot (*Rhizoctonia solani*) in beans, groundnut, cotton and radish, powdery mildew (*Sphaerotheca pannosa*) of roses or grey mould (*Botrytis cinerea*) of grapes (Mence and Hildebrandt, 1966; Gabler *et al.*, 2003). However, for some pathogen such as *Uncinula necator* causing powdery mildew of grape, the cuticle does not seem to be problematic for penetration, although spread of disease is reduced by preformed barrier at the cuticular region (Ficke *et al.*, 2004).

2. The waxes are hydrophobic and repel water, so pathogenic spores do not get water for germination on the plant surface. The spore structures of rust fungi do not germinate properly on wax-coated leaves of wheat and barley.

3. The wax is considered unpalatable by many insect pests, so wax deposition reduces insect attack.

The chemical composition of plant cuticular lipids influences the choice of the insect pests. Wild groundnut species exhibit different cuticular lipid composition than the cultivated groundnut, which provides a higher degree of resistance to fall armyworm and thrips (Yang *et al.*, 1993). Studies have shown that when corn earworm (*Helicoverpa zeae*) larvae are grown on diets having cuticular lipids extracted from silk of maize, the growth of the larvae was reduced. This suggests that the presence of cuticular lipids negatively affects the growth and feeding of earworms and provides resistance to the crop (Yang *et al.*, 1992).

Apart from serving as a barrier, plant cuticle has also been proposed to play a crucial role in systemic acquired resistance (SAR). SAR is activated in plants through signalling processes, which sense insect or pathogen attack in an attack site and induce defence response in distal tissues in the host. *Arabidopsis* mutants with defects in cuticle formation fail to transmit the SAR response, indicating that cuticles play an important role in SAR (Xia *et al.*, 2009).

Silicification

The role of silica in disease resistance in cereal crops has been established for quite a long time (Germar, 1935). Silicon uptake through roots increases insoluble silicon deposition, which acts as a physical barrier to the pathogen's entrance (Zeyen *et al.*, 1983). The amorphous silica located in the cell wall has a marked effect on the cellular wall physical properties. It induces formation of papilla in the inner surface of epidermal cell walls and the deposition of callose (Bélanger *et al.*, 2003). The papilla inhibits penetration of fungal hyphae and callose encases the fungal haustoria, resulting in resistance reaction. The system of silicon-mediated physical resistance works well in many host–pathogen interaction systems including rice–rice blast, wheat–powdery mildew (*Blumeria graminis* f. sp. *tritici*), barley–powdery mildew (*Erysiphe graminis* f. sp. *hordei*), oat–powdery mildew, coffee–*Cercospora* and sorghum–*Rhizoctonia*. Insoluble deposition of silica layers in the penetration sites increases the resistance response to powdery mildew pathogens.

Among the grass crops, rice shows the highest silicon uptake, which goes up to 10–15% of total dry weight. Silicon is deposited in all rice plant parts, the maximum being in stem and leaves. In leaves it is deposited as a glass-like coating in the epidermis, middle lamella, intercellular spaces, sclerenchyma, vascular bundle and bundle sheath, but not in cytoplasm (Kim *et al.*, 2002). The formation of such an impervious layer with cuticular deposition serves as a physical barrier for both insects and pathogens, including blast (*Magnaporthe grisea*), brown spot (*Bipolaris oryzae*) and sheath blight (*Rhizoctonia solani*).

Besides the physical barrier, soluble silicon also creates a physiological barrier through upregulation of plant defence systems (Koga *et al.*, 1988; Bélanger *et al.*,

2003). Due to the accumulation in the epidermis of leaves, soluble silicon activates genes involved in the production of secondary compounds of metabolism, such as polyphenols, and enzymes related to plants' defence mechanisms.

Lignification and suberization

Lignin is a complex phenolic substance resulting from the polymerization of *p*-coumaryl, coniferyl and sinapyl alcohols in the plant cell wall, which act as a physical barrier for pathogen infection (Ride, 1983). The accumulation of lignin and lignin-like materials in plant parts for resisting pathogen invasion is a very common structural defence phenomenon. Lignin is highly resistant to attack by microorganisms, and lignified cell walls are an effective barrier to pathogen entrance and spread. Lignification toughens the cell wall, which exhibits higher resistance to mechanical pressure applied during penetration by fungal appressoria. Due to this water-resistant barrier, other components of the cell wall are less accessible to cell wall-degrading enzymes.

There are many examples of pathogen inhibition in crop plants by lignification, and probably in the majority of host–pathogen interaction it serves as a major component of host defence system. In wheat, lignification seems to be a specific response mechanism for protection against filamentous fungi, but not for other physical stresses (Ride *et al.*, 1989). It has been observed that silencing of monolignol synthesizing genes makes wheat highly susceptible to powdery mildew (*Blumeria graminis* f. sp. *tritici*) in susceptible as well as in resistant lines by allowing penetration of fungal appressoria (Bhuiyan *et al.*, 2009).

Panama wilt is a serious disease of banana caused by *Fusarium oxysporum* f. sp. *cubense*, infecting roots of the crop and causing characteristic wilting symptoms. Comparison of root anatomy of resistant and susceptible cultivars shows that higher lignification and callose deposition in the roots provides a considerable level of resistance against this disease (De Ascensao and Dubery, 2000). Rapid building up of mechanical barrier in the root epidermal cells through cell wall esterification and lignification serves as the main defence system of banana against this disease.

Suberin deposition creates a barrier layer that prohibits progress of fungal pathogens during infection and the release and spread of fungal enzymes, and also creates toxicity to microbes due to the high proportion of phenolic compounds (Kolattukudy and Espelie, 1989). It provides structural resistance in many crops against pathogens, such as wilt of tomato and potato (*Vertcillium dahliae*) and root rot of soybean (*Phytophthora sojae*). Suberin formation is common when disease progresses in the vascular bundle. The vascular bundle develops a resistance response by heavily impregnating surrounding cells of the infection zone so that the progress of disease is halted. Suberization is a common feature in the barks of tree crops as a structural defence mechanism.

Cell wall esterification

Pathogenic fungi penetrate the plant cell wall by primarily localized enzymatic degradation of cell wall through a variety of cell wall-degrading enzymes. Cell wall esterification is a defence response initiated by the host plant to inhibit degradation of cell wall components. Esterification with phenolic compounds such as hydroxycinnamic acids modifies the polysaccharides of the cell wall, so that fungal enzymes cannot degrade the polysaccharides. The esters form a cross-linking platform, which also serves as a physical barrier. These cross-linking regions are later further strengthened by lignification.

10.2.2 Modification of pores and stomata

The majority of fungal pathogens use the pores and lenticels formed in the outer cellular layer of the host tissue as the entry point for infection. Consequently, the number of pores present in the plant surface is inversely related to the resistance. Grey mould-resistant grape cultivars have been found to bear a smaller number of pores and lenticels compared to susceptible plants (Eibach, 1994; Gabler *et al.*, 2003).

Stomata are yet other natural openings in the plant that are essential for physiological activities but serve as entry points for pathogen attack. It has been observed that rust fungi sense the relative distance of the stomatal guard cells and accordingly penetrate the stomata at a desirable height (Hoch *et al.*, 1987). When the fungus hyphae encounter a lip of the proper height, they undergo a developmental programme resulting in the formation of invasive structures that enter the stomata and begin colonization of the leaf interior. In mulberry plants, susceptibility to powdery mildew is related to higher stomatal density (Chattopadhyay *et al.*, 2011); therefore low stomatal density is a promising character for selection of resistant cultivars.

10.2.3 Barriers at inner layers

Apart from the outer epidermal layers, anatomical changes are also observed in the inner cellular layers including vascular bundles, parenchyma and sclerenchyma cells. These cells are thickened by esterification, lignification and suberin deposition. In many cases, cultivars having thickened inner cell walls exhibit resistance to the pathogen, as observed in case of head blight of wheat (*Fusarium graminearum*).

For viral pathogens that are transmitted from cell to cell by sap, thickening of cell walls of vascular bundles and development of structures inhibiting flow of cell sap reduce progress of the disease. In sugarcane, a slow flow rate generally indicated resistance to viral ratoon stunting disease (RSD) while a fast flow rate indicated susceptibility (Teakle *et al.*, 1978). The resistant clones with slow flow rates exhibit higher branching of the large metaxylem vessels in the nodes than susceptible cultivars.

Lignification of sclerenchyma cells surrounding vascular bundles contributes to thickening of the leaf sheath in grasses and serves as a mechanism for resistance to insect pests. In St Augustine grass (*Stenotaphrum secundatum*), a lawn grass, lignification does not contribute to epidermal thickening, but leads to thickening of sclerenchyma cells contributing to resistance against Southern chinch bug (*Blissus insularis*) (Rangasamy *et al.*, 2009).

10.3 Plant Trichome, a Specialized Multipurpose Structure for Insect Pest and Disease Resistance

10.3.1 Trichomes

Trichomes are single or multicellular outgrowths of the plant epidermis and collectively constitute the pubescence (hairiness) of the plant surface. These epidermal hairs in many crop species are specialized for defence against attack by insects and mites. The mode of defence used by trichomes is determined by its nature such as nonglandular or glandular, as well as their density, length, shape, and degree of erectness. Presence of trichomes and their special modifications have been preferred in a number of crops to develop genotypes resistant to insect pest damage.

Non-glandular trichome

This type of trichome does not secrete any chemical substance, but creates a physical hindrance for the insect pests or diseases attacking the plant tissues. When present on the plant surface at high densities, nonsecretory trichomes create a physical barrier to insect feeding on the underlying surface or internal tissues as well as a barrier against migration. Barrier defence is an important element of resistance to leafhoppers in cultivated crop plants such as lucerne (alfalfa), cotton and soybean. Beans have evolved fish-hook-shaped trichomes that help to anchor their climbing vines, but the hooked feature is also defensive because leafhopper and aphid pests are captured by these hairs.

The ability to impart resistance in crop plants by non-glandular trichomes depends on the number, size and shape of the trichomes. High trichome density on leaves reduces oviposition by the insects. In okra, genotypes having higher trichome density are not preferred by the leafhoppers (Singh, 2005) for laying eggs. Presence of dense trichomes is particularly useful for providing resistance to sucking insect pests. High number of trichomes in the underside of the leaf helps in providing resistance to hoppers in cotton. Presence of non-glandular trichomes at the abaxial surface inhibits feeding of young maggots thus increasing host resistance in sorghum. The wild species of sorghum have a high trichome density on the lower surface of the leaves and are nearly immune to shoot fly attack (Bapat and Mote, 1982). Density of non-glandular pointed trichomes is a reliable and very simple indicator for the selection of sorghum genotypes resistant to shoot fly (Maiti and Gibson, 1983).

The nature and occurrence of leaf trichomes in sorghum were studied at ICRISAT, India, following initial observations that many lines having field resistance to *Atherigona soccata* Rond. had trichomes on the leaves. Trichomes appeared to be of rather infrequent occurrence and varied in number per unit area of the leaf surface, and in length, angle and morphology, in those genotypes in which they occurred. The presence or absence of trichomes was a stable varietal characteristic. Trichome frequency on the leaf was, however, influenced by the plant and leaf sampled and the time of sampling. A field sampling methodology for trichome frequency, designed to minimize the variance of cultivar means, was presented. The presence of trichomes on the leaf surface was related to a lesser frequency of oviposition by adults of *A. soccata* and plants destroyed by the larvae. Neither the density of trichomes, trichome angle nor trichome length were related to differences in shoot-fly damage, however (Maiti and Bidinger, 1980).

Glandular trichome

The most elegant specializations of plant hairs for defence are glandular trichomes, which secrete adhesive materials that physically entrap and immobilize insects or contain toxic or deterrent substances. Trichome glands produce and secrete various types of exudates affecting insects, microbes and herbivores. Glandular trichomes are common in the nightshade family (Solanaceae). Selection for presence of trichomes has been helpful in developing new varieties of potato and tomato that resist insect pests because of glandular hairs on their leaves and stems. Other crop plants in which glandular trichomes are being used to breed for imparting pest resistance include lucerne, strawberry, sunflower and tobacco.

Several groups of chemicals including phenols, quinones, resins and volatiles are released from the glandular trichomes. These chemicals act in various ways to reduce insect damage such as by reducing insect population growth, creating traps to reduce mobilization of insects, making the host tissues unpalatable, or by repelling insects from coming near to the host. Many crop plants including potato and tomato release phenolic compounds through their trichome glands, producing sticky substances to trap small insects such as aphids. Type A glandular trichomes of wild potato *Solanum berthaultii* produce phenolic exudates (glucose ester of p-hydroxyphenylpropionic acid), which are oxidized by polyphenol oxidase (PPO), producing a cement-like substance resulting in entrapment and death of the aphids (Kowalski *et al.*, 1993). High PPO activity has also been correlated with resistance to potato tuber moth (*Phthorimaea operculella*). The glandular trichomes from trichomes of *S. berthaultii* also provide resistance to Colorado potato beetle (*Leptinotarsa decemlineata*) by exudate deposition on larvae and also providing a barrier to feeding (Neal *et al.*, 1991). A second type (type B) of glandular trichome has been identified in other wild potatoes, *Solanum sparsipilum* and *Solanum pinnatisectum*, which exhibits a strong repulsive effect on oviposition of tuber moth (Musmeci *et al.*, 1997).

Many of the exudates are secreted under the influence of trichome-specific gene expression. Alteration or suppression of endogenous genes involved in these secondary metabolic pathways provides opportunity for development of genotypes with higher resistance to insect pests. In tobacco, expression of a P450 hydroxylase gene specific to the trichome gland leads to formation of cembratriene-diol (CBT-diol) from its precursor cembratriene-ol (CBT-ol) (Wang *et al.*, 2001). CBT-diol, the main exudate, is less effective in aphidicidal activity than CBT-ol. Suppression of this gene increases concentration of CBT-ol with a remarkable increase in resistance to aphids.

Both glandular and non-glandular trichomes can coexist in the same genotype. In pigeonpea, three glandular (Types A, B, and E) and two non-glandular (Types C and D) trichome types have been detected that are present on leaves, pods and calyxes of both the cultivated and wild species (Romeis *et al.*, 1999). Presence of high density of non-glandular trichomes in pigeonpea helps in preventing attack of pod borer (*Helicoverpa armigera*), which is the most serious pest of this crop. The wild species having higher trichome density is more resistant to *H. armigera* infestation, which establishes the importance of trichomes in insect-pest resistance.

Studies on inheritance of trichomes

Studies have been undertaken on the inheritance of trichomes. Segregation ratios in successive generations of five single-cross matings between trichomed and trichomeless parents indicated that the presence of leaf trichomes is controlled by a single recessive gene. Inheritance of this character in three crosses among four trichomed parents involved the same locus, designated tr. Trichome density on the abaxial leaf surface varied among trichomed lines from single crosses. Heritability of trichome density between the F3 and F4, estimated in the cross IS1054 × B CK60, was 0.75 (Gibson and Maiti, 1983).

Trichomed, segregating and trichomeless F3 and F4 lines from four trichomed × trichomeless crosses and their parents were studied in the field in Patancheru.

Compared with trichomeless lines, trichomed lines had significantly lower percentages of plants with *Atherigona soccata* eggs 18 days after emergence and of dead hearts at both 18 and 23 or more days after emergence. The ratio of the difference between the means of trichomed and trichomeless lines for the percentage of dead hearts to the corresponding difference between the parents ranged from 0.16 to 0.92 and exceeded 0.32 in seven of nine comparisons, indicating that trichomes were a major factor in resistance. Means of parents and progenies were regressed on four possible genetic models and results indicated that, besides trichomes, at least two loci that interact with each other were involved in resistance (Maiti and Gibson, 1983).

In a separate study, leaf surfaces of seven genotypes of *Sorghum bicolor*, two of maize, *Zea mays*, and two of pearl millet, *Pennisetum americanum*, were examined by scanning electron microscopy for possible morphological differences. Glossiness was determined by spraying water, which adhered to the glossy leaves. Cuticular transpiration of detached third and fifth leaves was estimated from the rate of water loss after abscisic acid induced stomatal closure. Glossy character was correlated with the reduction or absence of wax deposits on the leaf surfaces, while hairiness might occur in either glossy or non-glossy genotypes. Unlike sorghum and maize, in which all leaves after the fifth or seventh were glossy, pearl millet showed no glossiness through the ninth leaf. Measurements showed that cuticular transpiration of glossy leaves was often more than double that of non-glossy leaves. Comparisons among sorghums showed that non-glossy lines had higher stomatal densities than glossy lines (Traore *et al.*, 1989).

Further study may reveal that the density of non-glandular pointed trichomes oriented at an acute angle on leaf surfaces offers barriers to the movement of maggots/larvae to reach the shoot apex through the collar. When associated with non-glandular trichomes, waxy coatings impart glossiness on the leaf surface, which prevents insects from laying eggs. The presence of bicellular trichomes in non-glossy lines may also stimulate shoot fly attack.

Do all trichomes contribute to plant defence?

Not all trichomes are suitable for resistance to insect pests. The ability of trichomes to impart resistance to insect pests depends on the structure and the properties of the trichomes. It has been observed that there is great variation in the structure of trichomes, including length of trichome, body shape, tip shape and size, gland shape and size and number of cells in base and head. In tomato, seven types of trichomes have been described by Luckwill (1943). Some of these trichomes (for example type VI, which has a glandular head) are associated with higher mortality of larvae of potato tuber moth than other trichome types. Similarly, presence of dense type I trichomes, which is longer in size than other trichomes, is inversely related to feeding ability of the potato tuber moth larvae (Simmons *et al.*, 2006).

Presence of glandular trichomes that do not exhibit insect resistance properties may increase susceptibility to insect pests. In sorghum, genotypes having non-glandular trichomes are more tolerant to shoot fly than the genotypes with glandular trichomes (Maiti and Gibson, 1983). They proposed that the glandular trichomes of sorghum may produce some chemicals that favour insect growth. In pigeonpea, glandular trichomes (type A) on the calyxes and pods contribute to susceptibility to pod borer (*H. armigera*), while the non-glandular trichomes (trichome type C and D) are associated with resistance (Sharma *et al.*, 2009). Although most of the evidence on glandular trichome research suggests negative effects of exudates on insects, recently Weinhold and Baldwin (2011) found that *o*-acyl sugars, the most abundant metabolite of glandular trichomes in wild tobacco *Nicotiana attenuata*, is readily consumed by the larvae of Lepidopteran insect pests *Spodoptera littoralis*, *Spodoptera exigua* and *Manduca sexta*. These findings suggest that trichome exudates may also attract insect pests thereby increasing the susceptibility of a genotype.

References

Bapat, D.R. and Mote, U.N. (1982) Sources of shoot fly resistance in sorghum. *Journal of Maharashtra Agricultural University* 7, 238–240.

Bélanger, R.R., Benhamou, N. and Menzies, J.G. (2003) Cytological evidence of an active role of silicon in wheat resistance to powdery mildew (*Blumeria graminis* f. sp. *tritici*). *Phytopathology* 93, 402–412.

Bhuiyan, N.H., Selvaraj, G., Wei, Y. and King, J. (2009) Role of lignification in plant defense. *Plant Signal Behaviour* 4(2), 158–159.

Chattopadhyay, S., Ali, K.A., Doss, S.G., Das, N.K., Aggarwal, R.K., Bandopadhyay, T.K., Sarkar, A. and Bajpai, A.K. (2011) Association of leaf micromorphological characters with powdery mildew resistance in field grown mulberry (*Morus* spp.) germplasm. AOB Plants. Available at: http://www.aobpla.oxfordjournals.org/content/2011/plr002.full (accessed 1 March 2012).

De Ascensao, A.R.D.C.F. and Dubery, I.A. (2000) Panama disease: cell wall reinforcement in banana roots in response to elicitors from *Fusarium oxysporum* f. sp. *cubense* race four. *Phytopathology* 90, 1173–1180.

Dhaliwal, G.S., Dhawan, A.K. and Singh, R. (2007) Biodiversity and ecological agriculture: Issues and perspectives. *Indian Journal of Ecology* 34(2), 100–109.

Eibach, R. (1994) Defense mechanisms of the grapevine to fungus disease. *American Vineyard* 1, 8–10.

Ficke, A., Gadoury, D.M., Seem, R.C., Godfrey, D. and Ian, B.D. (2004) Host barriers and responses to *Uncinula necator* in developing grape berries. *Phytopathology* 94, 438–445.

Gabler, F.M., Smilanick, J.L., Mansour, M., Ramming, D.W. and Mackey, B.E. (2003) Correlations of morphological, anatomical, and chemical features of grape berries with resistance to *Botrytis cinerea*. *Phytopathology* 93, 1263–1273.

Gibson, P.T. and Maiti, R.K. (1983) Trichomes in segregating generations of sorghum matings. I. Inheritance of presence and density. *Crop Science* 23(1), 73–75.

Hoch, H.C., Staples, R.C., Whitehead, B., Comeau, J. and Wolf, E.D. (1987) Signaling for growth orientation and cell differentiation by surface topography in Uromyces. *Science* 235, 1659–1662.

Kim, S.G., Kim, K.W., Park, E.W. and Choi, D. (2002) Silicon-induced cell wall fortification of rice leaves: a possible cellular mechanism of enhanced host resistance to blast. *Phytopathology* 92, 1095–1103.

Koga, H., Zeyen, R.J., Bushnell, W.R. and Ahlstrand, G.G. (1988) Hypersensitive cell death, autofluorescence, and insoluble silicon accumulation in barley leaf epidermis cells under attack by *Erysiphe graminis* f. sp. *hordei*. *Physiological and Molecular Plant Pathology* 32, 395–409.

Kolattukudy, P.E. and Espelie, K.E. (1989) Chemistry, biochemistry and functions of suberin-associated waxes. In: Rowe, J.W. (ed.) *Natural Products of Woody Plants*. Springer-Verlag, New York, pp. 235–287.

Kowalski, S.P., Plaisted, R.L. and Steffens, J.C. (1993) Immunodetection of polyphenol oxidase in glandular trichomes of *S. berthaultii, S. tuberosum* and their hybrids. *American Journal of Potato Research* 70, 185–199.

Kunst, L. and Samuels, L. (2009) Plant cuticles shine: advances in wax biosynthesis and export. *Current Opinion in Plant Biology* 12, 721–727.

Luckwill, L. (1943) *The Genus* Lycopersicon: *historical, biological, and taxonomic survey of the wild and cultivated tomatoes*. Aberdeen University Press, Aberdeen, UK.

Maiti, R.K. and Bidinger, F.R. (1980) A simple approach to the identification of shoot fly tolerance in sorghum. *Indian Journal of Plant Protection* VII (20), 135–140.

Maiti, R.K. and Gibson, P.T. (1983) Trichomes in segregating generations II. Association with shoot fly resistance. *Crop Science* 23, 76–79.

Mence, M.J. and Hildebrandt, A.C. (1966) Resistance to powdery mildew in rose. *Annals of Applied Biology* 58, 309–321.

Musmeci, S., Ciccoli, R., Di Gioia, V., Sonnino, A. and Arnone, S. (1997) Leaf effects of wild species of *Solanum* and interspecific hybrids on growth and behaviour of the potato tuber moth, *Phthorimaea operculella* Zeller. *Potato Research* 40, 417–430.

Neal, J.J., Plaisted, R.L. and Tingey, W.M. (1991) Feeding behavior and survival of Colorado potato beetle, *Leptinotarsa decemlineata* (Say), larvae on *Solanum berthaultii* Hawkes and an F6 *S. tuberosum* L. X *S. berthaultii* hybrid. *American Journal of Potato Research* 68, 649–658.

Oerke, E.-C., Dehne, H.-W., Schonbeck, F. and Weber, A. (1994) Crop production and crop protection. In: *Estimated Losses in Major Food and Cash Crops*. Elsevier, Amsterdam and New York, 808 pp.

Rangasamy, M., Rathinasabapathi, B., McAuslane, H.J., Cherry, R.H. and Nagata, R.T. (2009) Role of leaf sheath lignification and anatomy in resistance against southern chinch bug (Hemiptera: Blissidae) in St. Augustinegrass. *Journal of Economic Entomology* 102(1), 432–439.

Ride, J.P. (1983) Cell walls and other structural barriers in defense. In: Callow, J.A. (ed.) *Biochemical Plant Pathology*. Wiley Publishers, Chichester, UK, pp. 215–236.

Ride, J.P., Barber, M.S. and Bertram, R.E. (1989) Infection-induced lignification in wheat. In: *Plant Cell Wall Polymers*. ACS Symposium Series, 399, 361–369.

Riederer, M. (2006) Introduction: biology of the plant cuticle. In: Riederer, M. and Müller, C. (eds) *Biology of the Plant Cuticle*. Blackwell Publishing, Oxford, UK pp. 1–10.

Romeis, J., Shanower, T.G. and Peter, A.J. (1999) Trichomes on pigeonpea (*Cajanus cajan* (L.) Millsp.) and two wild *Cajanus* spp. *Crop Science* 39, 564–569.

Sharma, H.C., Sujana, G. and Manohar Rao, D. (2009) Morphological and chemical components of resistance to pod borer, *Helicoverpa armigera* in wild relatives of pigeonpea. *Arthropod-Plant Interactions* 3, 151–161.

Simmons, A.T., Nicol, H.I. and Gurr, G.M. (2006) Resistance of wild *Lycopersicon* species to the potato moth, *Phthorimaea operculella* (Zeller) (Lepidoptera: Gelechiidae). *Australian Journal of Entomology* 45, 81–86.

Singh, R.L. (2005) Influence of leaf vein morphology in okra genotypes (Malvaceae) on the oviposition of the leafhopper species *Amrasca biguttula* (Hemiptera: Cicadellidae). *Entomologia Generalis* 28, 103–114.

Teakle, D.S., Appleton, J.M. and Steindl, D.R.L. (1978) An anatomical basis for resistance of sugar cane to ratoon stunting disease. *Physiological Plant Pathology* 12, 83–88.

Traore, M., Sullivan, C.Y., Rosowski, J.R. and Lee, K.W. (1989) Comparative leaf surface morphology and the glossy characteristic of sorghum, maize and pearl millet. *Annals of Botany* 64, 447–453.

Wang, E., Wang, R., De Parasis, J., Loughrin, J.H., Gan, S. and Wagner, G.J. (2001) Suppression of a P450 hydroxylase gene in plant trichome glands enhances natural-product-based aphid resistance. *Nature Biotechnology* 19, 371–374.

Weinhold, A. and Baldwin, I.T. (2011) Trichome-derived O-acyl sugars are a first meal for caterpillars that tags them for predation. *Proceedings of National Academy of Sciences USA* 108, 7855–7859.

Xia, Y., Gao, Q.M., Yu, K., Lapchyk, L., Navarre, D., Hildebrand, D., Kachroo, A. and Kachroo, P. (2009). An intact cuticle in distal tissues is essential for the induction of systemic acquired resistance in plants. *Cell Host & Microbe* 5, 151–165.

Yang, G., Wiseman, B.R. and Espelie, K.E. (1992) Cuticular lipids from silks of seven corn genotypes and their effect on development of corn earworm larvae (*Helicoverpa zea* (Boddie)). *Journal of Agricultural and Food Chemistry* 40, 1058–1061.

Yang, G., Espelie, K.E., Todd, J.W., Culbreath, A.K., Pittman, R.N. and Demski, J.W. (1993) Cuticular lipids from wild and cultivated peanuts and the relative resistance of these peanut species to fall armyworm and thrips. *Journal of Agricultural and Food Chemistry* 41, 814–818.

Zeyen, R.J., Carver, T.L.W. and Ahlstrand, G.G. (1983) Relating cytoplasmic detail of powdery mildew infection to presence of insoluble silicon by sequential use of light microscopy, SEM and X-ray microanalysis. *Physiological Plant Pathology* 22, 101–108.

11

Anatomical Adaptation for Drought and Waterlogging Stress Tolerance

11.1 Introduction

Abiotic stresses include adverse conditions imposed by environmental factors such as drought, salinity, cold, high temperature, degraded soil conditions such as toxicity of mineral elements and heavy metals as well as pollution caused by various agents. The rapid expansion of the human population in the last century and the shift in social and cultural behaviour is primarily responsible for the emergence of many abiotic stress factors. Soil salinity, for example, is a direct result of indiscriminate use of chemical fertilizer and intensive farming practices. Several soil-related problems such as of iron and aluminium toxicity are a result of acidification of soil, loss of soil organic carbon and fixation of phosphates from chemical fertilizers.

In previous sections discussing the anatomy of different crops (Chapters 5–9) we have cited many examples establishing that crop plants have modified their structural anatomy in response to different biotic and abiotic stresses for adaptation to the stressed environment. These modifications have helped the crop plants to survive in odd environments. In spite of the stress conditions, the crop plants have not only survived, but also developed mechanisms to continue reserves of storage materials providing food and other economic products

to us. Discussing such adaptations and their significance again will be a repetition, so in the present chapter we will concentrate on general anatomical mechanisms for adaptation under two major abiotic stresses, i.e. drought and waterlogging (hypoxia/anoxia), citing specific examples pertaining to taxonomically similar crop species.

11.2 Anatomical Adaptations for Drought Tolerance

Drought is considered to be the most threatening globally distributed abiotic stress to agriculture. The incidence of drought has been estimated to have more than doubled within the very short span of the last 30 years (Isendahl and Schmidt, 2006). Many drought-tolerant crops have developed special structural mechanisms for maintaining the water requirement. One of the major objectives of such avoidance mechanisms is to maintain the leaf water potential by reducing transpiration. Plants that have developed means to reduce transpiration are commonly known as 'water savers'. However, there are other anatomical modifications in plants that help to economize the water utilization of plants under drought stress. The selection for adaptation of crop plants for drought-prone agriculture has led to accumulation of both these mechanisms for increasing water use efficiency under drought conditions.

11.2.1 Mechanisms of avoidance of drought through anatomical modifications

Reduction of transpiration by modulating stomatal activity and density

Stomatal activity is more influenced by drought stress than photosynthetic activities in plants. Opening and closing of stomata is controlled by the guard cells surrounding the stomatal pore. Under water sufficiency conditions, water influx from surrounding cells to guard cells causes the guard cells to swell and rise, leading to opening of stomata. This is regulated by a cellular network involving ion channels and transporter proteins in the plasma membrane, and is influenced by many factors such as concentrations of K^+, Ca^{2+}, sugars and CO_2 and also on the spectrum of light. The stomata can sense drought conditions through the increase in abscisic acid (ABA) concentration, which causes efflux of solute from guard cells resulting in closure of stomata. It can sense the rise in ABA concentration in nearer leaf tissues as well as in distant regions such as roots through xylem-mediated ABA transport. Rapid stomatal sensitivity under drought stress is a desirable mechanism to close the stomata and reduce water loss. Mutations affecting stomatal sensitivity readily reduce water availability, resulting in rapid wilting of plants.

Stomatal density is directly correlated with stomatal conductance and thus crop productivity, by helping to maintain the photosynthetic activities in leaves. Sustenance of stomatal conductance during the crucial drought cycles helps to increase productivity under drought stress.

Cuticular resistance

Along with stomata, plant cuticle is involved in transpiration. When stomata are closed, transpiration takes place through cuticle. Thus the ability to conserve water under drought stress depends on the thickness and deposition of cuticle. If cuticle is thin with no deposition, stomatal resistance becomes ineffective under drought stress.

Cuticle is a comparatively thin layer (0.1–10 µm), so the resistance to water loss is primarily determined not by the thickness of cuticle, but by the wax deposition (see Chapter 9, this volume) on the cuticle (epicuticular wax). Mechanical removal of epicuticular wax results in higher transpiration. The wax deposition benefits the plants under drought stress in two ways: by reducing transpiration and also by increasing reflectance of the leaf (Maiti, 1996). Studies in maize show that cuticular deposition is inversely related to epidermal water loss. However, some studies show that the amount of cuticular wax has no correlation with transpirational loss (Ristic and Jenks, 2002). Thus rather than the amount, the chemical composition may be more pertinent to water loss through the epidermis.

Anatomical modification in root characters

Drought stress increases root growth compared to shoot growth as plants try to harvest higher soil moisture. The root–shoot ratio in almost all crop plants increases under drought stress, indicating root growth is a common adaptation across species under water-limited environments. This growth is activated by and associated with a number of signaling processes, including ABA accumulation, concentration of reserve carbohydrates in the apical zone of root, accumulation of proline, cytokinins and expansion proteins. A characteristic symptom of drought stress is activation of primary root growth and cessation of secondary root growth, which is driven by cytokinins-mediated blocking of auxin accumulation. Primary root growth enables the plant to penetrate deeper in the soil to extract soil moisture. In most of the landraces of crop plants adapted to arid region, the deep root system has been observed to be the major factor contributing to drought tolerance (Trethowan and Mujeeb-Kazi, 2008).

Under drought conditions, root hydraulic conductivity also plays a key role for adaptation to stress, which is further determined by anatomical features of the root. For example, hydraulic conductivity of rice root is lower than maize root due to barriers created by endodermis and aerenchyma (Miyamoto *et al.*, 2001). During water shortage, cavitations in the water transport channel are created, resulting in loss of connection between root and shoot, making rice less adapted to drought-prone environments than maize.

Hardening of cell wall

Under drought stress, plants tend to impregnate the cell wall with various lignocellulosic and ferrulic depositions to prevent cellular dehydration. This leads to hardening of the cell wall, which helps to maintain turgor pressure in the cell. But the hardening also reduces growth of the cell. Depending on the extent of stress, the plants try to maintain a balance between cellular growth and hardening. The process of hardening starts with signals received through various drought-induced biochemical pathways, in which ABA plays a significant role.

11.2.2   Adaptations in drought-tolerant cereals

In cereals, drought conditions during vegetative stages reduce growth and development of vegetative tissues and induces cellular death by disturbing osmotic potential. If the onset of drought coincides with the pre-flowering or flowering stage, fertilization is hampered as a result of which grains become chaffy. However, during the grain filling period, drought seems to be beneficial, as common farming practice is to withhold water supply to the field during the period of maturity. In comparison to rice, sorghum, millets and wheat are more tolerant to drought. Although all these crops belong to same Gramineae family, the

differences in response to drought stress in these crop species are partly manifested by anatomical differences in plant structure, although several other cellular mechanisms contribute to the differential response. However, striking differences in certain anatomical features are clearly observable among these species. It must also be noted that a number of non-crop grass species, which are mostly perennial in nature, are well adapted to arid environments.

Grasses roll their leaves in response to drought stress, which is a common phenotypic signal of drought stress in the field. The leaf rolling is manifested by loss of turgor in bulliform cells present between the veins along the leaf axis. Leaf rolling under drought stress helps the plant to reduce transpiration and protect the leaf cells from direct sunlight, creating a microclimate inside the leaf where the temperature is less (Nar *et al.*, 2009). Under high stress the leaves are rolled completely. Drought-tolerant genotypes exhibit delayed or less leaf rolling than the drought-sensitive genotypes, hence it is a very important trait for selection against drought stress in breeding programmes.

A distinct mechanism helping cereals to avoid drought stress is by deposition of epicuticular wax. It contributes towards higher tolerance to drought stress in sorghum, wheat and millets and is absent in rice. Crops such as sorghum and wheat have higher wax deposition on leaves, which is the primary reason behind the glossy appearance of several sorghum and wheat genotypes. Rice leaves, on the other hand, have little wax deposition. Increase in leaf glossiness has been suggested as a major breeding objective for improvement of rice performance in water-limited environments (Nguyen *et al.*, 1997).

The stay green trait in cereals, which results from delayed senescence of leaves due to slower rate of chlorophyll breakdown, is also considered an important character for drought tolerance (Borrell *et al.*, 2000). It has been observed that the stay green genotypes of sorghum, maize and wheat perform better under water-limited environments than senescent types. This phenotype contributes to better grain filling in cereals during post-anthesis period, since the greener leaves produce more photosynthates during this period and translocate it to the endosperm of developing seeds. The stay green genotypes have developed internal structural modifications for better assimilation of water and mineral nutrients.

At metabolite level, increase in proline and soluble concentration is a common phenomenon among the grass crops. Proline provides tolerance to abiotic stress via osmo-protection as well as by increasing the antioxidant enzyme activities. The soluble sugars also contribute to stress tolerance as signalling agents, osmo-protectants and as a pool for metabolic activities. Studies have shown that accumulation of these metabolites is much higher in drought-tolerant cultivars of rice, wheat and sorghum than the susceptible cultivars. In the case of rice, the soluble sugars are accumulated more in leaf sheaths than other plant parts, as sheaths are more exposed to drought stress compared to other plant parts (Cabuslay *et al.*, 2002). Leaf sheaths also provide mechanical protection to actively dividing meristematic tissues. The drought-tolerant rice cultivars adapt to the water-limited environment by accumulating more soluble sugars and proline for higher osmo-protection in the leaf sheaths than leaf blades.

11.2.3 Adaptations in arid legumes

The semi-arid and arid regions of the world are characterized by low agricultural productivity and higher percentage of population under extreme poverty with nutritional deficiency. The legume crops are important diets for poor people in this region; hence they are grown in intermittent or terminal drought conditions mostly utilizing residual soil moisture. The arid and semi-arid legumes such as pigeonpea (*C. cajan*), cowpea (*V. unguiculata*), chickpea (*C. arietinum*, desi type) and black gram (*V. mungo*) are drought-tolerant species. Some other legumes that have moderate drought tolerance are common bean (*P. vulgaris*), chickpea

(Kabuli type), mungbean (*V. radiata*), lentil (*L. culinaris*) and soybean (*G. max*). Only a few legume species such as faba bean (*Vicia faba*) are comparatively more susceptible to drought.

The anatomical features contributing to drought tolerance in arid legumes are similar to general features of drought tolerance or avoidance (see Chapter 6, this volume), which attempt to maximize water conservation by modifying stomatal conductance and cuticular structures, maintenance of photosynthetic activities, economization of water distribution, efficient mobilization of storage reserve and enhanced water uptake through development of a deep primary root system (White and Castillo, 1989; Ramirez-Vallejo and Kelly, 1998). Although epicuticular wax deposition is not commonly found in legumes, black gram genotypes deposit epicuticular wax under drought stress to reduce water loss. Similarly, osmotic adjustment, which plays a major role in drought tolerance in cereals, seems to be of less importance in the drought tolerance mechanism of legumes (Likoswe and Lawn, 2008).

Pigeonpea is one of the most drought-tolerant tropical legume crops. The ability to withstand high drought in this crop species is primarily attributed to the deep root system, better osmotic adjustment, higher photosynthetic ability and multiple flowering flushes (Odeny, 2007). However, there is considerable within and between species variability in utilizing these mechanisms for drought tolerance among arid legumes. For example, stomatal conductance contributes to drought tolerance in some common bean genotypes, but it does not appear as a primary mechanism for drought tolerance in other common bean genotypes or drought-tolerant genotypes of faba bean (Ludlow and Muchow, 1990; Ehleringer *et al.*, 1991; Grzesiak *et al.*, 1999). In some drought-tolerant genotypes, stomatal closure is faster in response to drought. Common bean cultivars that are susceptible to drought exhibit reduction in thickness of epidermis, protoderm and area of parenchymatic cells along with plasmolysis and death of root cells (Peña-Valdivia *et al.*, 2010). Stratification of root epidermis and protoderm starts under drought stress. The cortex and xylem cell wall thickens in response to drought stress, resulting in reduction in diameter of xylem vessels, which limits water movement. These responses are also observed in non-legume crops under stress, such as maize. Studies on maize shows that in the drought-tolerant cultivars the protoderm thickness is less affected than in susceptible cultivars (Peña-Valdivia *et al.*, 2005). Wild relatives of common bean show higher drought tolerance and anatomical features better adapted to drought stress, which suggests that during domestication, some traits contributing to drought tolerance were lost, which makes the present day common bean genotypes less adapted to arid regions compared to their wild relatives (Peña-Valdivia *et al.*, 2010).

Optimal resource allocation is a critical issue in arid legumes. Plants with large vegetative biomass and larger leaf area lose more water through transpiration, which is the main hindrance for the development of a drought-tolerant mungbean cultivar with high productivity. The mungbean leaves are large and distributed horizontally, resulting in higher transpirational loss. Moreover, under drought stress root–shoot ratio increases considerably in all legumes, since root growth is less affected than shoot growth under drought.

11.3   Tolerance to Waterlogging (Submergence)

Waterlogging and soil flooding results in submergence of below-ground parts of the plant. This not only reduces the oxygen availability in the soil, creating hypoxia or anoxia, it also changes the soil chemical properties resulting in difficulties for absorbance of chemical nutrients from the soil. A prominent outcome of flooding in soil is the iron toxicity, which results in the conversion of available iron in the soil into non-available forms, creating deposition of iron oxides and sulfides in the root

zone. This symptom is very common in rice fields of Asian countries. It has been long recognized as 'bronzing' of rice plants.

The rice plant under submerged conditions exhibits two major anatomical adaptations: (i) the underground root zone by formation of aerenchyma; and (ii) enhanced growth and shoot elongation under deep water. Both these mechanisms are very unique among crop species, as no other crop plant grown on soil shows such extreme anatomical adaptations under submerged condition.

11.3.1   Aerenchyma formation in rice as a response to waterlogging conditions

Although some crop species can survive in waterlogged condition for a few days, rice is a remarkable crop whose life cycle is completed mostly in the waterlogged anaerobic environment. It also grows well under aerobic conditions (upland rice), which makes it a very good candidate for studying various anatomical, physiological and cellular adaptations under aerobic as well as anaerobic environmental conditions. Rice is a semi-aquatic plant with a very high adaptability to different environmental conditions. The most striking feature of rice anatomical adaptation for waterlogging stress tolerance is the formation of aerenchyma. Aerenchyma are formed in many other plants under waterlogging condition, including wheat, barley, maize, jute, tomato and many forage grasses. However, while in other plants, development of aerenchyma is induced only in response to waterlogging stress, rice has a special genetic mechanism for constitutive aerenchyma formation even in aerobic condition (Kawai *et al.*, 1998). However, flooding enhances aerenchyma formation in rice.

Aerenchyma is a cellular space filled with air, which provides oxygen to the respiring cells under anoxic and hypoxic condition by accumulating oxygen from the environment by diffusion (Colmer, 2003). Aerenchyma create a diffusion path for transport of oxygen from aerial parts to the waterlogged tissues as well as helping in diffusion of other volatile compounds including CO_2 and ethylene, both of which are crucial for survival of the rice plant under waterlogged conditions (Vartapetian and Jackson, 1997). Ethylene serves as a signalling agent for aerenchyma formation in maize; however, it plays little role in development of aerenchyma in rice. Underwater CO_2 concentration reduces in the daytime due to photosynthesis while it increases in the night time by respiration. Inadequate supply of CO_2 during photosynthesis results in reduced submergence tolerance (Setter *et al.*, 1989).

However, formation of aerenchyma leads to weakening of root structure, thus its penetration ability and mechanical support to the plant is reduced. The rice plant maintains a balance between aerenchyma formation and functional ability of root to provide these mechanical supports. Rice is commonly grown in environments in alternate aerobic and anaerobic environments. Maintaining stagnant water in the field for days or weeks is a general practice in farmers' fields, which increases the rate of formation of aerenchyma in the roots, so that more oxygen can be harvested under anaerobic condition. Due to less rainfall or withholding irrigation, water is often depleted from the soil, resulting in hardening of soil. This exerts high mechanical pressure on the rice roots, resulting in collapse of the aerenchyma structure. The rice plant then again shifts to the formation of vascular tissues in place of aerenchyma. Anatomical studies reveal that rice roots develop a smaller amount of aerenchyma under irrigated cultivation conditions (alternate wetting and drying) than complete submerged conditions (Mostajeran and Rahimi-Eichi, 2008). The inherent plasticity of rice root to develop aerenchyma under submerged conditions and degrade it under aerobic conditions is a key anatomical adaptation for its semi-aquatic nature. The cost of maintaining plasticity obviously affects the productivity of rice. New rice cultivation techniques

(System Rice Intensification or SRI) minimize the anaerobic environment by maintaining non-flooded soil conditions. This management practice has led to increased rice productivity by increasing tiller number and better plant growth as well as reducing the cost of cultivation (Sato and Uphoff, 2007).

11.3.2 Shoot elongation in deep water rice

Certain rice genotypes adapted to complete submergence (deep water rice) have the unique capability to shoot elongation and float over the water level for growth and development. The ability to elongate under such an environment depends on several plant and environmental factors, such as genotype, age of plant, root and shoot anatomical characters, light perception through water, temperature of the water etc.

Shoot elongation is a primary response in rice and in many wetland plants under submerged condition. The internodes of shoots elongate vertically rapidly through the water until they contact the atmosphere (Jackson, 1985). As in roots, aerenchyma plays a crucial role in elongating shoots by providing oxygen and other gaseous elements to the shoot tissues. More internode air spaces develop under submerged condition in rice. Development of internode air spaces is higher in submergence-tolerant cultivars than susceptible cultivars (Datta and Banerji, 1974). The air spaces also reduce the culm weight and help to maintain buoyancy in the water.

In deep water rice, the air layers formed in the leaves serve two purposes: (i) to accumulate oxygen in the air chambers through diffusion, which is used by the leaf for respiration (Matsukura *et al.*, 2000); and (ii) the air chambers increase floating capability of the leaves above water, so that part of the leaves may directly come in contact with the atmosphere for photosynthesis and respiration.

References

Borrell, A.K., Hammer, G.L. and Henzell, R.G. (2000) Does maintaining green leaf area in sorghum improve yield under drought? II. Dry matter production and yield. *Crop Science* 40, 1037–1048.

Cabuslay, G.S., Ito, O. and Alejar, A.A. (2002) Physiological evaluation of responses of rice (*Oryza sativa* L.) to water deficit. *Plant Science* 163(4), 815–827.

Datta, S.K. and Banerji, B. (1974) Anatomical variations of flood-resistant and deep-water rices under field and deep-water conditions. *Phytomorphology* 24, 164–174.

Ehleringer, J.R., Klassen, S., Clayton, C., Sherrill, D., Fuller-Holbrook, M., Fu, Q. and Cooper, T.A. (1991) Carbon isotope discrimination and transpiration efficiency in common bean. *Crop Science* 31, 1611–1615.

Grzesiak, S., Grzesiak, M. and Hura, T. (1999) Effects of soil drought during the vegetative phase of seedling growth on the uptake of 14C and the accumulation and translocation of 14C in cultivars of field bean (*Vicia faba* L. var *minor*) and field pea (*Pisum sativum* L.) of different drought tolerance. *Journal of Agronomy and Crop Science* 183, 183–192.

Isendahl, N. and Schmidt, G. (2006) *Drought in the Mediterranean*. WWF policy proposals. WWF, Madrid.

Jackson, M.B. (1985) Ethylene and responses of plants to soil waterlogging and submergence. *Annual Review of Plant Physiology* 36, 145–174.

Kawai, M., Samarajeewa, P.K., Barrero, R.A., Nishiguchi, M. and Uchimiya, H. (1998) Cellular dissection of the degradation pattern of cortical cell death during aerenchyma formation of rice roots. *Planta* 204, 277–287.

Likoswe, A.A. and Lawn, R.J. (2008) Response to terminal water deficit stress of cowpea, pigeonpea, and soybean in pure stand and in competition. *Australian Journal of Agricultural Research* 59, 27–37.

Ludlow, M.M. and Muchow, R.C. (1990) Critical evaluation of the possibilities for modifying crops for high production per unit of precipitation. *Advances in Agronomy* 43, 107–153.

Maiti, R.K., Amaya, L.E.D., Cardena, S.I., Dimas, A.O., de la Rosa-Ibbara, M. and de Leon Castillo, H. (1996) Genotypic variability in maize cultivars (*Zea mays* L.) for resistance to drought and salinity at the seedling stage. *Journal of Plant Physiology* 148, 741–744.

Matsukura, C., Kawai, M., Toyofuku, K., Barrero, R.A., Uchimiya, H. and Yamaguchi, J. (2000) Transverse vein differentiation associated with gas space formation – fate of the middle cell layer in leaf sheath development of rice. *Annals of Botany* 85, 19–27.

Miyamoto, N., Steudle, E., Hirasawa, T. and Lafitte, R. (2001) Hydraulic conductivity of rice roots. *Journal of Experimental Botany* 52, 1835–1846.

Mostajeran, A. and Rahimi-Eichi, V. (2008) Drought stress effects on root anatomical characteristics of rice cultivars (*Oryza sativa* L.). *Pakistan Journal of Biological Sciences* 11, 2173–2183.

Nar, H., Saglam, A., Terzi, R. *et al.* (2009) Leaf rolling and photosystem II. Efficiency in *Ctenanthes etosa* exposed to drought stress. *Photosynthetica* 47, 429–436.

Nguyen, H.T., Babu, R.C. and Blum, A. (1997) Breeding for drought resistance in rice – physiology and molecular genetics considerations. *Crop Science* 37, 1426–1434.

Odeny, D.A. (2007) The potential of pigeonpea (*Cajanus cajan* (L.) Millsp.) in Africa. *Natural Resources Forum* 31, 297–305.

Peña-Valdivia, C.B., Sánchez-Urdaneta, A.B., Trejo, C., Aguirre, R.J.R. and Cárdenas, E. (2005) Root anatomy of drought sensitive and tolerant maize (*Zea mays* L.) seedlings under different water potentials. *Cereal Research Communications* 33, 705–712.

Peña-Valdivia, C.B., Sánchez-Urdaneta, A.B., Rangel, J.M., Muñoz, J.J., García-Nava, R. and Velázquez, R.C. (2010) Anatomical root variations in response to water deficit: wild and domesticated common bean (*Phaseolus vulgaris* L.) *Biological Research* 43, 417–427.

Ramirez-Vallejo, P. and Kelly, J.D. (1998) Traits related to drought resistance in common bean. *Euphytica* 99, 127–136.

Ristic, Z. and Jenks, M.A. (2002) Leaf cuticle and water loss in maize lines differing in dehydration avoidance. *Journal of Plant Physiology* 159, 645–651.

Sato, S. and Uphoff, N. (2007) Raising factor productivity. In: *Irrigated Rice Production: Opportunities with the System of Rice Intensification*. CAB International, Wallingford, UK.

Setter, T.L., Waters, I., Wallace, A., Bhekasut, P. and Greenway, H. (1989) Submergence of Rice II: Growth and photosynthetic response to CO_2 enrichment of floodwater. *Australian Journal of Plant Physiology* 16, 251–263.

Trethowan, R.M. and Mujeeb-Kazi, A. (2008) Novel germplasm resources for improving environmental stress tolerance of hexaploid wheat. *Crop Science* 48, 1255–1265.

Vartapetian, B.B. and Jackson, M.B. (1997) Plant adaptation to anaerobic stress. *Annals of Botany* 79, 3–20.

White, J.W. and Castillo, J.A. (1989) Relative effect of root and shoot genotypes and yield on common bean under drought stress. *Crop Science* 29, 360–362.

12

Anatomical Adaptation in Crop Plants to Harvest Higher Energy

12.1 Introduction

Crop plants are distinct in features from non-crop plants for the directional selection towards higher productivity, which has accelerated the rate of evolution in these plants. However, due to continuous selection, the genetic resources have reduced drastically. The diversity of modern-day crop gene pool is much less than that of their wild and weedy relatives. Two distinct changes in the domestication process are more evident. The first is the reduction in fitness of crop plants in the natural environment. Modern maize or wheat varieties are unable to survive in the natural environment without the help of human intervention. In other words, these crop plants have become completely dependent on the human population for continuation as a species. The second change is the tremendous increase in storage reserves in the organs which are harvested as the product by us. This is particularly important for all the crop plants, except where the whole plant biomass is the economic product (fodder crops). In such cases, human selection has targeted for increase in overall vegetative growth. Increase in amount of reserve compounds in a crop plant may be achieved in two ways. In the first case, total reserve can be increased by the number of units

storing the reserve food. In most of the grain crops, increase in grain number per plant is the main strategy for increasing productivity. The second case is the increase in amount of reserve compounds per unit of reserve tissue. In grain crops, grain weight is considered to be an important factor for increasing yield. However, it is easier to increase grain number than grain weight, since the structural boundaries of a grain limits the amount of reserve compounds that can be stored. However, continuous selection has improved the grain size considerably in many crops including wheat, rice and pea. However, in many other crops remarkable increases in grain size have not been observed even after continuous genetic improvement, such as in maize, lentil, sorghum and millets. This is also valid for vegetative storage organs, where reserve food compounds are stored for supporting asexual propagation of the plant. In potato, genotypes with high productivity produce many tubers of medium size. The genotypes with few tubers of large size generally exhibit lower productivity.

To store more reserve compounds, the crop plant has to produce more photosynthate, which will be converted to the storage compounds after meeting the need of the growing plant. Thus selection for higher productivity has essentially been the quest for genotypes that can harvest higher solar energy and convert this into chemical energy efficiently. The second issue is to channel the storage compounds to storage organs efficiently (from source to sink). The structural anatomy of crop plants has thus been modified accordingly to harvest higher energy, conversion of energy to storage product, minimization of vegetative growth activities during reproductive maturity phase, better assimilation of mineral nutrients from the environment (soil and air) and reduction in unnecessary losses of water from the plant. We should remember that human selection has not brought all the changes in plants to be required as crop. During the domestication phases of crop plants, the human population has employed opportunistic selection for crop species from natural variability according to their choice. The process of selection was primarily based on domestication-related traits including higher productivity, suitability of consumption and feasibility of cultivation as a crop. Per unit productivity was not very important to early agrarian populations due to smaller consumption requirement and availability of land for cultivation. Crop yield has become the primary and sometimes the sole concern with the increase in human population size and competition for resources and is the main concern for modern day plant breeding. The changes in structural anatomy of crop plants are clear indications of this evolutionary process, which is being changed continuously depending on the adaptation conditions and demand for higher productivity.

During evolution, several anatomical changes in the plant kingdom have been found to be beneficial for humans, making these plants suitable as a crop. A leading example is the evolution of C_4 plants from C_3 plants. The higher photosynthetic ability with desirable features of higher harvestable produce for consumption have made a good number of C_4 species crop plants, such as maize, sorghum, sugarcane and pineapple. However, several other grass species having the C_4 photosynthesis mechanism were not selected as a crop,

due to a lack in other important desirable characters, such as higher harvestable product and usability as food, fuel or fodder. A second example of structural change and adaptation is the ability to develop a symbiotic relationship with microorganisms for fixation of nitrogen. The specially developed root nodules in legumes are factories for converting nitrogen into nitrogenous compounds, which are supplied to the plant for growth and development. Conservation of energy for increasing the amount of storage reserves and efficient channelling of reserve foods toward storage tissues have also been accomplished by a number of structural modifications in crop species.

12.2   Higher Photosynthesis

Over 90% of the total biomass of the crop is derived from photosynthetic activities (Zelitch, 1982). Thus adaptation for higher photosynthesis is a major driving force for selection of crop plants. During the process of domestication, selection led to improvement of the photosynthetic capacity of crop plants. Comparison of photosynthetic ability of modern crop plants with that of their wild relatives reveals that the crop plants exhibit higher photosynthesis (Bhagsari and Brown, 1976; Saitoh *et al.*, 2004). Rice and wheat, two principal C_3 crops, exhibit higher photosynthesis than other C_3 crops, which contributes to the high yield of these two crops (Murata, 1981).

However, until recently, net photosynthesis was considered not to be related to crop yield (Makino, 2011). Researchers found no correlation between net CO_2 exchange per unit leaf area and yield per unit area, which is considered paradoxical because the reserve food expressed as yield would not have been produced if there was no photosynthesis (Evans, 1975; Zelitch, 1982). But experiments with elevated CO_2 concentration proved that yield is increased when the rate of photosynthesis was higher, provided the other factors are unaltered (Long *et al.*, 2006).

12.2.1 C_3 crops versus C_4 crops

The most dramatic evolution for higher photosynthesis is the appearance of C_4 plants. It was discovered that in crop plants, increased photosynthetic activity of C_4 plants increases total biomass by fixing more CO_2 and trapping higher solar energy. However, the ability to produce photosynthates alone does not ensure higher productivity. Several other factors, including mineral nutrition supply, effective trapping of solar energy (crop canopy) and balanced partitioning of assimilates to the economic storage body of the crop are required to ensure high yield. The structural anatomy of the major C_3 crops and C_4 crops (maize, sorghum) have been described in detail in this book in earlier sections with their functional significance, which will not be repeated here.

12.2.2 Feasibility of conversion of C_3 to C_4 crops – structural difficulties

Rice, a C_3 crop, is the most important crop in tropical countries, feeding the majority of the world's population. Productivity of rice has not increased significantly in the last decades in the majority of the rice-growing countries. Conversion of the C_3 photosynthesis system in rice to a C_4 system is being considered a major strategy for increasing productivity. However, simple introduction of the C_4 enzymes into rice by genetic engineering has not been very successful because of the inherent anatomical differences of the C_4 and C_3 photosynthesis systems. The CO_2 enrichment mechanism in C_4 plants by the special bundle sheath cells with Krantz tissue allow RUBISCO to function efficiently under high CO_2 concentration. In rice leaves, such compartmentalization is absent. Thus overexpression of pyruvate orthophosphate dikinase (PPDK) of maize, NADP-malate dehydrogenase (MDH) of sorghum or NADP-malic enzyme (ME) of rice individually or in combination did not increase photosynthesis in rice under normal CO_2 concentration (Taniguchi *et al.*, 2008). When the CO_2 concentration was higher, the transgenic plants containing all the C_4 specific enzymes exhibited higher photosynthesis. This emphasizes that structural compartmentalization is essential for increasing CO_2 concentration and rate of photosynthesis. It has been suggested that the C_4 plants originated from C_3 plants at different periods of evolution, which have been often associated with increase in temperature. The structural anatomy of bundle sheath cells of leaves of C_4 plants show that cell wall thickening of bundle sheath cells helps in concentrating CO_2 in these cells. The coexistence of C_3 and C_4 systems in the same plant (*Eleocharis vivipara*, vascular tissue in tobacco) indicate that the transition from C_3 to C_4 system during evolution did take place via intermediate structures combining both the systems. Sage (2004) suggested that this transition occurred gradually by accumulation of one feature at a time (incremental hypothesis). However, the repeatability of C_3 to C_4 conversion (50 times, 19 families) also suggests that the mutation mechanisms for evolution of C_4 plants have been common, which indicates a primary mutation event followed by other changes. The genetic control of Kranz anatomy is still not well understood, which is a major bottleneck of converting a C_3 crop like rice to a C_4 crop. Thus conversion of a C_3 crop species to C_4 crop would obviously require conversion of a good proportion of mesophyll cells of the vascular bundle into bundle sheath cells in a particular orientation, so that the non-converted mesophyll cells remain enclosed by the bundle sheath cells for trapping more CO_2 during photosynthesis. The repeatability of the C_4 mutations and universality of the Kranz anatomy in several families suggest that some genetic switch may be involved in transformation of C_3 to C_4 species.

12.2.3 Higher partitioning of photosynthates

The harvest index

Productivity in economic terms is designated by the ratio of economic product to the non-harvestable biomass. Although almost every part of crop species has certain utility, productivity is mainly concerned with the principal economic product. In the case of grain crops, seeds are the principal product, where the stored material in seed is used either as food (cereals and pulses) or for extraction of oil (rapeseed, mustard, soybean, groundnut). In these crops, the productivity is determined by harvest index (ratio of grains produced and total biomass). The cereal crops have occupied the bulk of the agricultural area not only for the supply of essential carbohydrates, but also for their high harvest index (HI) compared to other crops. The HI of intensively cultivated varieties of major crops falls in the range of 0.4–0.6 (Hay, 1995).

During the early selection periods, productivity enhancement was mainly achieved by increase in biomass (Kawano, 1990). This led to the development of locally adapted genotypes, which have a higher vegetative growth under the low input conditions. Since early agriculture was not input intensive, genotypes responsive for high input condition having higher harvest index were not paid much attention; productivity was enhanced primarily through higher crop growth, resulting in both vegetative and reproductive yield. However, certain historical evidence suggests that the HI of crops grown by the ancient civilization was not very low (Nagato *et al.*, 1988; Amir and Sinclair, 1994). HI of major cereal crops decreased in the middle ages with the increased alternative use of vegetative parts of cereal crops as animal feed.

The concept of HI in modern plant breeding came in the early 20th century, when Beaven proposed the term 'migration coefficient' in 1914 to describe the ratio of grain weight to total plant weight (Donald and Hamblin, 1976). The concept gained importance in plant breeding only in the 1960s when Donald (1962) provided mathematical expressions for HI determination. In the second half of the 20th century, modern plant breeding has led to a substantial increase in the HI of grain crops (Sinclair, 1998). However, in certain crops such as maize, the HI was already high, and major genetic improvement has taken

place in this crop through increase in biomass (Hay, 1995).

Partitioning of biomass through anatomical adaptations

According to Evans (1975), sink capacity and yield tend to increase in parallel until they reach a particular limit set by the photosynthetic capacity. The photosynthetic and storage capacities of improved genotypes of major crops are likely to maintain a close balance. Thus improving one or the other will have a coordinated effect on productivity. In certain crops, the sink capacity is limited by structural boundaries, such as grain size in maize. Thus even with a higher rate of photosynthesis as a C_4 plant, maize productivity is not higher than rice, a C_3 plant, which has a lower rate of photosynthesis but higher sink capacity. Hybrid maize has increased the productivity through higher vegetative growth as well as increase in size of cobs, but the HI did not increase significantly.

Structural limitations of source and sink are primary bottlenecks of productivity enhancement. Zelitch (1982) proposed that the limitation of sink capacity in crop plants is the major limiting factor for realizing high yield despite adequate photosynthesis. Crop ideotypes (described in Chapter 15) have been proposed by many workers that can break the barrier of source or sink limitation. Remarkable achievements have been made in rice improvement using the ideotype concept.

12.3   Biological Nitrogen Fixation

The capability of fixing atmospheric nitrogen in symbiosis with microbial fauna is a remarkable feature of plant species to meet the nitrogen demand. Higher nitrogen concentration is a characteristic feature of plants that exhibit biological nitrogen fixation. The additional nitrogen is required to meet high nitrogen demand of the tissues as well as to develop a number of nitrogenous secondary metabolites that provide protection from predation and pathogenesis. For examples, the leguminous nitrogen-fixing species produce a variety of polyphenols, which protect them from pest and disease attack.

The leguminous plants add substantial amounts of nitrogen to the soil. It has been observed that the amount of nitrogen fixed by leguminous crops is about 40 Mt/year (Vitousek *et al.*, 1997). Although the ability of fixing nitrogen by different legume species varies considerably (soybean fixes more nitrogen than chickpea), on an average, the major legume crops fix about 9–245 kg nitrogen/ha (Unkovich and Pate, 2000).

12.3.1 Plant-bacteria specificity for N-fixation

Nodule formation in plants is induced by two groups of soil bacteria, rhizobia and actinobacteria. A new group of beta-rhizobia has recently been identified, which can induce nodule formation in legumes. Rhizobial bacteria comprise the genera *Rhizobium, Sinorhizobium, Bradyrhizobium, Azorhizobium, Allorhizobium* and *Mesorhizobium*. Rhizobial nodulation is limited to the plant families Leguminosae and Ulmaceae. The interactions of rhizobia and the plant species are specific, and each rhizobial species has a particular host range. The actinobacteria (hyphae-forming bacteria) are the members of genus *Frankia*, which induce nodulation in many plant families including Casurianaceae, Myricaceae, Rhamnaceae, Rosaceae, Betulaceae, Datiscaceae, Eleagnaceae and Coriariaceae. However, for crop productivity, rhizobial-leguminous crop species interaction is most important. The specificity of legume crops and rhizobia is described in Table 12.1.

Apart from the food crop species of Leguminosae, a large number of tree species of agroforestry systems (*Leucaena leucocephala, Calliandra calothyrsus, Gliricidia sepium* and *Sesbania sesban*) are capable of biological nitrogen fixation.

Recent discoveries show that several non-rhizobial microbes can induce nodulation

Table 12.1. Legume crop-rhizobial species/strain specificity.

Legume crop host	Rhizobial symbiont
Pea	*Rhizobium leguminosarum* bv. *viciae*
Soybean	*Bradyrhizobium japonicum* *Sinorhizobium fredii*
Bean	*Rhizobium leguminosarum* bv. *phaseoli* *Rhizobium etli*
Clover	*Rhizobium leguminosarum* bv. *trifolii*

and fix nitrogen in legumes, such as *Blastobacter, Devosia, Methylobacterium, Agrobacterium, Phyllobacterium, Ochrobactrium, Cupriavidus, Herbaspirillum* and *Burkholderia* (Balachandar *et al.*, 2007). This group is known as beta-rhizobia to differentiate it from the rhizobial bacteria. These species in association with *Rhizobium, Mesorhizobium* and *Bradyrhizobium* add substantial amounts of nitrogen to the soil through leaf senescence. Nitrogen fixation by symbiosis is also observed in other non-leguminous crops (sugarcane, sweet potato), where rhizobia exist as endophytes in plants and fix nitrogen, although nodules are not formed (Dong *et al.*, 1994; Terakado-Tonooka *et al.*, 2008). These recent discoveries clearly suggest that the impact of biological nitrogen fixation by crop plants in symbiosis with microorganisms is much more than anticipated one or two decades ago.

12.3.2 Mechanism of nodulation

What is the mechanism behind the specificity of host–rhizobium interaction? The legumes release a group of isoflavonoid and betaine compounds through their root, which serve as chemo-attractants to these bacterial species (Cooper, 2007). The recognition of a particular host by the rhizobia is specified by the structure of the flavonoid compounds released. A group of genes in the bacterial genome, known as *nod* (*nodulation*) genes are activated, releasing nod factors. Three of these genes are common to all rhizobial species (*nodA, nodB* and *nodC*, develop the basic structure of Nod factors), a few provide host specificity (*nodE, nodF, nodP, nodQ* etc.), and one acts as regulatory gene (*nodD*). The flavonoid compounds released by the host plant are recognized by the *NodD* protein, inducing transcription of other *nod* genes. The Nod factors are lipochitin oligosaccharides, a group of signal molecules. The host root, sensing the presence of nodulating rhizobia, bends in a characteristic fashion to enclose the rhizobial population, known as 'Shepherd's crook'. The rhizobial bacteria move to the root hair region or in lateral root regions, colonize and induce nodule formation. Structural alteration in the root hair cells and the epidermal region lead to curling of the roots and development of an infection thread. A small group of cells of root cortex opposite to the protoxylem region, known as nodule primordium, starts dividing rapidly under the influence of auxin influx and differentiates into a nodule.

12.3.3   Nodule anatomy

A nodule is a modified root, which attains spherical or ovoid shape due to cell division of root cortex cells. In some cases pericycle cells are also involved in nodule formation (actinorrhizal nodule, indeterminate legume nodule). The anatomy of the nodule is typical of the root with a vascular system. However, the vascularization is peripheral as in the stem, with a central void which is a reserve for the rhizobial population. The rhizobia penetrates the nodule by forming an infection thread, which enters the nodule primordium through root epidermal transfer cells (Lin *et al.*, 2008). It lives there as a bacteroid, where it is encircled by plant membrane called peribacteroid membrane. Depending on the growth characteristics, the nodules are grouped into determinate nodules (groundnut, soybean) and indeterminate nodules (pea). In determinate nodules, growth is controlled and the shape of the nodule is spherical. In the case of indeterminate nodules, the cell division continues

almost throughout the period of nodule activity; as a consequence, they are oval in shape. An apical meristem region is present in the indeterminate types, which leads to continuous growth. In both the nodules, the cortex region is surrounded by a periderm. The characteristic pink colour is due to presence of high concentration of leghaemoglobin, which plays a pivotal role in nitrogen fixation.

12.3.4   Nitrogen fixation

Nitrogen fixation is performed by the following biochemical reaction catalysed by the bacterial nitrogenase enzyme complex.

$$N_2 + 8e^- + 8H^+ + 16ATP \rightarrow 2NH_3 + H_2 + 16ADP + 16P_i$$

The fixed ammonia is rapidly transferred to surrounding cells and is converted to amides (asparagines, glutamine) or ureides (allantoin, allantoic acid, citrulline) and is transported to shoots via xylem. The energy for nitrogen fixation comes from oxidative phosphorylation, which requires high amount of oxygen influx in the nodules. However, the reaction itself requires a reducing environment. The exclusion of oxygen from the zones of nitrogen fixation is performed by leghaemoglobin, which has a very high affinity for oxygen (ten times more than blood haemoglobin). In many legumes the hydrogen produced is oxidized by hydrogenase enzyme. In other legumes the hydrogen produced diffuses from the nodules into the soil, where it is taken up by microorganisms (Peoples *et al.*, 2008).

12.4   Higher Nutrient Assimilation

12.4.1   Anatomical modifications for nutrient assimilation

Nutrient is essential for optimum growth of a crop plant through efficient xylem for assimilation of minerals through root and its translocation to the leaves and stems, associated with the translocation of photosynthates from leaves downwards through phloem.

12.4.2 Exclusion of toxic mineral elements – the role of root anatomy

Aluminium (Al) toxicity is the most common form of soil toxicity caused by minerals. About 15% of the world's land area is affected by Al toxicity (Bot *et al.*, 2000). At acidic pH (<5.5), Al toxicity stunts root growth and reduces crop productivity substantially. Roots have mechanisms for the exclusion of Al by chelation, ROS scavenging, lipid peroxidation, or by modification of cell wall properties. Among the cereal crops wheat exhibits good levels of tolerance to Al toxicity, whereas barley is susceptible. In wheat, a group of malate transporter genes is activated by sensing higher Al concentration in the root zone (Hoekenga and Magalhaes, 2011). This switches on malate transporter systems, which release a large quantity of malate in the rhizosphere for chelation to Al ion.

Anatomical modification by reducing cell wall extensibility is another important mechanism to decrease the concentration of Al in the xylem sap. Al is deposited in the root cell wall. It has been suggested that in the Al-tolerant genotypes, de-esterification of pectins in the cell wall can cause greater Al-loading in the cell wall, thereby reducing Al uptake by roots (Jones *et al.*, 2006). It also reduces the root growth under an Al-toxic environment, Al^{3+} ions bind to the cell wall expansin molecules, preventing extensibility and growth of roots.

Although iron is an essential element for plant growth, excessive accumulation of iron in plants causes several toxic symptoms, including yellowing and bronzing of seedlings. It is a major soil mineral stress of rice plants grown under submerged conditions causing yield loss of about 35–45% (Audebert and Sahrawat, 2000). The toxicity leads to stunted growth of the tillers, reduction in tillering ability and panicle numbers, reduced photosynthesis, spikelet sterility, and accumulation of dark coloured $Fe(OH)_3$ in the root zone. The accumulation of insoluble iron (Fe^{3+}) in the root zone under anaerobic conditions inhibits absorption of other minerals, leading to complex mineral deficiency symptoms. In the rice plant, iron forms insoluble oxides or phosphates or binds to phytoferritin and accumulates in the lower leaves, blocking mineral transport. Even under this condition, the young leaves suffer from iron deficiency due to the low mobility of iron. As iron is an important member of the photosynthetic machinery, chlorosis of young leaves with bronzing of older leaves is a characteristic symptom of iron toxicity. The bronzing due to accumulation of iron oxides in cells and reduction of chloroplast size and number in young leaves are observed. Iron accumulation also blocks the vascular system of the root. The plant cells store excess iron in vacuoles, mitochondria, plastids and also in apoplastic space. Manipulation of size and shape of these compartments and the iron transport channels is expected to provide higher tolerance to iron toxicity.

References

Amir, J. and Sinclair, T.R. (1994) Cereal grain yield: Biblical aspirations and modern experience in the Middle East. *Agronomy Journal* 86, 362–364.

Audebert, A. and Sahrawat, K.L. (2000) Mechanisms for iron toxicity tolerance in lowland rice. *Journal of Plant Nutrition* 23, 1877–1885.

Balachandar, D., Raja, P., Kumar, K. and Sundaram, S.P. (2007) Non-rhizobial nodulation in legumes. *Biotechnology and Molecular Biology Reviews* 2, 049–057.

Bhagsari, A.S. and Brown, R.H. (1976) Photosynthesis in peanut (*Arachis*) genotypes. *Peanut Science* 3, 1–5.

Bot, A.J., Nachtergaele, F.O. and Young, A. (eds) (2000) Land resource potential and constraints at regional and country level. FAO Land and Water Development Division, FAO, Rome. Available at: http://www.fao.org/AG/agl/agll/terrastat/ (accessed on 22 August 2011).

Cooper, J.E. (2007) Early interactions between legumes and rhizobia: disclosing complexity in a molecular dialogue. *Journal of Applied Microbiology* 103, 1355–1365.

Donald, C.M. (1962) In search of yield. *Journal of Australian Institute of Agricultural Science* 28, 171–178.

Donald, C.M. and Hamblin, J. (1976) The biological yield and harvest index of cereals as agronomic and plant breeding criteria. *Advances in Agronomy* 28, 361–405.

Dong, Z., Canny, M.J., McCully, M.E., Roboredo, M.R., Cabadilla, C.E., Ortega, E. and Rodes, R. (1994) A nitrogen fixing endophyte of sugarcane stems: a new role for the apoplast. *Plant Physiology* 105, 1139–1147.

Evans, L.T. (1975) The physiological basis of crop yield. In: Evans, L.T. (ed.) *Crop Physiology*. Cambridge University Press, Cambridge, UK, pp. 327–335.

Hay, R.K.M. (1995) Harvest index: a review of its use in plant breeding and crop physiology. *Annals of Applied Biology* 126, 197–216.

Hoekenga, O.A. and Magalhaes, J.V. (2011) Mechanism of aluminum tolerance. In: De Oliviera, A.C. and Varshney, R.K. (eds) *Roo Genomics*. Springer, New York, pp. 133–153.

Jones, D.L., Blancaflor, E.B., Kochian, L.V. and Gilroy, S. (2006) Spatial coordination of aluminium uptake, production of reactive oxygen species, callose production and wall rigidification in maize roots. *Plant Cell and Environment* 29, 1309–1318.

Kawano, K. (1990) Harvest index and evolution of major food crop cultivars in the tropics. *Euphytica* 46, 195–202.

Lin, S.Y., Wu, J.Z. and Han, S.F. (2008) Inducement of root hair and the root epidermal transfer cells during the formation of root nodule in leguminous plants. *Acta Botanica Boreali Occidentalia Sinica* 28, 697–703.

Long, S.P., Zhu, X.G., Naidu, S.L. and Ort, D.R. (2006) Can improvement in photosynthesis increase crop yields? *Plant Cell and Environment* 29, 315–330.

Makino, A. (2011) Photosynthesis, grain yield, and nitrogen utilization in rice and wheat. *Plant Physiology* 155, 125–129.

Murata, Y. (1981) Dependence of potential productivity and efficiency for solar energy utilization on leaf photosynthetic capacity in crop species. *Japanese Journal of Crop Science* 50, 223–232.

Nagato, Y., Inanga, S. and Suzuki, H. (1988) Yield of cultivars, their wild ancestors and relatives in some crops. *Japanese Journal of Breeding* 38, 414–422.

Peoples, M.B., McLennan, P.D. and Brockwell, J. (2008) Hydrogen emission from nodulated soybeans [*Glycine max* (L.) Merr.] and consequences for the productivity of a subsequent maize (*Zea mays* L.) crop. *Plant and Soil* 307, 67–82.

Sage, R.F. (2004) The evolution of C-4 photosynthesis. *New Phytologist* 161, 341–370.

Saitoh, K., Nishimura, K. and Kuroda, T. (2004) Comparison of leaf photosynthesis between wild and cultivated types of soybean. *Plant Production Science* 7, 277–279.

Sinclair, T.R. (1998) Historical changes in harvest index and crop nitrogen accumulation. *Crop Science* 38, 638–643.

Taniguchi, Y., Ohkawa, H., Masumoto, C., Fukuda, T., Tamai, T., Lee, K., Sudoh, S., Tsuchida, H., Sasaki, H., Fukayama, H. and Miyao, M. (2008) Overproduction of C-4 photosynthetic enzymes in transgenic rice plants: an approach to introduce the C-4-like photosynthetic pathway into rice. *Journal of Experimental Botany* 59, 1799–1809.

Terakado-Tonooka, J., Ohwaki, Y., Yamakawa, H., Tanaka, F., Yoneyama, T. and Fujihara, S. (2008) Expressed nifH genes of endophytic bacteria detected in field-grown sweet potatoes (*Ipomoea batatas* L.). *Microbes and Environments* 23, 89–93.

Unkovich, M.J. and Pate, J.S. (2000) An appraisal of recent field measurements of symbiotic N_2 fixation by annual legumes. *Field Crops Research* 65, 211–228.

Vitousek, P.M., Aber, J.D., Howarth, R.W., Likens, G.E., Matson, P.A., Schindler, D.W., Schlesinger, W.H. and Tilman, D.G. (1997) Human alteration of the global nitrogen cycle: sources and consequences. *Ecological Applications* 7, 737–750.

Zelitch, I. (1982) The close relationship between net photosynthesis and crop yield. *BioScience* 32, 796–802.

13

Anatomical Adaptation for Better Reproduction Efficiency

13.1 Introduction

Changes in floral anatomy have directed substantial evolutionary consequences in the flowering plants as well as in the evolution of crop plants. About 70% of the present day angiosperm species are cross-pollinated, while the rest are self-pollinated. However, the majority of crop plants typically show the opposite trend. Most of the cereal and pulse crops that provide the bulk of the food are strictly or predominantly self-pollinated (Table 13.1), with the exception of maize and pearl millet, which are cross-pollinated crops. The area under self-pollinated crops and the total production outnumbers the area and production of strictly cross-pollinated crops. However, a great number of vegetable crops and fruit crops are cross-pollinated (Table 13.1).

Self-pollination has evolved as a mechanism from obligate outcrossing, mostly in species where a small group of individual genotypes have spread to new habitats. In many cases, the changes have little impact on flower anatomy. For example, self-pollination in grasses has evolved through reduction in anther size (Stebbins, 1970). Since assisted pollination may not be available in a new habitat, the outcrossing plants have developed mechanisms to ensure receipt of self-pollen through minimal changes in floral anatomy and morphology.

13.2 Modifications for Better Reproduction

13.2.1 Strategies for better pollination

Pollination and fertilization are the two key mechanisms for successful reproduction in plants. Evolution of the floral structure in angiosperms from the non-flowering gymnosperms is a key step for increasing the efficiency of pollination and thereby ensuring fertilization. The gymnosperms lack flower structure, hence depend on the wind for pollination. This is a less efficient mechanism to ensure reproduction. Although it might be a good reason behind the tall structures of gymnosperm trees so that they can shed pollen better, the mechanism is less competent in terms of fertilization events per unit biomass. In the present day, angiosperm species (>200,000) have outnumbered the gymnosperm species (<1000) in abundance, efficiency in pollination and fertilization, protection of seeds produced and provision of nutrients to the emerging seedlings for initial establishment as well as efficiency in vascular transport.

Table 13.1. Pollination mechanism of major crops.

Self-pollinated	Cross-pollinated	Intermediate pollination
Rice (*Oryza sativa*)	Maize (*Zea mays*)	Sorghum (*Sorghum bicolor*)
Wheat (*Triticum aestivum*)	Pearl millet (*Pennisetum glaucum*)	Amaranth (*Amaranthus* spp.)
Oat (*Avena sativa*)	Buckwheat (*Fagopyrum esculentum*)	Common millet (*Panicum miliaceum*)
Barley (*Hordeum vulgare*)	Kidney bean	Foxtail millet (*Setaria italica*)
Soybean (*Glycine max*)	Sunflower (*Helianthus annuus*)	Cowpea (*Vigna unguiculata*)
Lentil (*Lens culinaris*)	Carrot (*Daucas carota*)	Cotton (*Gossypium* spp.)
Chickpea (*Cicer arietinum*)	Coriander (*Coriandrum sativum*)	Pigeonpea (*Cajanus cajan*)
Pea (*Pisum sativum*)	Cucumber (*Cucumis sativus*)	Broad bean (*Vicia faba*)
Groundnut (*Arachis hypogea*)	Rapeseed (*Brassica napus*)	Lima bean (*Phaseolus lunatus*)
Green gram (*Vigna mungo*)	Pumpkin (*Cucurbita moschata, C. maxima*)	Chilli, Capsicum (*Capsicum* spp.)
Adzuki bean (*Phaseolus vulgaris*)	Radish (*Raphanus sativus*)	Brinjal (*Solanum melongena*)
Lettuce (*Lactuca sativa*)	Spinach (*Spinacea oleracea*)	Okra (*Abelmoschus esculentus*)
Tomato (*Lycopersicon esculentum*)	Sugar beet (*Beta vulgaris*)	
Garlic (*Allium sativum*)	Muskmelon (*Cucumis melo*)	
Jute (*Corchorus* spp.)	Watermelon (*Citrullus lanatus*)	
Flax (*Linum usitatisimum*)	Onion (*Allium cepa*)	
Kenaf (*Hibiscus cannabinus*)	Squash (*Cucurbita pepo*)	
Safflower (*Carthamus tinctorius*)	Cabbage (*Brassica oleracea var. capitata*)	
Sesame (*Sesamum indicum*)	Cauliflower (*Brassica oleracea var. botrytis*)	
Lupin (*Lupinus* spp.)	Apple (*Malus domestica*)	
Citrus (*Citrus sinensis*)	Apricot (*Prunus armeniaca*)	
Grapefruit (*Citrus* × *paradisea, Citrus maxima* var. *racemosa*)	Mango (*Mangifera indica*)	
Lemon (*Citrus* × *limon, Citrus limonum*)	Cocoa (*Theobroma cacao*)	

High population density

Abundance of a species in a particular geographic area is a key determinant of successful pollination. Crop species are predominantly cultivated in defined areas with much higher population density under human protection and management. Reproduction in a typical self-pollinated crop species does not depend on the number of plants per unit area until this factor affects the resource utilizing mechanism (high plant density will increase competition among plants for nutrients and sunlight, which will reduce reproductive efficiency), because the pollination and fertilization take place in an environment where sufficient amounts of self-pollen are available. However, this is an important factor for survival of cross-pollinated species in a natural environment, because population density is a key factor for reproductive efficiency.

The major cross-pollinated crops are planted at a much smaller plant to plant distance (higher plant density) than required for pollination. The maize plant, which is the most important cereal crop, is pollinated by wind. Maize pollen is heavy $(2.5 \times 10^{-7}$ g) and the largest in size among the grass species (Miller, 1985). It is also one of the heaviest airborne particles. Several experiments have shown that the percentage of pollen concentrations at 50–60 m downwind compared with those at 1 m from the source is about 2%, while at 200 m distance this reduces to 1%. About 50% of the pollination of any individual plant is caused by pollen of plants within a radius of about 12 m (Emberlin, 1999). The population density of cultivated maize is usually 65,000–70,000/ha, which is more than sufficient for ensuring pollination and fertilization.

The maize inflorescence (tassel) continues shedding pollen for 5–10 days, producing about 2–5 million pollen grains. With 70,000 maize plants per hectare, the total pollen numbers of a maize field can be estimated to be $1.4–3.5 \times 10^5$ million, which is enormous for an angiosperm species. Such a large pollen load resembles more closely to gymnosperm species, which are also wind pollinated.

Flower anatomy favouring pollination by insects

Insect pollination has a tremendous impact on global food production. About 75% of the crops grown in different countries are pollinated by insects, which contribute to 35% of the total world production (Klein *et al.*, 2007) (Table 13.2). Insect pollination has specially evolved as a mechanism of assisted cross-pollination in angiosperms. A large group of insects including bees, butterflies, moths, flies and beetles assist in cross-pollination of crop plants. Among these, honeybee (*Apis* spp.), bumble bee (*Bombus impatiens*), solitary bee, stingless bee (*Melipona* spp.) and wild bees are major insect pollinators.

The insects visit flowers for the collection of nectar and pollen for conversion to honey. For this reason, the flowers of the insect-pollinated species have evolved to store nectar in the flower and have developed attractive flower morphology and anatomy. Sunflower, lucerne, clover, cotton, apple, cherry, litchi etc. are considered a very good source for honey and are visited frequently by insects. Insect-pollinated flowers are generally large, opening sufficiently to allow entry and exit of bees and other insects, are extremely

Table 13.2. Insect pollinated crops and their pollinators.

Crop	Pollinator	Dependence level
Cotton (*Gossypium* spp.)	Bumble bee, honeybee	Modest
Rapeseed (*Brassica napus*)	Honeybee, solitary bee	Modest
Sunflower (*Helianthus annuus*)	Wild bees	Modest
Buckwheat (*Fagopyrum esculentum*)	Honeybee	High
Berseem/Egyptian clover (*Trifolium alexandrinum*)	Honeybee	
Alfalfa/Lucerne (*Medicago sativa*)	Honeybee, bumble bee, *Halictus, Megachile, Melissodes*	
Cowpea (*Vigna unguiculata*)	Bumble bee	Little
Broad bean (*Vicia faba*)	Bumble bee	Modest
Lima bean (*Phaseolus lunatus*)	Honeybee, bumble bee, thrips	Little
Carrot (*Daucus carota*)	Honeybee	High for seed production Not required for harvest
Cauliflower (*Brassica oleracea* var. *botrytis*), Cabbage (*Brassica oleracea* var. *capitata*)	Honeybee, wild bee, solitary bee	High for seed production Not required for harvest
Coriander (*Coriandrum sativum*)	Honeybee	High
Cucumber (*Cucumis sativus*)	Honeybee, bumble bee, squash bee	High
Pumpkin (*Cucurbita moschata, C. maxima*)	Honeybee, bumble bee	Essential
Watermelon (*Citrullus lanatus*)	Honeybee, bumble bee	Essential
Apple (*Malus domestica*)	Wild bees, honey bee	High
Apricot (*Prunus armeniaca*)	Honeybee, wasp, flies	High
Cocoa (*Theobroma cacao*)	Midges	Essential
Cherry (*Prunus* spp.)	Honeybee	High
Fig (*Ficus* spp.)	Wasp	High
Mango (*Mangifera indica*)	Honey bee	High

Source: McGregor, 1976; Chacoff *et al.*, 2011

colourful to attract insects, and often contain nectary glands.

The size of the insect pollinator corresponding to the size of the flower largely influences the extent of pollen attached to the insect body and the amount of the pollen shedding on stigma. This character along with the frequency of visits made by the pollinator charts the success of cross-pollination. In cotton, the bumble bee is considered to be a more efficient pollinator than the honeybee. Because of the relatively larger size of the bumblebee corresponding to honeybee and the small size of the cotton flower, the bumble bee cannot collect honey without touching the anthers, thereby collecting pollen and depositing it on the stigma of self-flower or another flower. The honeybees on the other hand are interested in extrafloral nectary and seldom enter the cotton flower. Thus the cotton genotypes having extrafloral nectary are preferred more by the honeybee for pollination. In contrast, the rape flower anatomy fits well to the honeybee for the collection of pollen and nectar. The positioning of anther and stigma helps the stigma to be in touch with the surface of the bee carrying foreign pollen, thereby aiding pollination. The anatomy of florets of sunflower also is well suited for honeybees, where it is the principal pollinator. However, in this crop, the numbers of flower units are quite high compared to the average bee population. Thus the seed yield of a sunflower crop largely depends on the size of the bee population visiting the crop.

An interesting flower anatomy and pollination mechanism is observed in fig, a tropical fruit tree (Ramirez, 1969). The pollination is assisted by a group of wasps belonging to family Agaonidae. The wasp completes its life cycle within the fig inflorescence (syconium); males never emerge from the inflorescence, but females after emerging from one inflorescence invade another receptive inflorescence through a small hole called an ostiole. It carries pollen from one syconium to another and during the process of oviposition helps in pollination of female flowers within the syconium. In monoecious fig, some female flowers within the syconium

are used for oviposition and gall formation for hatching, while others are left for seed production. In dioecious fig, different syconiums are used for seed production and gall formation (Moe *et al.*, 2011).

Flower anatomy favouring pollination by other agents

Wind pollination is the next most important pollination mechanism after insect pollination (Table 13.3). The mechanism of wind pollination evolved much earlier than insect pollination, which is the primary mode of pollination of gymnosperms. The transition from wind pollination towards insect pollination is considered a major landmark in evolution of angiosperms (Hu *et al.*, 2008). This change has been intrinsically associated with the evolution of flower anatomy, where several features have been added, such as a large flower structure and organs to produce sticky and sugary substances. However, during recent evolution within angiosperms, a number of species have changed their pollination mechanism from animal to wind pollination, suggesting some common, recurrent mutations may be involved in such transition. The evolution of insect pollination from wind pollination is more prominent in primitive families of angiosperms, such as Cyperaceae, Caryophyllaceae and Moraceae (Wragg and Johnson, 2011).

The anatomy of the inflorescence of wind-pollinated crops is characterized by smaller size of flowers with increase in flower number. It is observed primarily in monoecious (maize) and dioecious (coconut) crop plants. The wind-pollinated flowers

Table 13.3. Crops pollinated by wind and other mechanisms.

Crop	Pollination mechanism
Maize	Wind
Coconut	Wind, insects
Date palm	Manual by hand pollination, wind, insect
Beet	Wind
Vanilla orchid (*Vanilla planifotus*)	Hand pollination

generally do not produce any nectar and are of unattractive colour or shape.

Position and timing of pollen landing

The position of landing of pollen on stigmatic surface and the timing of pollen dispersal largely determine the success of pollination and fertilization. For example, cotton stigma remains receptive from morning to noon, but receptivity drops sharply after noon. The cotton ovule contains about 50 seeds, thus at least 50 pollen grains must germinate, reach to the ovary and release male gametophyte for successful fertilization. The position of pollen adhesion to the stigmatic surface is crucial for pollen hydration and germination. The pollen that land on the base of the stigma germinate poorly and receive more mechanical hindrance for germination and pollen tube growth. Abundance of pollen throughout the stigmatic surface results in better seed set and fibre yield. This is the reason for yield improvement in cotton fields where bee visits are frequent, which helps in better pollen dispersal and landing during the peak receptivity period of stigma.

13.3   Reduction in Cost of Seed Production

13.3.1   Reduction in direct cost towards seed development

Biologically, the direct cost for seed production is defined by the amount of energy spent for development of the seed. However, in broader agronomic terms, this cost also includes the cost for whole fruit development where fruit is the economic product (vegetables, fruit crops). In most of the grain crops, productivity improvement has been achieved through the increase in seed size and number of seeds per plant. For increasing the number of seeds per plant energy has to be distributed towards a higher number of fertilization events, which is ensured by more ovule and pollen production. Both of these events require expenditure of more energy towards

meiotic events, including chromosome doubling and distribution in egg cells and companion reproductive cells (synergids and antipodal cells in the ovule). The metabolic activities for these events draw the required energy from assimilates developed in the vegetative tissues. Consequently, deposition of food reserve in the endosperm sink is reduced. Plant breeders often experience negative correlation between seed number and seed size. Simultaneous improvement in both the components is a challenging task, which requires high photosynthetic activity as well as balanced distribution of assimilates between metabolic activity spent for reproduction and accumulation of food reserves in the endosperm.

13.3.2   Reduction in accessory cost

Accessory costs of reproduction include energy expense towards development and maturity of seed, excluding the direct energy used for embryo and endosperm development (Thompson and Stewart, 1981). Such costs include energy utilized for development of support structures for pollination, fertilization and seed maturity as well as energy spent towards aborted ovules. It is further classified in two broad categories, pre-pollination cost and post-pollination cost (Haig and Westoby, 1991). The pre-pollination accessory costs include investment made in development of flower structure and peduncles, which is required for pollen capture and fertilization. Development of protective structures of the seed, cost of ovule packaging and structures needed for dispersal of seed (wings, tufts, hairs etc.) are included in the post-pollination accessory costs. Investigations on 14 diclinous non-crop plant species revealed that the accessory costs comprise 33–96% of the total reproductive cost (Lord and Westoby, 2006). The economics of seed development largely depends on efficient management in balancing these costs, since a larger amount of energy is consumed for development of these structures compared to the direct cost involved in seed development. The grain crops, adapted for better translocation of food reserve towards endosperm or cotyledon, are expected to have reduced accessory costs. However, studies on accessory cost determination in crops are limited. More information is required for understanding the role of domestication and selection in economizing floral anatomy in crop species.

Economization of pollen capture

The structural anatomy of the flower determines the cost required for development of floral organs. Reduction in cost of pollen capture has been achieved by many plant species by enclosing the stamens with petal structures for ensuring self-pollination. The legume flower structure is a prominent example of closing the flower so that pollen of self-flower is not wasted outside and entry of foreign pollen is restricted. In many cross-pollinated species, pollen capture is economized by grouping the pollen-receiving flowers. Maize is a prominent example where female flowers are grouped together in a cob so that pollen capture efficiency can be increased. However, being a wind-pollinated crop, the cost spent towards pollen development is higher (2–5 million pollen grains, equivalent to 0.5–1.25 million meiotic divisions) compared to self-pollinated cereal crops.

Reduction in cost of packaging, protection and dispersal of seed

A large proportion of accessory cost is spent towards packaging of ovules and providing protection to the seed. In many cereal crops, the seed is encased by a number of protective structures, such as lemma, palea, glumes and awn. Construction of these structures requires considerable energy. However, these structures not only provide protection and hold the grain, but also contribute significantly towards photosynthesis and assimilation of food reserve in grains. Thus selection in the natural environment as well as under cultivated environments has been directed towards balancing the floral anatomy so that the accessory costs for seed production are optimized.

In wheat, the photosynthetic contribution made by these components of ear can go up to 45% (Kriedemann, 1966; Wang et al., 2001). The awn of wheat is particularly significant as a protective structure prohibiting entry of pests and pathogens and as a light-capturing photosynthetic organ contributing significantly particularly at the grain-filling stages (Li et al., 2006). Domestication of wheat has resulted in modification of awn structure and number. In wild wheat, two prominent awns are observed, which act additionally as a balancing organ during seed dispersal as well as provide motility for dispersion in air and soil (Elbaum et al., 2007). Since self-dispersal is not a requirement in the cultivated wheat species, domestication has led to the reduction and sometimes the elimination of awns (awnless varieties). The reduction in awn number is a good example of economization of floral anatomy of wheat, reducing the cost of awn development.

References

Chacoff, N.P., Morales, C.L., Garibaldi, L.A., Ashworth, L. and Aizen, M.A. (2011) *The Americas Journal of Plant Science and Biotechnology* 3, 106–111.

Elbaum, R., Zaltzman, L., Burgert, I. and Fratzl, P. (2007) The role of wheat awns in the seed dispersal unit. *Science* 316, 884–886.

Emberlin, J. (1999) A Report on the Dispersal of Maize Pollen. Available at: http://www.mindfully.org/GE/Dispersal-Maize-Pollen-UK.htm (accessed 18 July 2011).

Haig, D. and Westoby, M. (1991) Seed size, pollination costs and angiosperm success. *Evolutionary Ecology* 5, 231–247.

Hu, S.S., Dilcher, D.L., Jarzen, D.M. and Taylor, D.W. (2008) Early steps of angiosperm–pollinator coevolution. *Proceedings of the National Academy of Sciences USA* 105, 240–245.

Klein, M.A., Vaissière, E.B., Cane, H.J., Steffan-Dewenter, I., Cunningham, S.A., Kremen, C. and Tscharntke, T. (2007) Importance of pollinators in changing landscapes for world crops. *Proceedings of Royal Society of London (B)* 274, 303–313.

Kriedemann, P. (1966) The photosynthetic activity of the wheat ear. *Annals of Botany* 30, 349–363.

Li, X., Wang, H., Li, H., Zhang, L., Teng, N., Lin, Q., Wang, J., Kuang, T., Li, Z., Li, B., Zhang, A. and Lin, J. (2006) Awns play a dominant role in carbohydrate production during the grain-filling stages in wheat (*Triticum aestivum*). *Physiologia Plantarum* 127, 701–709.

Lord, J.M. and Westoby, M. (2006) Accessory costs of seed production. *Oecologia* 150, 310–317.

McGregor, S.E. (1976) Insect Pollination of Cultivated Crop Plants. Available at: afrsweb.usda.gov/SP2UserFiles/Place/.../OnlinePollinationHandbook.pdf (accessed 13 May 2011).

Miller, P.D. (1985) Maize pollen: collection and enzymology. In: Sheridan, W.F. (ed.) *Maize for Biological Research*. A Special Publication of the Plant Molecular Biology Association, USA, pp. 279–282.

Moe, A.M., Rossi, D.R. and Weiblen, G.D. (2011) Pollinator sharing in dioecious figs (*Ficus*: Moraceae). *Biological Journal of the Linnean Society* 103, 546–558.

Ramirez, W.B. (1969) Fig wasps: mechanism of pollen transfer. *Science* 163, 580–581.

Stebbins, G.L. (1970) Adaptive radiation of reproductive characteristics in Angiosperms I: pollination mechanisms. *Annual Review of Ecology, Evolution and Systematics* 1, 307–326.

Thompson, K. and Stewart, A.J.A. (1981) The measurement and meaning of reproductive effort in plants. *American Naturalist* 117, 205–211.

Wang, Z.M., Wei, A.L. and Zheng, D.M. (2001) Photosynthetic characteristics of non-leaf organs of winter wheat cultivars differing in ear type and their relationship with grain mass per ear. *Photosynthetica* 39, 239–244.

Wragg, P.D. and Johnson, S.D. (2011) Transition from wind pollination to insect pollination in sedges: experimental evidence and functional traits. *New Phytologist*, doi: 10.1111/j.1469-8137.2011.03762.x.

14

Anatomical Basis of Crop Ideotype for Higher Productivity

14.1   Introduction

Crop plants have been selectively domesticated and cultivated by the human population in order to provide food, fuel, fibre or energy. The shift from the migratory nature of early hominids to colonist society is largely attributed to the introduction of agriculture. Recent excavation studies show that domestication of major crop plants was initiated in different parts of the world 15,000–6,000 years ago. During this period, the crop genotypes having higher utility to humans have been selected for, while the genotypes with characters less fit or undesirable to agriculture have been eliminated and lost. Compared to the natural evolution of plants, the selection process has been very rapid towards higher productivity. In modern agricultural systems, the rates of replacement of genotypes are quite high for many food crops that increased the productivity significantly, owing to the tremendous efforts made by the plant breeders of the 20th century.

The aims of early domestication and modern plant breeding are essentially same, to identify an ideal plant type having maximized productivity under a particular system. The term ideotype (ideal type) was first coined by Donald (1968), who defined it as 'a plant with model characteristics known to influence photosynthesis, growth and grain (for cereals) production'. Donald proposed ideotypes for wheat and barley, which will have erect large ears, a strong stem producing few tillers with smaller leaves. As most of the selection criteria used by plant breeders are morphological observations, anatomical studies have not been given much importance in ideotype development. However, the morphological features proposed for most of the crop ideotypes are manifested by changes in anatomical structures. A strong stem is the result of strengthening of tissues of the vascular bundle, and the increase in number of cellular layers or cellular thickness. Similarly, higher photosynthesis results from a greater number of mesophyll cells, changes in chloroplast structure, number and orientation. Several features of the plant at structural anatomy level have been observed to be changed in different crop ideotypes resulting in higher yield.

14.2   Changes in Crop Anatomy for Higher Harvest Index

Harvest index (HI) is the ratio of mass of the principal economic harvestable product to the total biomass of the crop. It is a principal indicator of productivity, particularly in

©R. Maiti, P. Satya, D. Rajkumar and A. Ramaswamy 2012. *Crop Plant Anatomy*
(R. Maiti *et al.*)

the case of food crops. The higher the HI, the higher is the channelling of the food reserve from non-harvestable to harvestable produce. Where the economic product is seed, HI is the ratio of total seed yield to biomass yield.

Three prominent examples where replacement of old genotypes by new ideotypes has been very rapid in the 20th century are maize, wheat and rice, the principal food crops of the world. There are several other examples, where plant types have considerably been changed to suit the agricultural system or to adapt to higher environments. Needless to say, the structural anatomy of the ideotypes has also changed considerably, and played a significant role in adaptation of new genotypes in the production system. In some cases, ideotype construction has been directed by selective anatomical features, while in other cases anatomy has played an underlying but important role in formation and selection of plant ideotype. The criterion for this ideal plant type was fixed by Donald as a plant type that would 'make a minimum demand on resources per unit of dry matter produced'. This led to an important conclusion that the sink must be large enough to store all the photosynthetic reserves produced by the plant. Although Donald proposed ideotypes for wheat and barley that will contain few tillers, erect leaves, small canopy size and higher seed number, the ideotype breeding has been most successful in another important cereal crop, rice (Virk *et al.*, 2004).

14.2.1 Ideotype for increasing productivity in rice

Rice is probably the only crop to be grown in diverse ecological conditions, starting from above 2000 m in small villages of the Himalayas in India and Nepal to below sea level in southern India and South-east Asia. As a crop it withstands extreme dry conditions and high temperature in western India and Pakistan. On the other hand, it is also grown under moderately low temperatures in parts of China and Japan. However, it cannot withstand chilling temperature as much as another grain crop, wheat. Nevertheless, it can tolerate acidic soil (pH <4.0) and alkalinity (pH >11.0) higher than other cereal crops such as wheat, maize and sorghum. This remarkable plasticity makes rice one of the most interesting crop species for studying crop adaptation. This also makes the job of defining a rice ideotype more difficult, as the ideal type of cultivars will vary according to the environment. To augment anatomical structures to these ideotype descriptions is a mammoth task requiring in-depth study of structural changes of major plant parts under such environments.

Rice was predominantly cultivated in Asian countries as photosensitive, tall plants producing few tillers. The Asian cultivated rice is differentiated in three eco-geographic races, *indica, japonica* and *javanica*. The race differentiation culminated from climatic adaptation and human selection. *Indica* rice is adapted to the humid, tropical to subtropical environment of South Asia, particularly the Indian subcontinent. It is characterized by tall plants with many tillers spreading from the base. The grains are borne on tall panicles, are medium to long and are easily threshed. The *japonica* rice is short and adapted to a temperate climate, bears a high number of tillers and short panicles with short, bold grains. The *javanica* rice is adapted to the tropical climate of Indonesia, being taller but bearing fewer panicles than *indica* rice. The anatomical differences among these three types are prominently observed in culm and leaves.

Early rice ideotype in the first half of 20th century could be described by genotypes that matured late, as the predominant cropping system was single crop (rice followed by fallow land). As the reproductive growth phase in rice is fixed at around 25–35 days, late maturity types had higher vegetative growth and photosynthesis, this being channelled into grain development for higher productivity. Longer vegetative growth also led to higher cellular growth and maturation, providing better strength to the tillers. The anatomical structure of stem and leaves of these types reveals higher maturity of vascular bundle tissue and higher deposition of lignin, suberin and silica in the stem. The leaves of these genotypes are also thicker with a higher number of cell layers and increased thickness of cell wall. Resistance to blast disease (causal organism – *Pyricularia oryzae*) is also considered as a character for rice ideotype in this period, although the formal concept of ideotype was developed much later.

A second ideotype was developed by combining the desirable features of *indica* and *japonica* rice. Both these ecotypes have their inherent problems of low productivity. Development of an ideal type combining the features of both of these ecotypes (long panicle with higher grains per panicle, semi-dwarf non-lodging stature with high tillering) was targeted by the breeders of many countries to improve productivity, including Korea, Indonesia and India. Since the two ecotypes exhibit high sterility in intercrosses, limited success was obtained, except in Korea. At structural anatomy level, strong, well-developed vascular bundle was combined with higher panicle extensibility, rapid cellular growth, better translocation of soil nitrogen to leaves and structures favouring higher translocation of food reserves from source to sink.

Development of semi-dwarf, high fertilizer-responsive rice varieties in the 1960s in the International Rice Research Institute (IRRI) in the Philippines transformed the rice production system, uplifting the yield potential of rice from 6 t/ha to 10 t/ha (Virk *et al.*, 2004). The plant features of these new semi-dwarf varieties were short stature, sturdy stem with profuse tillering, dark green and erect leaves (Jennings, 1964). The semi-dwarf rice types could support rapid cellular growth and development by the uptake of more nitrogen from the soil and channelling it for further vegetative growth. This led to an increase in the number and thickness of cell layers in the stem and leaves, higher cellular growth, and more cell layer formation in the root. The leaves of semi-dwarf rice exhibit better developed chloroplasts with a higher capacity to trap solar energy. Although anatomical features were not the basis of selection of semi-dwarf rice, they served to develop the ideotype, which was selected by the plant breeders at morphological level.

A new plant type (NPT) was developed in the late 1980s by modifying the features of the semi-dwarf varieties (Khush, 1995). The semi-dwarf varieties produce many unproductive tillers with small panicles, limiting the sink size as well as contributing to unproductive vegetative growth (low HI). The NPT for irrigated tropical *indica* rice, which is capable of yielding 10–14 t/ha, consists of a semi-dwarf variety with 8–10 tillers bearing heavy panicles with a high number of grains per panicle (200–250), a plant height of 90–100 cm, high grain weight (25 g/1000 grains) and increased HI (Virk *et al.*, 2004). The NPT is designed to have higher leaf growth at the vegetative stage and higher translocation of food reserves leading to more grain filling with a longer reproductive phase. To sustain such heavy panicles with high grains the stem strength should be very high with vigorous root growth. Thus the morpho-anatomical structure of NPT rice should bear dense vascular bundles, with higher cell wall thickness. Better translocation of food reserve during reproductive growth phase depends on the activity of phloem tissues, which should be well developed in NPT rice.

14.3 Changes in Crop Anatomy in Vegetative Structures of Economic Importance

Apart from the grain crops, a substantial amount of food requirement is supplied by starchy and sugary crops, where seed is not the economic product. These include potato (stem tuber), sweet potato (root tuber) and cassava (root tuber), which form the staple food crop of millions of people in different countries. There are two other major crops, sugarcane (stem) and sugarbeet (root tuber), where vegetative structures are the principal source of sugar. An interesting distinction of these species from the seed crop species is that in most of these crops, seeds are formed through sexual reproduction, but the food reserve is channelled in specialized vegetative structures. As these vegetative structures are economic products to humans, selection has been practised for the enhancement of the size and shape of these vegetative structures. In natural populations, these structures have primarily been evolved for asexual reproduction, where new clones are developed from specialized tissues (e.g. scale or eye in potato, which is a dormant bud). The new plants derive their initial food requirement from the food reserve in these storage tissues. Efficient channelling of the starch and sugar in these reserve tissues depends primarily on the metabolic processes leading to food reserve development and structural anatomy of the plant for translocation of the reserve to storage tissues.

14.3.1 Potato

Potato (*Solanum tuberosum* subsp. *tuberosum*) is by far the most important tuber crop and is grown in about 150 countries. A related species *S. tuberosum* subsp. *andigena* is cultivated to some extent in Chile. Although the crop originated in South America, its domestication is one of the most recent events in the history of crop domestication. Although potato cultivation dates back to 13,000 BP in Chile (Ugent *et al.*, 1982), the tetraploid *S. tuberosum* subsp. *andigena* was brought to Europe by the Spanish invaders during the discovery of the New World. Only during the mid-1800s, when all the cultivated clones of *S. tuberosum* subsp. *andigena* were devastated by late blight disease (causal organism – *Phytophthora infestans*), a resistant *S. tuberosum* subsp. *tuberosum* clone (Rough Purple Chili) was introduced to save the European countries from food crisis. The present-day potato cultivation spread very quickly in the rest of the world within a very short period of time.

The potato tuber is a shortened fleshy stem. Botanically it is an enlarged branch where raised nodes with buds are present. It is attached to the main stem with a narrow attachment known as a stolon. The growth and size of the tuber depends on the number and size of the stolons. Genotypes with many stolons produce numerous small tubers. This leads to low productivity, as the plant's energy and food reserve is divided among many translocation paths.

Cellular density in the tuber is yet another important anatomical feature. Large cells pack larger starch granules, while in small cells the size of the starch grains are smaller. The rate of enlargement of starch grains in the tuber cells is also important to increase productivity without extending the period of tuberization and maturation. A high rate of starch granule formation is thus a selection criterion for early maturing varieties without sacrificing productivity. The late maturing varieties have higher cellular density. The rate of cell division is higher at the pith than the cortical region in the initial stages of tuber development. In the later stages, extensive growth takes place between the region of cortex and the pith.

14.3.2 Sugarbeet

About 25% of the total sugar produced in the world comes from sugarbeet (*Beta vulgaris* L. ssp. *vulgaris*). It is a crop that came under cultivation only in the 19th century and consequently, the domestication process of sugarbeet as a crop is very recent and the genetic base of present-day beet varieties is narrow. However, it is possibly the only crop where scientific breeding started as early as the process of domestication. The process of mass selection and progeny testing was first carried out in this crop by Louis de Vilmorin during the 1850s, which is considered as a pioneer work in scientific plant breeding.

A significant change in crop architecture of sugarbeet has been obtained by inducing autopolyploidy by colchicine. The tetraploid varieties ($2n = 4x = 36$) produced better root shape, larger leaves with stronger petioles with better photosynthesis.

An autotriploid (3*x*) sugarbeet population was developed by crossing tetraploid and diploid genotypes to provide a higher sugar yield and better root shapes than the diploid varieties.

Root anatomical characters influence the sugar yield, sugar percentage and shape of the tuber. Deeper vertical grooves in beet roots are considered as an undesirable character and has been removed by mass selection (Saunders *et al.*, 1999). The number of vascular rings in the root is directly related to the root yield.

14.3.3   Cassava

The economic product of cassava (*Manihot esculenta* Cranz, also known as manioc or yucca) is the root tuber. Botanically it is a root swollen by deposition of starch (>80% starch on dry weight basis). It is a staple food source of tropical South American countries. The root tuber is high in carbohydrate and low in protein (1–3%). The starch content of the tuber depends on the maturity of the tubers. Young tubers that are harvested within 6 months have a lower amount of starch than 1-year-old tubers. However, when tubers are in the field for more than 18 months, they become lignified, woody and unsuitable for consumption or starch preparation. An inspection of tuber anatomy reveals three distinct regions in the tuber, outer periderm region, middle cortex region and central pith region. Starch content is primarily dependent on the diameter of the central pith region. Some cassava genotypes taste bitter due to the presence of cyanogenic glucosides in the periderm and outer cortex regions. For extraction of commercial starch, these regions are removed during processing by peeling the outer layer and boiling.

The cassava roots after harvest cannot be kept in storage for an extended period due to post-harvest physiological deterioration (PPD), which makes the roots unpalatable (Reilly *et al.*, 2003). It is caused by reactive oxygen species-mediated discoloration of xylem parenchyma, similar to wounding response in many plants. Screening for low PPD cassava genotypes is a major target of cassava breeding.

Stem lignification is a determining character for usability of a stem as vegetative propagating material. A low lignified stem when used as planting material dehydrates rapidly and gives poor germination. On the other hand highly lignified stem would have less storage reserve and will not support growth of germinated seedlings. Although this feature is not directly linked to yield per plant, it is a contributing character for higher productivity per unit area, since selection of immature or over-mature planting material will reduce the number of plants per unit area, poor crop stand and susceptibility to insect pest and disease attack, all of which deteriorate cassava productivity (Ceballos *et al.*, 2010).

Dry matter partitioning is a crucial factor for increasing productivity of cassava. Carbohydrate is stored in roots as well as stems during the growth period. This vegetative competition is absent in grain crops, where the vegetative phases slow down after the reproductive phase is initiated and all the reserve is channelled to grains. In cassava, the roots, stem and leaves compete for partitioning of dry matter (Cock, 1984). Since photo-assimilates are also required for growth of the plant, translocation of food towards the aboveground part is higher during the first 4–5 months, leading to stem and leaf growth and development. The rate of storage in roots increase starting from about 5–6 months onwards with concomitant reduction in aboveground translocation (leaf senescence and stem lignification) and continues until harvest. The senescence process limits further leaf development, while the lignification and sclerification of the stems inhibits further accumulation of reserve food in stem tissues. Thus the structural anatomy and physiological processes determine the amount of the starch to be stored in root tubers, and affect productivity. Genotypes that are able to transfer more reserve food towards the root earlier without hampering the crop growth (early bulking tubers) are considered to produce more starch.

References

allos, H., Okogbenin, E., Pérez, J.C., López-Valle, L.A.B. and Debouck, D. (2010) Cassava. In: Bradshaw, J.E. (ed.) *Handbook of Plant Breeding: Root and Tuber Crops*. Springer, New York, pp. 53–96.

k, J.H. (1984) Cassava. In: Goldsworthy, P.R. and Fisher, N.M. (eds) *The Physiology of Tropical Field Crops*. Wiley, New York, pp. 529–549.

1ald, C.M. (1968) The breeding of crop ideotypes. *Euphytica* 17, 385–403.

)STAT (2012) Available at: http://faostat.fao.org/site/567/DesktopDefault.aspx?PageID=567#ancor (accessed 9 March 2012).

inzen, F., Ramos, J. and Tivano, J.C. (2002) Comparative quantitative anatomy of grass species in Santa Fe province. *Revista Fave Seccion Ciencias Agrarias* 1, 57–63.

nings, P.R. (1964) Plant type as a rice breeding objective. *Crop Science* 4, 13–15.

ush, G.S. (1995) Breaking the yield frontier of rice. *GeoJournal* 35, 329–332.

illy, K., Gomez-Vasquez, R., Buschman, H., Tohme, J. and Beeching, J.R. (2003) Oxidative stress responses during cassava post-harvest physiological deterioration. *Plant Molecular Biology* 53, 669–685.

unders, J.W., McGrath, J.M., Halloin, J.M. and Theurer, J.C. (1999) Registration of SR94 sugar beet germplasm with smooth root. *Crop Science* 1, 297.

gent, D., Pozorski, S. and Pozorski, T. (1982) Archeological potato tuber remains from the Casma Valley of Peru. *Economic Botany* 36, 182–192.

rk, P.S., Khush, G.H. and Peng, S. (2004) Breeding to enhance yield potential of rice at IRRI: the ideotype approach. *International Rice Research Notes* 29, 5–9.

14.3.4 Sugarcane

Sugarcane (*Saccharum* spp.) is the world's most important sugar-producing crop, meeting over 70% of the total sugar requirement of the world. Unlike other vegetative crops, it does not produce any specialized structure to conserve reserve food material, but stores sugar in the stems (tillers). Until the technology for extraction of sugar was developed in India about 3000 years ago, sugarcane was considered to be a garden grass plant. Today it is cultivated in over 100 countries covering an area of 23.8 million ha and production of 1685.44 Mt (FAOSTAT, 2012). It is also a major source of ethanol. The by-products are used as fuel, and also for the generation of electricity.

Sugarcane belongs to the grass family. Several cane-producing species of the genus *Saccharum* are cultivated for production of sugar, which are genetically related and polyploid having various chromosome numbers. Being a C_4 plant and polyploid, the crop produces high biomass.

High sugar yield has been the major target for selection in sugarcane during the phases of domestication and assisted breeding efforts. The adaptation of different cane species during the domestication periods depended on the climatic conditions, productivity of the species and extent of cultivation. Although sugarcane was domesticated in New Guinea about 13,000 years ago, it was predominantly cultivated in the Indian subcontinent. The thin canes of India and China (*Saccharum barberi* and *Saccharum sinense*) were adapted to the subtropical climatic conditions of northern India and southern China. Although the tillers were thin, these species were selected due to high sugar content in the juice. *Saccharum spontaneum*, a low sugar-yielding species has wide adaptability to both tropical and subtropical conditions, acquiring several stress tolerance traits. The stem anatomy of *S. spontaneum* reveals a hollow centre and pithy regions, which is the main reason for low sugar content. This species has been adapted to both tropical and subtropical conditions, but has not been preferred much for cultivation. *Saccharum officinarum*, the most widely cultivated

species of sugarcane, ha: canes with thicker pith, sugar yield. This species i cal conditions, where it l other cane species.

A considerable propo is used for biofuel prodι about 50% of the sugarc used for production of etha biofuel production requires growth, the ideotype for b should produce high biom good amount of sugar.

14.4 Changes in Cro; for Better Forage (

The forage and the fodder cro; number of crop species prim; to Poaceae (Gramineae) and and some species of Brassicac der beet). Apart from the straw cereal food crops, animal fee obtained from these forage crc vegetative biomass is consume mal either raw or in a process the structural anatomy of plaι sumed largely determines the fι in these species. The ratio of dige and non-digestible tissues is a indicator of forage quality, whicl dicted from anatomical observatic *et al.*, 2002). Selection for forage primarily targeted two key char ductivity as expressed by increase(growth and suitability as animal fe mined by palatability and digestib

14.4.1 Forage grasses

Apart from fodder maize and sorgl of the forage grasses are perenι major fodder grasses are *Panicum n* (Guinea grass) and other *Panicu, Pennisetum purpureum* (Napier or grass), hybrid Napier (Pearl millet × *Brachiaria mutica* (Para grass), *Loliu dinaceum* (tall fescue) and *Sorghum ense* (Sudan grass).

14.3.4 Sugarcane

Sugarcane (*Saccharum* spp.) is the world's most important sugar-producing crop, meeting over 70% of the total sugar requirement of the world. Unlike other vegetative crops, it does not produce any specialized structure to conserve reserve food material, but stores sugar in the stems (tillers). Until the technology for extraction of sugar was developed in India about 3000 years ago, sugarcane was considered to be a garden grass plant. Today it is cultivated in over 100 countries covering an area of 23.8 million ha and production of 1685.44 Mt (FAOSTAT, 2012). It is also a major source of ethanol. The by-products are used as fuel, and also for the generation of electricity.

Sugarcane belongs to the grass family. Several cane-producing species of the genus *Saccharum* are cultivated for production of sugar, which are genetically related and polyploid having various chromosome numbers. Being a C_4 plant and polyploid, the crop produces high biomass.

High sugar yield has been the major target for selection in sugarcane during the phases of domestication and assisted breeding efforts. The adaptation of different cane species during the domestication periods depended on the climatic conditions, productivity of the species and extent of cultivation. Although sugarcane was domesticated in New Guinea about 13,000 years ago, it was predominantly cultivated in the Indian subcontinent. The thin canes of India and China (*Saccharum barberi* and *Saccharum sinense*) were adapted to the subtropical climatic conditions of northern India and southern China. Although the tillers were thin, these species were selected due to high sugar content in the juice. *Saccharum spontaneum*, a low sugar-yielding species has wide adaptability to both tropical and subtropical conditions, acquiring several stress tolerance traits. The stem anatomy of *S. spontaneum* reveals a hollow centre and pithy regions, which is the main reason for low sugar content. This species has been adapted to both tropical and subtropical conditions, but has not been preferred much for cultivation. *Saccharum officinarum*, the most widely cultivated species of sugarcane, has been selected for canes with thicker pith, resulting in higher sugar yield. This species is adapted to tropical conditions, where it has out-yielded all other cane species.

A considerable proportion of sugarcane is used for biofuel production. In Brazil, about 50% of the sugarcane produced is used for production of ethanol. Since higher biofuel production requires more vegetative growth, the ideotype for biofuel sugarcane should produce high biomass as well as a good amount of sugar.

14.4 Changes in Crop Anatomy for Better Forage Crops

The forage and the fodder crops include large number of crop species primarily belonging to Poaceae (Gramineae) and Leguminosae, and some species of Brassicaceae (kale, fodder beet). Apart from the straw component of cereal food crops, animal feed is primarily obtained from these forage crops. Since the vegetative biomass is consumed by the animal either raw or in a processed condition, the structural anatomy of plant parts consumed largely determines the forage quality in these species. The ratio of digestible tissues and non-digestible tissues is an important indicator of forage quality, which can be predicted from anatomical observations (Heinzen *et al.*, 2002). Selection for forage grasses has primarily targeted two key characters, productivity as expressed by increased vegetative growth and suitability as animal feed as determined by palatability and digestibility.

14.4.1 Forage grasses

Apart from fodder maize and sorghum most of the forage grasses are perennial. The major fodder grasses are *Panicum maximum* (Guinea grass) and other *Panicum* spp., *Pennisetum purpureum* (Napier or elephant grass), hybrid Napier (Pearl millet × Napier), *Brachiaria mutica* (Para grass), *Lolium arundinaceum* (tall fescue) and *Sorghum sudanense* (Sudan grass).

References

Ceballos, H., Okogbenin, E., Pérez, J.C., López-Valle, L.A.B. and Debouck, D. (2010) Cassava. In: Bradshaw, J.E. (ed.) *Handbook of Plant Breeding: Root and Tuber Crops*. Springer, New York, pp. 53–96.

Cock, J.H. (1984) Cassava. In: Goldsworthy, P.R. and Fisher, N.M. (eds) *The Physiology of Tropical Field Crops*. Wiley, New York, pp. 529–549.

Donald, C.M. (1968) The breeding of crop ideotypes. *Euphytica* 17, 385–403.

FAOSTAT (2012) Available at: http://faostat.fao.org/site/567/DesktopDefault.aspx?PageID=567#ancor (accessed 9 March 2012).

Heinzen, F., Ramos, J. and Tivano, J.C. (2002) Comparative quantitative anatomy of grass species in Santa Fe province. *Revista Fave Seccion Ciencias Agrarias* 1, 57–63.

Jennings, P.R. (1964) Plant type as a rice breeding objective. *Crop Science* 4, 13–15.

Khush, G.S. (1995) Breaking the yield frontier of rice. *GeoJournal* 35, 329–332.

Reilly, K., Gomez-Vasquez, R., Buschman, H., Tohme, J. and Beeching, J.R. (2003) Oxidative stress responses during cassava post-harvest physiological deterioration. *Plant Molecular Biology* 53, 669–685.

Saunders, J.W., McGrath, J.M., Halloin, J.M. and Theurer, J.C. (1999) Registration of SR94 sugar beet germplasm with smooth root. *Crop Science* 1, 297.

Ugent, D., Pozorski, S. and Pozorski, T. (1982) Archeological potato tuber remains from the Casma Valley of Peru. *Economic Botany* 36, 182–192.

Virk, P.S., Khush, G.H. and Peng, S. (2004) Breeding to enhance yield potential of rice at IRRI: the ideotype approach. *International Rice Research Notes* 29, 5–9.

Index

Page numbers in **bold** type refer to figures and tables